Communications
in Computer and Information Science   2977

Series Editors

Gang Li, *School of Information Technology, Deakin University, Burwood, VIC, Australia*
Joaquim Filipe, *Polytechnic Institute of Setúbal, Setúbal, Portugal*
Zhiwei Xu, *Chinese Academy of Sciences, Beijing, China*

Rationale

The CCIS series is devoted to the publication of proceedings of computer science conferences. Its aim is to efficiently disseminate original research results in informatics in printed and electronic form. While the focus is on publication of peer-reviewed full papers presenting mature work, inclusion of reviewed short papers reporting on work in progress is welcome, too. Besides globally relevant meetings with internationally representative program committees guaranteeing a strict peer-reviewing and paper selection process, conferences run by societies or of high regional or national relevance are also considered for publication.

Topics

The topical scope of CCIS spans the entire spectrum of informatics ranging from foundational topics in the theory of computing to information and communications science and technology and a broad variety of interdisciplinary application fields.

Information for Volume Editors and Authors

Publication in CCIS is free of charge. No royalties are paid, however, we offer registered conference participants temporary free access to the online version of the conference proceedings on SpringerLink (http://link.springer.com) by means of an http referrer from the conference website and/or a number of complimentary printed copies, as specified in the official acceptance email of the event.

CCIS proceedings can be published in time for distribution at conferences or as post-proceedings, and delivered in the form of printed books and/or electronically as USBs and/or e-content licenses for accessing proceedings at SpringerLink. Furthermore, CCIS proceedings are included in the CCIS electronic book series hosted in the SpringerLink digital library at http://link.springer.com/bookseries/7899. Conferences publishing in CCIS are allowed to use our online conference service (Meteor) for managing the whole proceedings lifecycle (from submission and reviewing to preparing for publication) free of charge.

Publication process

The language of publication is exclusively English. Authors publishing in CCIS have to sign the Springer CCIS copyright transfer form, however, they are free to use their material published in CCIS for substantially changed, more elaborate subsequent publications elsewhere. For the preparation of the camera-ready papers/files, authors have to strictly adhere to the Springer CCIS Authors' Instructions and are strongly encouraged to use the CCIS LaTeX style files or templates.

Abstracting/Indexing

CCIS is abstracted/indexed in DBLP, Google Scholar, EI-Compendex, Mathematical Reviews, SCImago, Scopus. CCIS volumes are also submitted for the inclusion in ISI Proceedings.

How to start

To start the evaluation of your proposal for inclusion in the CCIS series, please send an e-mail to ccis@springer.com

Kazi Lutful Kabir · Hisham Al-Mubaid ·
Oliver Eulenstein
Editors

Bioinformatics and Computational Biology

18th International Conference, BICOB 2026
Honolulu, Hawaii, USA, March 23–25, 2026
Proceedings

 Springer

Editors
Kazi Lutful Kabir (ID)
George Mason University
Fairfax, VA, USA

Hisham Al-Mubaid
University of Houston-Clear Lake
Houston, TX, USA

Oliver Eulenstein
Iowa State University
Ames, IA, USA

ISSN 1865-0929   ISSN 1865-0937 (electronic)
Communications in Computer and Information Science
ISBN 978-3-032-26027-7   ISBN 978-3-032-26028-4 (eBook)
https://doi.org/10.1007/978-3-032-26028-4

This Springer imprint is published by the registered company Springer Nature Switzerland AG
The registered company address is: Gewerbestrasse 11, 6330 Cham, Switzerland

If disposing of this product, please recycle the paper.

Preface

We as the Organizing Committee of the 18th International Conference on Bioinformatics and Computational Biology (BICOB 2026) would like to express our sincere gratitude and extend a warm welcome to all authors, presenters, attendees, and organizers of BICOB 2026, sponsored by the International Society for Computers and their Applications (ISCA).

For the past eighteen years, the BICOB conference has enjoyed excellent participation and attendance from scientists and practitioners in the fields of bioinformatics and computational biology. BICOB has been establishing itself, and growing, as one of the leading events for bioinformatics researchers and practitioners. BICOB offers a unique platform for scientists and researchers to present and discuss their research results, methods, and findings with peers who share similar interests. This year, the BICOB conference was held in Honolulu, Hawaii, USA, on March 23–25, 2026.

BICOB 2026 featured keynote talks by well-known scientists and highly competitive contributed talks. The conference covered a wide range of bioinformatics and computational biology topics, including bioinformatics algorithms, computational proteomics, genomics, structural bioinformatics, graph and machine learning applications in bioinformatics, as well as medical and healthcare informatics. BICOB 2026 brought together researchers, scientists, practitioners, and attendees from multiple countries worldwide. Participants came from various research institutions, including universities, corporations, and government agencies.

Through a thorough review process, BICOB 2026 maintained high standards and quality. This year, each paper submitted to BICOB 2026 underwent single-blind peer review by three international program committee members. The papers were evaluated based on their originality, significance, technical content, application content, and presentation style. BICOB 2026 received 38 papers and through a comprehensive and rigorous review process, 21 papers were accepted, giving an acceptance rate of 55%. Also, BICOB 2026 featured a Best Paper Award, which was presented during the conference. The online conference management system, *Meteor-Springer*, was used to automate the submission, reviewing, and refereeing processes. The BICOB 2026 Proceedings are published by Springer. All the BICOB 2026 papers will be available online via the *Springer Nature* site, making them easily retrievable through Google and Google Scholar, and they will also be indexed in Scopus.

The professional work of the international program committee and the sub-reviewers is greatly appreciated. We also want to thank the ISCA president, the board of directors, and the ISCA executive director for their continuing support. Furthermore, we extend our gratitude to all the presenters and attendees for their valuable contributions to the success of BICOB 2026.

We enjoyed engaging in presentations and discussions at the conference. All partici-pants were invited to connect and establish new networks within the BICOB family. We sincerely hope that every participant benefited from the conference.

March 2026   Hisham Al-Mubaid
                                                                    Kazi Lutful Kabir
                                                                    Oliver Eulenstein

Organization

General Chair

Hisham Al-Mubaid   University of Houston – Clear Lake, USA

Program Committee Chairs

Kazi Lutful Kabir   George Mason University, USA
Oliver Eulenstein   Iowa State University, USA

International Program Committee

Nasrin Akhter   University at Buffalo, USA
Fardina Alam   University of Maryland, College Park, USA
Tamer Aldwairi   Temple University, USA
Hisham Al-Mubaid   University of Houston - Clear Lake, USA
Abdullah Arslan   Texas A & M University – Commerce, USA
Evaldo Costa   CESAR Innovation Center, Brazil
Scott Emrich   University of Tennessee, USA
Oliver Eulenstein   Iowa State University, USA
Negin Forouzesh   California State University - Los Angeles, USA
Irina Hashmi   George Mason University, USA
Nurit Haspel   University of Massachusetts, Boston, USA
Filip Jagodzinski   Western Washington University, USA
Kazi Lutful Kabir   George Mason University, USA
Anowarul Kabir   University of South Florida, USA
Danny Krizanc   Wesleyan University, USA
Ion Mandoiu   University of Connecticut, USA
Tin Nguyen   Auburn University, USA
Yuri Pirola   University of Milano-Bicocca, Italy
Letu Qingge   North Carolina Agricultural and Technical State
                University, USA
Mustafijur Rahman   Texas A&M University - San Antonio, USA
Salim Sajjed   Georgia Southern University, USA
Sing-Hoi Sze   Texas A&M University, USA
Ugo Vaccaro   University of Salerno, Italy

| Jianxin Wang | Central South University, China |
| Ka-Chun Wong | City University of Hong Kong, China |

Additional Reviewers

Zanoni Dias	University of Campinas, Brazil
Fathalla Rihan	United Arab Emirates University, UAE
Alexey Sadovski	Texas A&M University, Corpus Christi, USA

Contents

Modeling Virus–Tumor Interactions and Control in Oncolytic Virotherapy

Fathalla A. Rihan[✉]

Department of Mathematical Sciences, College of Science, UAE University, Al Ain 15551, UAE
frihan@uaeu.ac.ae

Abstract. A combination of biological delays, tumor-microenvironment transport barriers, and dosing cadence determines the success of oncolytic virotherapy. In this study, we develop a delay-differential model incorporating virus-tumor-immune dynamics with explicit infection, lysis, and immune-priming delays, as well as a tunable transport-impedance field that modulates viral contact and diffusion. A tractable viral invasion metric relates barrier severity to eradication thresholds. Meanwhile, stability and Hopf bifurcation analyses demonstrate that cumulative delays can induce oscillations that promote relapse. A safety-aware optimal control formulation, combined with learning-based model predictive control (MPC), enables adaptive, biomarker-guided dosing under toxicity constraints. A spatial reaction-diffusion delay extension reveals stacked traveling fronts under severe transport impedance, which motivates the normalization of transport and optimized therapy sequencing. Based on this framework, personalized oncolytic virotherapy can be designed.

Keywords: Oncolytic virotherapy · Delay differential equations · Optimal control · Hopf bifurcation · Reaction-diffusion · Model predictive control

1 Introduction

Oncolytic viruses replicate preferentially in malignant cells, directly lysing tumor while releasing antigens and danger signals that recruit adaptive immunity [1–4]. Despite promising responses, outcomes vary widely. Three system-level mechanisms repeatedly surface: (i) intrinsic delays (time from virion entry to productive lysis; antigen presentation to effector expansion; and trafficking delays) [5–7]; (ii) transport barriers in the tumor microenvironment (TME) dense extracellular matrix, high interstitial pressure, hypoxia that throttle diffusion and cell contact [8,9]; and (iii) timing of multi-agent dosing, where cadence and intensity interact with intrinsic timescales and transport impediments [10,11].

To reason about and *control* these mechanisms, we develop a compact, interpretable *delay-differential* model with an explicit barrier field, amenable to stability analysis, bifurcation mapping, and constrained optimal control. We connect the ordinary-delay model to a spatial reaction–diffusion counterpart that

K. L. Kabir et al. (Eds.): BICOB 2026, CCIS 2977, pp. 1–12, 2026.
https://doi.org/10.1007/978-3-032-26028-4_1

captures stacked traveling fronts (separated tumor, virus, and immune waves) arising under severe transport impedance. We then formulate a safety-aware optimal control problem, derive first-order conditions, and outline a learning-in-the-loop model predictive control (MPC) scheme for personalization.

Optimal control guides therapy timing, dose, and schedule to maximize tumor kill with minimal toxicity. Many OVT design questions balancing antiviral vs. antitumor immunity, choosing combinations and doses, tailoring to tumor/virus type, and accounting for spatial heterogeneity are costly or infeasible to resolve experimentally. Mathematical modeling, especially delay differential equations, pinpoints key parameters, predicts outcomes in silico, and captures time-lagged tumor-immune-therapy dynamics to optimize regimens [7]. The proposed model provides clinically actionable guidelines for scheduling virotherapy, explains how identical cumulative doses can result in divergent outcomes, and supports precision treatment planning using routinely measurable biomarkers. The results are directly relevant to the optimization of oncolytic virus protocols in solid tumors, where transport barriers and immune delays limit the efficacy of therapeutics.

2   Mathematical Model

Delay-differential equation (DDE) models have played a central role in describing tumor-immune dynamics under therapeutic intervention. In a series of foundational studies, Rihan and collaborators demonstrated that immune response delays can destabilize otherwise effective tumor control strategies [12]. Subsequent extensions incorporating chemo–immunotherapy and optimal control further emphasized the critical role of treatment timing and scheduling [13, 14]. However, these models did not account for virus-mediated oncolysis nor for transport limitations imposed by the tumor microenvironment. The present work extends this framework by explicitly incorporating oncolytic virus dynamics, transport impedance, and learning-in-the-loop control, yielding a unified and more realistic model for optimizing oncolytic virotherapy.

Herein, we track viable tumor cells $T(t)$, infected cells $I(t)$, free virus $V(t)$, and effector immune cells $E(t)$. Intrinsic biological lags in infection, lysis, and immune priming are represented as fixed delays $\tau_{\mathrm{inf}}, \tau_{\mathrm{lys}}, \tau_{\mathrm{prim}}$, respectively [5,6]. To account for transport limitations imposed by the tumor microenvironment, we introduce a scalar barrier field $B \in [0,1]$ that captures extracellular matrix density, interstitial pressure, and hypoxia [8,9]. Barrier severity down-modulates effective viral contact through the factor $\phi(B) \in (0,1]$, for which we assume

$$\phi(B) = e^{-\eta B}, \qquad \eta > 0,$$

where η is a dimensionless barrier-sensitivity parameter. Thus, increasing B (or η) reduces effective infection and contact rates.

Control inputs $u_V(t)$ and $u_I(t)$ represent exogenous viral dosing and immunomodulatory therapy, respectively, subject to clinical bounds and cumulative-dose constraints [10]. The governing delay-differential system is

$$\dot{T}(t) = r\,T\left(1 - \frac{T}{K}\right) - \phi(B)\,\beta\,V\,T - \kappa_E\,E\,T,$$

$$\dot{I}(t) = \phi(B)\,\beta\,V(t - \tau_{\text{inf}})\,T(t - \tau_{\text{inf}})e^{-\delta\,\tau_{\text{inf}}} - (\delta_I + \kappa_I E)\,I,$$

$$\dot{V}(t) = p\,I(t - \tau_{\text{lys}})\,e^{-\delta_V\,\tau_{\text{lys}}} - c\,V - \phi(B)\,\beta\,V\,T + u_V(t), \tag{1}$$

$$\dot{E}(t) = s + \alpha\,\frac{I(t - \tau_{\text{prim}})}{h + I(t - \tau_{\text{prim}})} - d_E\,E + u_I(t).$$

The parameters r and K are tumor growth parameters; β is infection rate; p burst size; c viral clearance; s basal effector influx; α priming strength with half-saturation h; κ_E, κ_I are immune killing coefficients. The survival factors $e^{-\delta\,\tau_{\text{inf}}}$ and $e^{-\delta_V\,\tau_{\text{lys}}}$ account for loss during latent periods.

Bolus pulses can be modeled as Dirac sums or finite-rate infusions via box/trapezoid profiles, with bounds $0 \leq u_*(t) \leq \bar{u}_*$ (where $u_*(t) \in \{u_V(t), u_I(t)\}$) and cumulative limits an approach standard in biological optimal-control formulations [10].

2.1 Nondimensionalization

We adopt a standard nondimensionalization to reduce parameter count and reveal effective groups (burst/clearance, immune kill, transport factor), following common practice in tumor-virus-immune modeling [15, 16].

Scale $T = K\,\hat{T}$, $I = K\,\hat{I}$, $V = \frac{r}{\beta}\,\hat{V}$, $E = \frac{r}{\kappa_E}\,\hat{E}$, $t = \hat{t}/r$. Dropping hats:

$$\dot{T} = T(1 - T) - \underbrace{\phi(B)}_{\in(0,1]}VT - \gamma ET,$$

$$\dot{I} = \phi(B)\,V(t - \tau_1)\,T(t - \tau_1)\,e^{-\mu\tau_1} - (\mu_I + \gamma_I E)I,$$

$$\dot{V} = \pi\,I(t - \tau_2)\,e^{-\mu_V \tau_2} - \chi V - \phi(B)VT + v_V(t), \tag{2}$$

$$\dot{E} = \sigma + \alpha\,\frac{I(t - \tau_3)}{h + I(t - \tau_3)} - \delta_E E + v_I(t),$$

with dimensionless parameters

$$\gamma = \frac{\kappa_E K}{r},\ \gamma_I = \frac{\kappa_I K}{r},\ \pi = \frac{p\beta K}{r},\ \chi = \frac{c}{r},\ \mu = \frac{\delta}{r},\ \mu_V = \frac{\delta_V}{r},\ \delta_E = \frac{d_E}{r},\ \sigma = \frac{s}{rK},\ \tau_{1,2,3} = r\,\tau_{\text{inf,lys,prim}}.$$

2.2 Equilibria and Invasion Metrics

We summarize biologically relevant steady states at constant dosing $u. \equiv 0$.

Tumor-only (no virus) State. Setting $V = I = 0$, we have $E^* = \sigma/\delta_E$ and $T^* > 0$ satisfying

$$T^*\left(1 - T^*\right) - \gamma E^* T^* = 0 \;\Rightarrow\; T^* = 1 - \gamma E^*,$$

requiring $1 - \gamma E^* > 0$.

Heuristic Virus Invasion Number. Linearization about the tumor-only state yields a heuristic viral invasion number $\mathcal{R}_V$ that captures infection before clearance and burst after delays; its use as a design proxy is consistent with delay-system threshold analyses [5]. Linearizing the system (I, V) subsystem about (T^*, E^*), a single virion successfully infects at rate $\phi(B)T^*$ before clearance χ. A standard two-step heuristic yields the expected number of secondary virions produced by one virion:

$$\mathcal{R}_V \approx \underbrace{\frac{\phi(B)T^*}{\chi + \phi(B)T^*}}_{\text{infection before clearance}} \cdot \underbrace{\pi\, e^{-\mu\tau_1}\, e^{-\mu_V \tau_2}}_{\text{burst after delays}}. \tag{3}$$

Thus, higher barrier B (smaller $\phi(B)$) and longer delays depress $\mathcal{R}_V$. When $\mathcal{R}_V < 1$, virus fails to invade the tumor-only state.

Remark 1 (Strict Thresholds). A full next-generation calculation for the retarded system yields a threshold consistent with (3) up to factors accounting for concomitant loss to target depletion and immune lysis of I; we use (3) as a design metric.

3   Local Stability and Hopf Bifurcation

Linearizing the system (1) about an equilibrium $x^* = (T^*, I^*, V^*, E^*)$ yields the retarded characteristic equation

$$\det\left[\lambda I - A_0 - A_1\, e^{-\lambda\tau_1} - A_2\, e^{-\lambda\tau_2} - A_3\, e^{-\lambda\tau_3}\right] = 0. \tag{4}$$

The retarded characteristic equation admits delay-induced Hopf bifurcations as $\tau_{1,2,3}$ or B increase, generating oscillations classical in DDEs and central to our cadence rules [5,6].

Proposition 1 (Existence of Hopf). *Assume the tumor-only equilibrium is locally asymptotically stable at $\tau_{1,2,3} = 0$ with $\mathcal{R}_V < 1$. There exists a critical $\bar{\tau} > 0$ such that for $\tau := \tau_1 + \tau_2 + \tau_3 \in (0, \bar{\tau})$ the equilibrium remains stable; at $\tau = \bar{\tau}$ a simple pair of roots crosses the imaginary axis transversally, generating a Hopf bifurcation.*

Remark 2 (Design Implication). Delay and impedance $(B \uparrow)$ shift eradication thresholds, open relapse windows, and may induce dosing-locked oscillations; cadence must account for intrinsic periods near Hopf.

4   Safety-Aware Optimal Control

We pose a quadratic regularized OCP with input bounds, cumulative-dose limits, and state-safety constraints, following [10], and use these weights as tunable surrogates for toxicity and effort. We minimize tumor burden and toxicity while regularizing dose over horizon $[0, T_f]$:

$$\min_{u_V, u_I} \; J = \int_0^{T_f} \left[w_T\, T(t) + w_{\text{tox}}\, \Psi\big(E(t), V(t)\big) + \rho_V u_V(t)^2 + \rho_I u_I(t)^2 \right] dt, \quad (5)$$

subject to (1), bounds $0 \leq u_{\cdot}(t) \leq \bar{u}_{\cdot}$, cumulative dose limits $\int_0^{T_f} u_{\cdot}\, dt \leq D_{\cdot,\max}$, and *state constraints* (e.g. $E(t) \leq E_{\text{safe}}$) for clinical safety.

Pontryagin conditions (quadratic regularization)

Let $\lambda = (\lambda_T, \lambda_I, \lambda_V, \lambda_E)$ be the adjoint. The Hamiltonian is

$$\mathcal{H} = w_T T + w_{\text{tox}}\Psi + \rho_V u_V^2 + \rho_I u_I^2 + \lambda_T \dot{T} + \lambda_I \dot{I} + \lambda_V \dot{V} + \lambda_E \dot{E}.$$

Stationarity yields

$$u_V^*(t) = \Pi_{[0, \bar{u}_V]}\left(-\frac{\lambda_V(t)}{2\rho_V} \right), \qquad u_I^*(t) = \Pi_{[0, \bar{u}_I]}\left(-\frac{\lambda_E(t)}{2\rho_I} \right), \qquad (6)$$

with the usual delayed-costate equations (omitted for brevity) integrated backward with terminal conditions $\lambda(T_f) = 0$. In practice we solve (5) in a receding-horizon fashion (MPC).

Learning-in-the-loop MPC

We combine online estimation with constrained MPC over a receding horizon [11], enforcing X_{safe} and re-timing pulses from biomarkers.

Algorithm 1. Safety-aware learning-in-the-loop MPC

1: **Inputs:** horizon H, sample time Δ, safety sets $\mathcal{X}_{\text{safe}}$, bounds $\bar{u}_{\cdot\cdot}$
2: Initialize parameter prior θ_0, state x_0.
3: **for** $k = 0, 1, 2, \ldots$ **do**
4:      Acquire biomarkers y_k (tumor volume, viremia, immune counts).
5:      Update posterior $\theta_k | y_{0:k}$ and state estimate x_k via EnKF/UKF (delay-aware augmentation).
6:      Solve OCP (5) over $[t_k, t_k + H]$ with dynamics at θ_k, constraints $x \in \mathcal{X}_{\text{safe}}$.
7:      Apply the first control segment $u^*(t)$ for $t \in [t_k, t_k + \Delta)$.
8: **end for**

---

5 Spatial Extension and Stacked Traveling Fronts

The reaction-diffusion-delay extension produces stacked traveling fronts, separate tumor, virus, and immune waves with ordered speeds; this organization mirrors classic traveling-wave theory in reaction-diffusion systems [16,17] and explains sequencing effects under transport impedance.

Let $x \in \Omega \subset \mathbb{R}^n$. We extend the variables T, I, V, E to spatially distributed fields and model transport limitations by allowing their diffusion coefficients to depend on the local barrier field $B(x) \in [0, 1]$. For concreteness, we assume

$$D_\bullet(B(x)) = D_\bullet^0 e^{-\eta B(x)}, \qquad \eta > 0,$$

so that increasing barrier severity reduces effective diffusivity, consistent with the contact factor $\phi(B) = e^{-\eta B}$.

$$\partial_t T = \nabla \cdot (D_T(B)\nabla T) + T(1 - T) - \phi(B)VT - \gamma ET,$$
$$\partial_t I = \nabla \cdot (D_I(B)\nabla I) + \phi(B)V(\cdot, t - \tau_1) T(\cdot, t - \tau_1) e^{-\mu \tau_1} - (\mu_I + \gamma_I E)I,$$
$$\partial_t V = \nabla \cdot (D_V(B)\nabla V) + \pi I(\cdot, t - \tau_2) e^{-\mu_V \tau_2} - \chi V - \phi(B)VT + \upsilon_V, \qquad (7)$$
$$\partial_t E = \nabla \cdot (D_E(B)\nabla E) + \sigma + \alpha \frac{I(\cdot, t - \tau_3)}{h + I(\cdot, t - \tau_3)} - \delta_E E + \upsilon_I.$$

For sufficiently strong transport impedance, the reaction–diffusion–delay system organizes into *stacked traveling fronts*, consisting of a leading tumor front followed by viral and immune fronts with ordered propagation speeds

$$c_T \gtrsim c_V \gtrsim c_E.$$

As the barrier field B decreases (transport normalization), both the effective diffusivities D_V, D_E and the contact factor $\phi(B)$ increase, thereby reducing the spatial separation between fronts and enlarging the basin of eradication. This structure naturally motivates a sequencing strategy: transport-normalizing pretreatment $\rightarrow$ OVT pulse $\rightarrow$ delayed immune boost.

6 Numerical Experiments

Stability maps reflect the right-shift of the $R_V = 1$ boundary and a delay-induced Hopf wedge [5]. Under toxicity weighting and dose limits, MPC selects front-loaded pulses when $\phi \approx 1$, transitioning to normalize, then pulse as barriers rise [10,11]. Severe B yields separated T, V, E fronts; increasing D_V, D_E or ϕ collapses the stack, widening the eradication basin [8,16].

We outline a simulation protocol to reproduce the qualitative regimes.

Parameter Baseline (dimensionless).

Symbol	Description	Value (baseline)
γ	immune kill of tumor	0.05
γ_I	immune kill of infected	0.05
π	burst factor	5.0
χ	viral clearance	1.0
μ, μ_V	latent/lysis loss	0.1, 0.1
σ	basal effector influx	0.02
α	priming strength	0.3
h	priming half-saturation	0.1
δ_E	effector decay	0.1
$\tau_{1,2,3}$	delays	0.5, 0.5, 1.0
η	barrier sensitivity	2.0

Figure 1 (**well-mixed Delay Model**). The time courses in Fig. 1 show three phases that repeat across parameter scans: (i) a target-rich phase in which virus expands and drives a sharp decline in T; (ii) a priming phase delayed by (τ_1, τ_2, τ_3), during which E rises in response to antigen from infected/lysed cells; and (iii) a regulation phase in which clearance and immune contraction decide between eradication, dormancy, or relapse. As the total delay $\tau_1 + \tau_2 + \tau_3$ grows, the trajectories exhibit dose-locked oscillations consistent with the Hopf mechanism predicted by the characteristic equation (4) and the design metric $\mathcal{R}_V$ in (3). Practically, identical cumulative dose distributed with different cadence produces distinct outcomes when the barrier factor $\phi(B)$ is small: front-loaded pulses can overcome clearance and latent loss; slow drips dissipate below the infection threshold. These behaviors motivate the cadence rule: *match pulse spacing to intrinsic periods to avoid reinforcing delay-driven oscillations.*

The barrier factor $\phi(B) \in (0, 1]$ compresses effective infection/contact, shifting the viral invasion metric $\mathcal{R}_V$ downward and moving the eradication threshold to higher dosing/burst. In practice, identical cumulative dose administered with different *cadence* produces distinct outcomes when $\phi(B)$ is small: fast front-loaded dosing can overcome clearance and latent loss, whereas slow drips dissipate below the infection threshold. The immuno-boost placement is likewise cadence-sensitive: boosting before antigen has accrued (i.e., before the lysis delay elapses) yields limited benefit, but boosting after the virus has created substrate accelerates contraction of residual disease. Collectively, Fig. 1 supports three design rules: (1) match pulse spacing to intrinsic periods to avoid reinforcing delay-driven oscillations; (2) pre-normalize transport (increase ϕ) when feasible to reduce the required viral intensity; and (3) sequence immune stimulation to follow, not precede, the wave of antigen release.

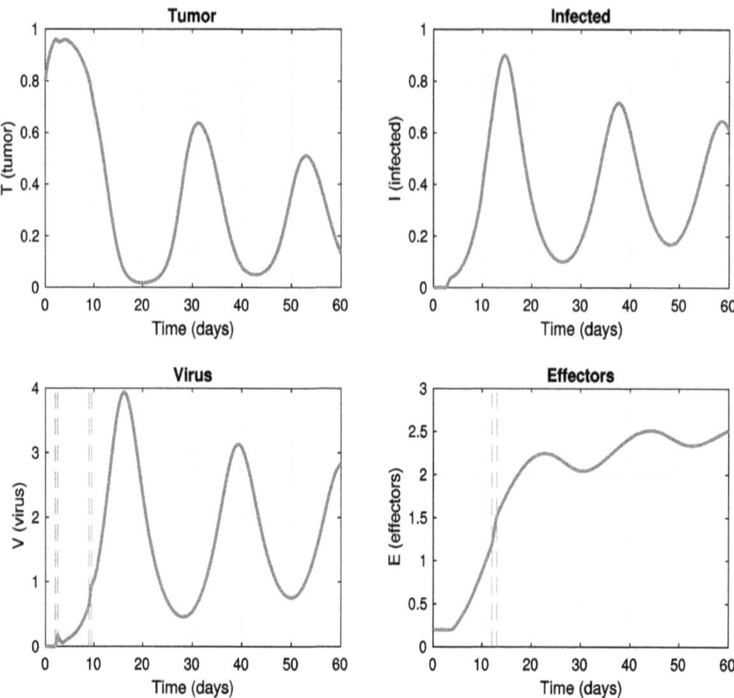

Fig. 1. Well-mixed dynamics under transport impedance and finite-rate dosing./Time courses of tumor T, infected cells I, virus V, and effector cells E. Dashed lines indicate dosing windows (virotherapy followed by immuno-boost). Intrinsic delays (τ_1, τ_2, τ_3) and the TME barrier factor $\phi(B)$ generate oscillatory transients and relapse windows; dosing cadence and intensity govern eradication, dormancy, or regrowth.

Figure 2 (Spatial Reaction–Diffusion–Delay). In the spatial model the solution organizes into stacked traveling fronts: a tumor front, a trailing viral front, and a slower immune front with ordered speeds $c_T \gtrsim c_V \gtrsim c_E$. Regions with higher $B(x)$ reduce both contact $(\phi(B))$ and diffusivities $D.(B)$, widening the gap between fronts and creating transiently cleared niches that can reseed after the viral wave passes. This explains why schedules that succeed in well-mixed settings can underperform in situ: the immune front may arrive too late to sterilize virus-cleared territory. Two actionable strategies follow: (i) transport-normalizing co-therapies (ECM remodeling, IFP reduction) that increase ϕ and (D_V, D_E) collapse the stack and widen the eradication basin; and (ii) sequencing—brief transport normalization $\rightarrow$ OVT pulse $\rightarrow$ delayed immune boost—aligns effector arrival with post-lysis antigen exposure.

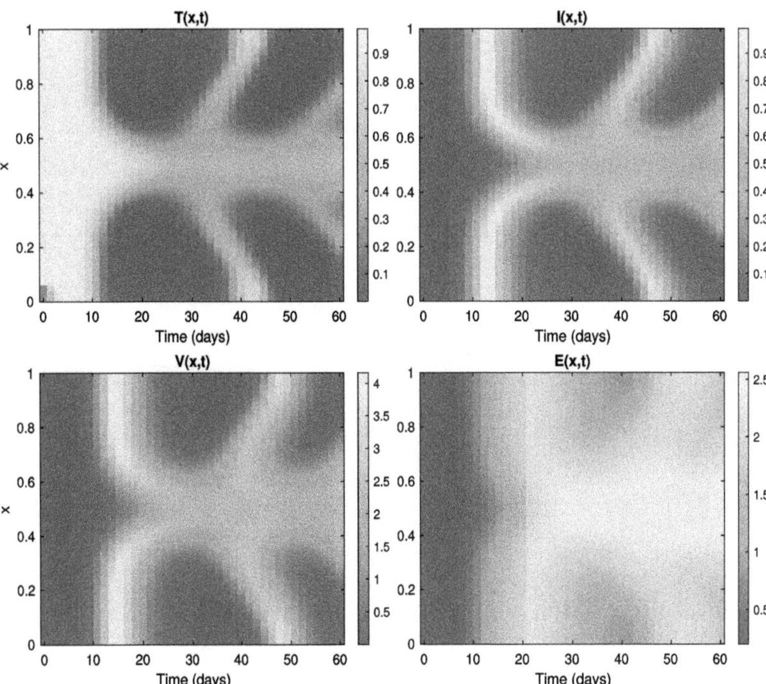

Fig. 2. Spatial (1D reaction-diffusion-delay) extension showing stacked traveling fronts. Heat maps of $T(x,t), I(x,t), V(x,t), E(x,t)$ in the presence of a heterogeneous barrier field $B(x)$. Transport impedance lowers $\phi(B)$ and effective diffusivities, separating tumor, virus, and immune waves with ordered speeds $c_T \gtrsim c_V \gtrsim c_E$. Transport-normalizing co-therapies (larger ϕ, higher D_V, D_E) are predicted to collapse the stack and widen the eradication basin.

The model predicts two actionable strategies. First, *transport-normalizing co-therapy* (e.g., ECM remodeling or pressure reduction) increases ϕ and D_V, D_E, collapsing the stack and widening the eradication basin. Second, *sequencing* matters: applying a brief transport-normalizing pretreatment, then an OVT pulse, then a delayed immune boost aligns the fronts so that effector arrival coincides with post-lysis antigen exposure. The heat maps also highlight measurable biomarkers for personalization: apparent front speeds (from serial imaging), the width of the V-E separation, and the rebound time of T behind barriers; each correlates with underlying parameters and can tune dosing in a learning-in-the-loop controller.

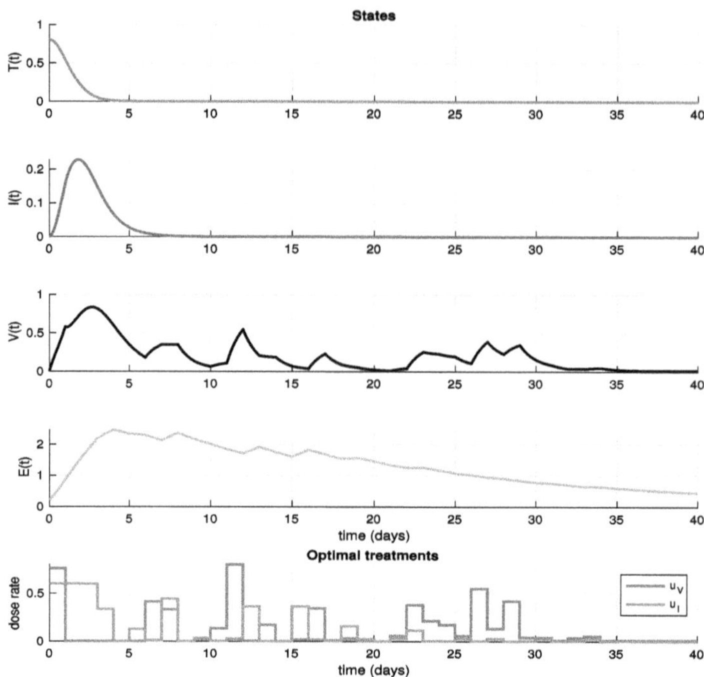

Fig. 3. Optimized dosing schedules eradicate the tumor over a shortened control horizon (0–40 days), consistent with the early-intervention objective of the optimal-control formulation. Panels show T, I, V, and E over time; bottom panel: virotherapy u_V and immune boost u_I. The terminal target is met: $T(T_f) \leq 10^{-4}$.

7   Conclusion

This paper presents a delay-differential model that integrates virus-tumor-immune dynamics, a tunable transport-impedance field, and explicit dosing inputs to model oncolytic virotherapy. The framework unifies intrinsic biological delays and tumor-microenvironment transport barriers, connects an interpretable viral invasion proxy R_V to eradication thresholds, and illustrates how delay-induced Hopf bifurcations shift stability, open relapse windows, and generate oscillatory dynamics, elevating dosing cadence to a first-class design variable.

Numerical simulations and stability maps yield three clinically relevant design rules: (i) match pulse spacing to intrinsic biological periods to avoid delay-driven oscillations, (ii) normalize transport to increase effective contact $\phi(B)$ before intensive virotherapy, and (iii) sequence immune stimulation after virus-mediated antigen release. These mechanisms are evident in both the well-mixed dynamics (Fig. 1) and the spatial reaction-diffusion-delay extension (Fig. 2), explaining why schedules successful in well-mixed models may fail *in situ* when the immune front lags behind the viral wave. A safety-aware optimal-control formulation combined with a learning-in-the-loop model predictive control (MPC) strategy translates these insights into optimized dosing schedules that achieve durable tumor control under dose and toxicity constraints (Fig. 3).

Future directions include distributed delays, richer immune structure, three-dimensional heterogeneous barriers, and robust or stochastic MPC formulations. By connecting delay dynamics, transport physics, and feedback dosing within a single tractable model, this work provides a practical foundation for designing personalized co-therapies and dosing cadences with improved translational potential.

Acknowledgment. This work has been supported by the UAE University, fund # 12S141.

Data Availability Statement. No datasets were generated or analyzed during the current study. All results were obtained from analytical derivations and numerical simulations of the proposed mathematical model.

A   Parameter Definitions and Units (dimensional)

Symbol	Meaning	Units	Role
r	tumor growth rate	day^{-1}	logistic growth
K	carrying capacity	cells (or vol)	logistic growth
β	infection rate	$(\text{virion cell day})^{-1}$	infection
p	burst size	virions/cell	lysis output
c	viral clearance	day^{-1}	loss of V
δ, δ_V	loss during delays	day^{-1}	survival factors
s	effector influx	cells/day	immune source
α, h	priming params	cells/day, cells	adaptive response
d_E	effector decay	day^{-1}	immune loss
κ_E, κ_I	kill coefficients	$(\text{cell day})^{-1}$	cytotoxicity
$\tau_{\text{inf}}, \tau_{\text{lys}}, \tau_{\text{prim}}$	delays	day	latent/lysis/priming
B, η	barrier & sensitivity	–	transport impedance

References

1. Russell, S.J., Peng, K.-W., Bell, J.C.: Oncolytic virotherapy. Nat. Biotechnol. **30**(7), 658–670 (2012). https://doi.org/10.1038/nbt.2287
2. Lichty, B.D., Power, A.T., Stojdl, D.F., Bell, J.C.: Going viral with cancer immunotherapy. Nat. Rev. Cancer **14**(8), 559–567 (2014). https://doi.org/10.1038/nrc3770
3. Fukuhara, H., Ino, Y., Todo, T.: Oncolytic virus therapy: a new era of cancer treatment at dawn. Cancer Sci. **107**(10), 1373–1379 (2016). https://doi.org/10.1111/cas.13027

4. Pol, J., Kroemer, G., Galluzzi, L.: First oncolytic virus approved for melanoma immunotherapy: talimogene laherparepvec. OncoImmunology **5**(1), e1115641 (2016). https://doi.org/10.1080/2162402X.2015.1115641

5. Hale, J.K., Verduyn Lunel, S.M.: Introduction to functional differential equations, vol. 99 of Applied Mathematical Sciences. Springer, New York (1993)

6. Bellen, A., Zennaro, M.: Numerical Methods for Delay Differential Equations. Oxford University Press, Oxford (2003)

7. Rihan, F.A.: Delay Differential Equations and Applications to Biology. Springer, 1st Ed., (2021). https://doi.org/10.1007/978-981-16-0626-7

8. Heldin, C.-H., Rubin, K., Pietras, K., Östman, A.: High interstitial fluid pressure-an obstacle in cancer therapy. Nat. Rev. Cancer **4**(10), 806–813 (2004). https://doi.org/10.1038/nrc1456

9. Jain, R.K.: Normalization of tumor vasculature: an emerging concept in antiangiogenic therapy. Sci. Transl. Med. **5**(185), 185ps9 (2013). https://doi.org/10.1126/scitranslmed.3005687

10. Lenhart, S., Workman, J.T.: Optimal Control Applied to Biological Models. Chapman and Hall/CRC, Boca Raton (2007)

11. Rawlings, J.B., Mayne, D.Q., Diehl, M.M.: Model Predictive Control: Theory, Computation, and Design. Nob Hill Publishing, 2nd edn. (2017)

12. Rihan, F.A., Rahman, D.A., Lakshmanan, S., Alkhajeh, A.S.: A time delay model of tumour-immune system interactions: global dynamics, parameter estimation, sensitivity analysis. Appl. Math. Comput. **232**, 606–623 (2014)

13. Rihan, F.A., Abdelrahman, D.H., Al-Maskari, F., Ibrahim, F., Abdeen, M.A.: Delay differential model for tumour?immune response with chemoimmunotherapy and optimal control. Comput. Math. Methods Med., 982978 (2014)

14. Rihan, F.A., Rihan, N.F.: Dynamics of cancer-immune system with external treatment and optimal control. J. Cancer Sci. Therapy **8**, 257–261 (2016)

15. Wodarz, D., Komarova, N.L.: Dynamics of Cancer: Mathematical Foundations of Oncology. World Scientific (2014)

16. Murray, J.D.: Mathematical Biology I: An Introduction, 3rd edn. Springer, New York (2003)

17. Sherratt, J.A., Murray, J.D.: Models of epidermal wound healing: the formation of excitable waves. J. Theor. Biol. **146**(4), 535–552 (1990). https://doi.org/10.1016/S0022-5193(05)80396-8

Fourier vs. Attention: A Re-look at Protein Sequence Generation Models

Rohann Rahul Shah[(✉)], Rishabh Kartik Jawagal, Sachin Rajpurohit, Nikhilesh Venkat, Namita Achyuthan, and Bhaskarjyoti Das

Department of CSE-AIML, PES University, Bengaluru, Karnataka, India
{pes1ug23am241,pes1ug23am235,pes1ug23am252,pes1ug23am192,
pes1202201119}@pesu.pes.edu, bhaskarjyotidas@pes.edu

Abstract. Protein sequence generation has traditionally relied on autoregressive and diffusion based models, while Fourier-based approaches remain largely unexplored in this domain. We apply FNet, a Fourier transform based architecture originally developed for natural language processing, to protein sequence generation and compare its performance against ESM2's direct masked sampling strategy and the autoregressive model ProGen. All three models were trained and evaluated on a curated kinase dataset under identical experimental conditions. FNet achieves substantial computational gains over ProGen, reducing generation time by 95% and memory consumption by 39%, while producing sequences with improved amino acid compositional accuracy and competitive structural plausibility scores. ESM2 generates sequences twice as rapidly as FNet, whereas ProGen shows stronger short-range motif preservation. These results indicate that Fourier-based token mixing provides a complementary alternative to attention based and autoregressive generation paradigms for protein sequence modeling, highlighting clear trade-offs between computational efficiency, global sequence statistics, and local motif fidelity. While attention-based models remain superior in overall performance, Fourier-based encoders demonstrate promise as efficient components within future hybrid or large scale protein modeling frameworks, with current limitations largely attributable to FFT implementation overhead rather than architectural constraints. **Code available here:** https://github.com/bashirgit/Fourier_vs._Attention.

Keywords: Generative protein modeling · Protein language models · Fourier networks

1 Introduction

Generative protein sequence models have emerged as important tools for rational biomolecular design in applications such as drug discovery and protein therapeutics. Computationally generating novel, functional proteins offers advantages over classical experimental screening by letting researchers explore sequence space far more rapidly [2,11]. Modern sequence models treat amino acid chains

K. L. Kabir et al. (Eds.): BICOB 2026, CCIS 2977, pp. 13–27, 2026.
https://doi.org/10.1007/978-3-032-26028-4_2

as one-dimensional strings similar to natural language [9] and use evolutionary information from large protein databases [29] to learn how sequence relates to structure and function. Two main approaches dominate current protein generation. Autoregressive models like ProGen [15] build sequences token-by-token, which captures local dependencies well but creates sequential bottlenecks that affect throughput for longer proteins. Masked language models like ESM2 [14] get around this through bidirectional self-attention that processes sequences in parallel. But attention mechanisms scale quadratically with sequence length [25], which becomes expensive for longer design targets. These efficiency problems have driven interest in alternative architectures that preserve global context more cheaply. FNet [13], originally built for natural language processing, swaps self-attention for Fourier transforms in token mixing, giving it O(n log n) complexity. FNet works well on NLP tasks, but has not been tested for protein generation. It's unclear whether Fourier-based mixing can handle the biochemical and structural constraints needed to design functional proteins. We compare FNet, ESM2, and ProGen for protein sequence generation under identical experimental conditions. Using a curated kinase dataset from UniProt [23], we measure computational efficiency, sequence-level statistics, and structural plausibility. Our goal is to determine whether Fourier-based token mixing represents a viable alternative to established autoregressive and attention-based approaches for generative protein modeling.

2   Related Work

2.1   Autoregressive Protein Language Models

Autoregressive models generate protein sequences token by token in a left-to-right manner. ProGen [15] demonstrated that causal language modeling can capture local dependencies and generate functional proteins. Later models such as ProGen2 and ProtGPT2 [6] scaled to billions of parameters and hundreds of millions of sequences, improving generative fluency. However, these models require extensive computational resources and specialized sampling strategies, hindering controlled comparison under identical hardware and data settings. Moreover, ProGen2's conditional token embeddings complicate architectural comparison with non-autoregressive models, while sequential decoding introduces computational bottlenecks that scale poorly with sequence length.

2.2   Masked Language Models for Protein Representation

Bidirectional masked language models have become a dominant paradigm for protein representation learning. The ESM family [14,21] and ProBERT [5] learn contextual representations via masked token prediction over large protein databases. ESM2, in particular, has shown strong performance in structure prediction and function annotation through bidirectional attention and parallel sequence processing [14]. However, ESM2 is primarily designed for representation learning, and its masked language modeling objective does not directly support de novo sequence generation without additional methodological adaptations.

2.3   Generative Applications of Protein Language Models

Protein language models have recently been extended to generative applications, primarily through diffusion-based frameworks. DiMA [18] performs diffusion over ESM2 embeddings, while Zhao et al. [30] and Hallee et al. [8] demonstrated biologically active and diverse sequences using similar approaches. Trinquier et al. [24] showed that MSA Transformer can generate sequences via iterative masking, indicating that masked language models possess inherent generative capacity. In contrast, the present work employs direct iterative masked-token resampling with ESM2, avoiding diffusion schedules or continuous embedding-space operations.

2.4   Efficiency and Alternative Models

As protein design targets grow in length, computational efficiency has become increasingly important [1, 25]. The quadratic complexity of self-attention has motivated alternative architectures. FNet [13] replaces self-attention with Fourier transforms for token mixing, achieving theoretical $O(n \log n)$ complexity and competitive performance on NLP benchmarks. Despite its efficiency advantages, the applicability of Fourier-based mixing to protein sequence generation and its ability to capture biochemical constraints remain unexplored.

2.5   Structure-Conditioned Generative Models

Beyond sequence-only models, structure-conditioned approaches have demonstrated strong design capabilities. RFdiffusion [26] generates three-dimensional protein backbones via diffusion, achieving high-quality designs but requiring precomputed structural priors and substantial computational resources. Chroma [10] similarly combines backbone, sequence, and side-chain generation within a diffusion framework, enabling precise structural control at high computational cost. ProteinMPNN [4] addresses conditional sequence design via inverse folding, differing fundamentally from unconditional sequence generation models.

3   Methodology

3.1   Dataset Preparation

ESM2, FNet, and ProGen were fine-tuned under matched optimization settings on 10,000 non-redundant kinase sequences and evaluated for runtime efficiency, generative fidelity, and biological quality (Fig. 1). All stages used deterministic seeds, version-locked dependencies, and automated logging to ensure reproducibility.

Kinase sequences were obtained from UniProt (2023 release; ~1.1M entries) and selected due to their functional diversity and therapeutic relevance.

Using BioPython (v1.85) [3], sequences with non-canonical residues, duplicates, and lengths outside the 1–1408 residue range were removed. From the

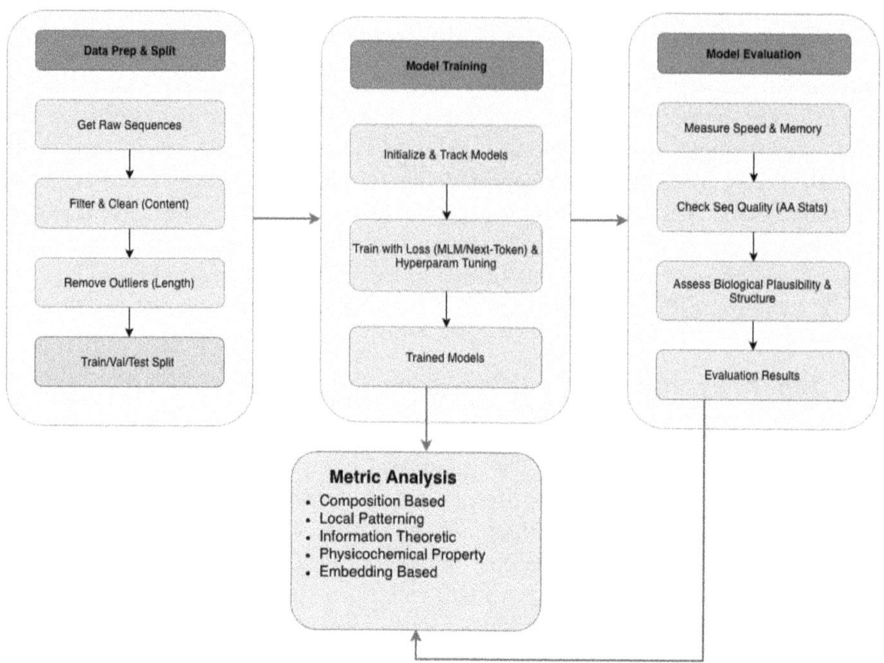

Fig. 1. Overview of the evaluation pipeline. Raw protein sequences are filtered and split into training and held-out sets. Models are then fine-tuned and monitored using their respective loss functions. Runtime efficiency, sequence-level quality, and biological plausibility metrics are assessed on generated sequences, and results are aggregated across evaluation dimensions.

cleaned set, 10,000 sequences were randomly sampled (seed − 42) and clustered with MMseqs2 (v14.7e284) [22] at 40% sequence identity and 80% coverage, yielding approximately 1,400 clusters. These thresholds follow standard protein de-redundancy practice and were chosen to ensure dataset de-redundancy rather than optimize downstream model performance; sensitivity analysis is deferred to future work [19].

Clusters were split into training, validation, and test sets using an 0.8/0.1/0.1 ratio (approximately 8,000 / 1,000 / 1,000 sequences), with no homolog overlap. All preprocessing was performed using BioPython, NumPy, and MMseqs2 on Ubuntu 22.04 LTS.

Labels were deterministically derived from UniProt hierarchical annotations. No manual annotation was performed; inter-annotator agreement is therefore not applicable. Consistency is ensured through the use of a single curated annotation source.

3.2 Model Architectures

Three models were evaluated, each representing a different approach to contextual encoding. **ProGen** (progen-small2; 151M parameters) is a causal Transformer that predicts each residue from its preceding context [15]. **ESM2** (facebook/esm2-t6-8M-UR50D) is an encoder-only Transformer that applies bidirectional self-attention [14]. **FNet** (google/fnet-base) replaces attention with two-dimensional Fast Fourier Transforms to achieve $O(n \log n)$ token mixing [13].

All models were selected at their respective base or smallest available checkpoint sizes to preserve architectural diversity while keeping model scale practical for single-GPU fine-tuning and controlled comparison. The ESM2–t6–8M checkpoint was used because it is the smallest encoder-only ESM variant with full bidirectional attention, allowing comparison without the overhead of larger versions (35M–650M parameters). The fnet-base checkpoint provides the standard Fourier-mixing baseline with a parameter scale comparable to ESM2–t6–8M, enabling a direct comparison between attention-based and FFT-based encoders. ProGen serves as the autoregressive reference model, reflecting the common left-to-right generation strategy in protein language modeling.

All sequences were standardized to a length of 512 residues by truncation or padding. ESM2 and FNet shared a character-level tokenizer and vocabulary, whereas ProGen used its native autoregressive tokenizer. A masking probability of $p = 0.15$ was applied for ESM2 and FNet. FNet used `float32` FFT operations to avoid type conflicts, and gradient checkpointing was enabled for all models to reduce GPU memory usage.

3.3 Training Setup

Fine-tuning was performed using the Hugging Face Trainer [27] (PyTorch 2.1) [20]. All models were trained for ten epochs using deterministic random seeds. Training was conducted on a RTX 4090 (24 GB) GPU, requiring batch sizes of 1 or 2 with gradient accumulation.

Hyperparameters were selected via a constrained grid search over values commonly reported in the protein language modeling literature, including learning rates $\{3 \times 10^{-5}, 5 \times 10^{-5}, 7 \times 10^{-5}\}$ $\{3 \times 10^{-5}, 5 \times 10^{-5}, 7 \times 10^{-5}\}$, weight decay $\{0.01, 0.05\}$ $\{0.01, 0.05\}$, batch sizes $\{1, 2\}$ $\{1, 2\}$, and gradient accumulation steps $\{4, 8\}$ $\{4, 8\}$, with a fixed maximum sequence length of 512.

Final settings were chosen from this grid and held constant across models wherever possible to avoid implicit tuning advantages. The focus was on architectural comparability under matched optimization effort rather than exhaustive hyperparameter optimization. All training logs, checkpoints, and hyperparameter configurations were recorded via centralized CSV-based tracking.

3.4 Generation Methodology

For the encoder only models (ESM2 and FNet), an iterative masked token resampling procedure was utilized to allow for sequence generation. A completely masked sequence of length 512 was provided as input to the model. At every refinement, a mask ratio of 15% of positions were remasked according to a uniform random distribution, and these positions are resampled using the model's token distribution. This was repeated for 40 refinement steps, a value selected empirically as a point at which perplexity improvements saturated, and further refinement provided negligible changes to sequence quality. The stopping criterion was therefore fixed at 40 iterations for all sequences to ensure consistent generation depth across models. For ProGen, sequences were generated autoregressively using nucleus sampling until a length of 510 residues was reached.

3.5   Evaluation

3.5.1   Training Evaluation

During training, model performance was tracked using the internal validation perplexity described above. For ESM2 and FNet, the corresponding loss was computed over the masked positions M as

$$ L_{\mathrm{MLM}} = -\frac{1}{|M|} \sum_{i \in M} \log P(x_i \mid x_{\backslash M}), \tag{1} $$

from which $\mathrm{PPL_{val}}$ was obtained using Eq. 4. For ProGen, the autoregressive objective directly aligns with its perplexity definition in Eq. 5.

3.5.2   Generation Evaluation

After fine-tuning, each model was evaluated on generation-time self-perplexity. For ESM2 and FNet, each token t in a sequence was masked individually and the pseudo log-likelihood (PLL) was computed as

$$ \mathrm{PLL_{ESM2}}(x_{1:T}) = -\frac{1}{T} \sum_{t=1}^{T} \log P_{\mathrm{ESM2}}\big(x_t \mid x_{1:T}^{(\backslash t)}\big). \tag{2} $$

For ProGen, generation-time perplexity follows the same autoregressive formulation as the validation metric:

$$ \mathrm{PPL_{ProGen}} = \exp\left(-\frac{1}{T} \sum_{t=1}^{T} \log P_{\mathrm{ProGen}}(x_t \mid x_{<t}) \right). \tag{3} $$

All metrics—including loss, perplexity, runtime, and training state—were logged throughout fine-tuning and generation. These logs were later used for statistical comparison across ESM2, FNet, and ProGen.

Perplexity values for ESM2 and FNet are computed using pseudo log-likelihood, while ProGen uses autoregressive likelihood, and these values are therefore not directly comparable across architectures.

3.6   Sequence Generation

Each of the fine tuned models generated 2,000 synthetic kinase like sequences (510 residues). ESM2/FNet: Iterative masked token resampling, temperature = 1.0, top-p = 1.0, 40 refinement steps). ProGen: Autoregressive nucleus sampling (temperature = 1.0) until 510 residues. Runtime, memory usage, and throughput were profiled using the PyTorch Profiler API [20]. Generated sequences and metadata (FASTA + CSV) were saved, and Sequence perplexity was calculated offline. All experiments used seed = 42 for full reproducibility.

3.7   Biological Evaluation Metrics

We evaluated sequences using standard metrics capturing amino acid composition [2,15,23,29], local residue motifs via 3-mer and 5-mer Jensen–Shannon divergence [29] and physicochemical properties such as hydrophobicity [12] charge and molecular weight [7]. We also assessed repetitive and low-complexity content [6,15,28]. Finally, embedding-based confidence scores derived from untrained models [5,14,16,17] provided an additional unsupervised signal of plausibility.

4   Results and Discussions

We compared two protein sequence generation models, the autoregressive **Pro-Gen** and the attention-based **ESM2**, to **FNet** using multiple quantitative metrics that capture both computational efficiency and sequence quality. We report results across three main dimensions:

- Computational efficiency and runtime behavior
- Biological plausibility and sequence-level quality metrics
- Robustness analyses via ablation studies over refinement and decoding strategies

   All values represent averages over three independent fine-tuning runs using identical random seeds and evaluation protocols.

4.1   Computational Efficiency

ESM2 was faster than both FNet and ProGen due to its parallel, non-autoregressive masked generation and highly optimized attention implementations. ProGen's sequential autoregressive decoding introduces inherent latency, while FNet, despite being theoretically fully parallelizable, suffers from practical inefficiencies. Profiling showed that FFT operations incur substantial overhead and suboptimal GPU utilization, limiting kernel fusion and increasing global memory movement. Consequently, FNet does not fully realize its theoretical parallel efficiency at the evaluated sequence length (Fig. 2).

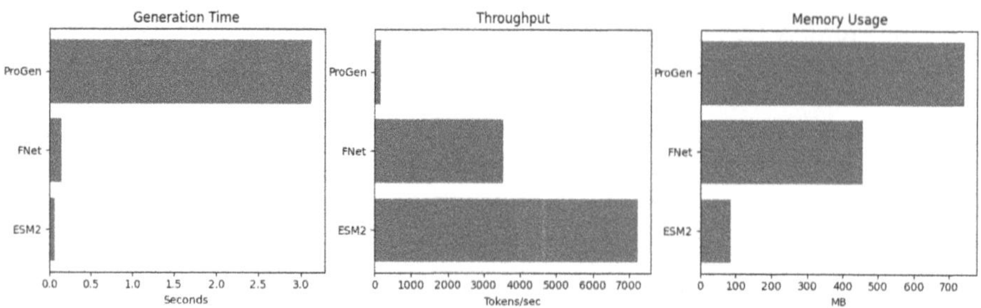

Fig. 2. Comparison of generation time, throughput, and memory use across the three models.

Relative to ProGen, FNet achieved a 95.4% reduction in generation time (0.14 s vs. 3.12 s per sequence) and a 38.7% reduction in GPU memory usage (456 MB vs. 745 MB). However, FNet remained approximately 2.1× slower than ESM2 (0.14 s vs. 0.07 s) and required about 5.4× more memory (456 MB vs. 85 MB) (Table 1).

Table 1. Performance metrics (mean ± SD).

Metric	ProGen	FNet	ESM2
Generation Time (s)	3.1235 ± 0.2756	0.1449 ± 0.0019	0.0705 ± 0.0015
Avg Token Latency (ms)	6.1649 ± 0.1468	3.4963 ± 0.0374	1.7191 ± 0.0373
Throughput (tokens/s)	162.2038 ± 4.7843	3533.9612 ± 45.7620	7232.0411 ± 114.7629
Sequence Length	505.3520 ± 43.6786	471.4865 ± 1.1893	509.9995 ± 0.0224
Self-Perplexity	4.2205 ± 11.1655	4.7887 ± 1.0289	18.4220 ± 1.2131
Memory Usage (MB)	744.5309 ± 12.7478	456.6045 ± 0.0000	85.1353 ± 0.0000

4.2   Sequence Quality

Because perplexity is defined differently for masked language models and autoregressive models, cross-model comparisons in this section rely on sequence-level statistics and efficiency metrics rather than absolute perplexity values. We evaluated the three models on a diverse set of sequence and structure level metrics to quantify their ability to generate naturalistic kinase-like sequences. Reference ("Train") sequences from the kinase cluster dataset were used as baselines (Fig. 3).

4.2.1   Sequence-Level Statistics

Across amino acid composition, entropy, and sequence complexity, ESM2 produced sequences most consistent with the training distribution, followed closely by FNet. ProGen showed the largest departures (Table 2).

Table 2. Profiling and runtime comparison (total sequence generation time).

Section	FNet Avg (ms)	ESM Avg (ms)	FNet (%)	ESM (%)
Sequence Generation Total	1348.59	1389.62	32.84	32.84
Iteration Total	33.02	34.17	32.43	32.30
Forward Pass	31.17	31.49	30.36	31.49
Perplexity Calculation	34.10	17.85	0.83	0.84

* Profiling was run for 500 sequences for both models.

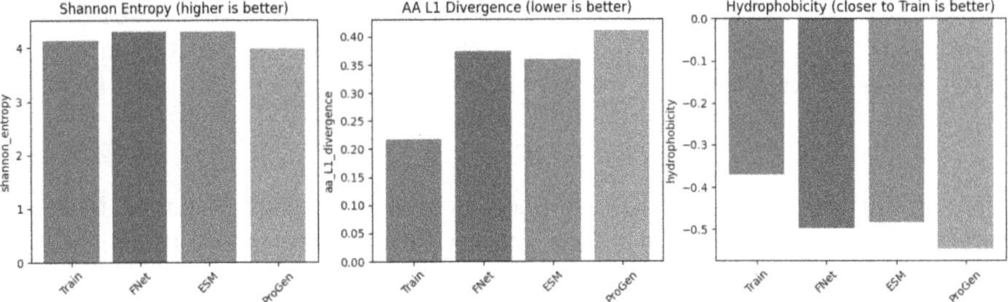

Fig. 3. Sequence-level metrics comparing generated proteins to the training set for entropy, amino acid divergence, and hydrophobicity.

Key Sequence Findings:

- **Amino acid L1 divergence**: ESM2 (0.3587) and FNet (0.3733) were closer to the training baseline (0.2167) than ProGen (0.4096).
- **Shannon entropy**: ESM2 and FNet (both approximately 4.29) were closer to the training value (4.1288) than ProGen (3.9701).
- **Complexity**: Repetitive content was lowest for ESM2 and FNet (approximately 0.002), significantly lower than ProGen (0.01699) and the training value (0.00758). Low- complexity fractions showed a similar pattern, with ESM2 and FNet being extremely low
- **Local Motif Similarity**: ProGen showed the best k-mer JSD scores (e.g., kmer3 JSD: 0.7805) compared to FNet (0.8531) and ESM2 (0.8423), suggesting ProGen preserved short motifs more effectively (Table 3).

4.2.2 Embedding-Based Plausibility Proxy

The sequences generated by ESM2 and FNet showed the highest, but most compact, confidence distributions, suggesting that these sequences fall into smooth, high likelihood regions of the respective embedding spaces of the language model. In contrast, reference kinase sequences exhibited lower scores and greater variability, reflecting genuine biological diversity and heterogeneity. Embedding-based confidence scores quantify the coherence of the internal language model rather than the biological correctness or functional validity. Elevated confidence values for generated sequences may therefore arise from reduced diversity and

Table 3. Biological plausibility assessment (mean ± SD).

Metric	Train	FNet	ESM	ProGen
Length	573.008 ± 258.487	471.487 ± 1.189	510.000 ± 0.022	505.352 ± 43.668
aa_L1_divergence	0.217 ± 0.072	0.373 ± 0.039	0.359 ± 0.039	0.410 ± 0.152
kmer3_JSD	0.753 ± 0.067	0.854 ± 0.007	0.842 ± 0.007	0.781 ± 0.022
kmer5_JSD	0.986 ± 0.007	0.999 ± 0.000	0.999 ± 0.000	0.997 ± 0.001
shannon_entropy	4.129 ± 0.066	4.293 ± 0.009	4.293 ± 0.009	3.970 ± 0.211
hydrophobicity	-0.370 ± 0.210	-0.497 ± 0.134	-0.484 ± 0.133	-0.546 ± 0.316
net_charge	0.027 ± 0.027	0.050 ± 0.023	0.049 ± 0.023	0.041 ± 0.035
molecular_weight	$(7.450\text{e}4 \pm 3.324\text{e}4)$	$(6.447\text{e}4 \pm 6.608\text{e}2)$	$(6.956\text{e}4 \pm 6.737\text{e}2)$	$(6.340\text{e}4 \pm 5.725\text{e}3)$
repetitive_content	0.008 ± 0.007	0.002 ± 0.002	0.003 ± 0.002	0.017 ± 0.012
low_complexity_fraction	0.008 ± 0.016	0.000 ± 0.000	0.000 ± 0.000	0.025 ± 0.029

over regularized representations, a phenomenon previously observed in generative language models that produce syntactically well-formed but semantically limited outputs [5, 17].

Table 4. Embedding-based structural confidence scores (mean ± SD).

Model	Kinases (Train)	FNet	ProGen	ESM
ProtBERT	0.7077 ± 0.1530	0.9670 ± 0.0170	0.9411 ± 0.0385	0.9753 ± 0.0120
ESM	0.8356 ± 0.0589	0.9718 ± 0.0090	0.9089 ± 0.0485	0.9741 ± 0.0090

4.3   Ablation Studies

We conducted targeted ablation studies to assess the robustness of masked protein sequence generation and to ensure that the reported quality and efficiency trends are not artifacts of specific refinement schedules or decoding choices. Unless otherwise stated, all experiments used fixed model checkpoints, a sequence length of 510–512 residues, identical random seeds, and a consistent GPU environment (RTX 4090) (Table 4).

4.3.1   Ablation of Iterative Refinement Parameters

We evaluated the sensitivity of masked sequence generation to the number of refinement steps and the per-iteration mask ratio. For each configuration, 500 sequences were generated per model, and mean values with standard deviations are reported.

The number of refinement iterations was varied in $\{10, 20, 40, 80\}$ while fixing the mask ratio at 0.15. FNet converges rapidly, with self-perplexity remaining approximately constant (~ 4.64) across all step counts, indicating that most gains are achieved in early iterations. Additional refinement primarily increases runtime, which scales linearly with the number of steps. In contrast, ESM2 benefits from deeper refinement, with self-perplexity decreasing monotonically from 17.93 at 10 steps to 16.87 at 80 steps, although improvements largely saturate beyond 40 iterations.

Varying the mask ratio in $\{0.10, 0.15, 0.20\}$ at a fixed refinement depth of 40 steps results in only modest changes for both models. GPU memory usage and output sequence length remain invariant across all configurations, confirming that refinement hyperparameters influence runtime but not memory footprint (Fig. 4).

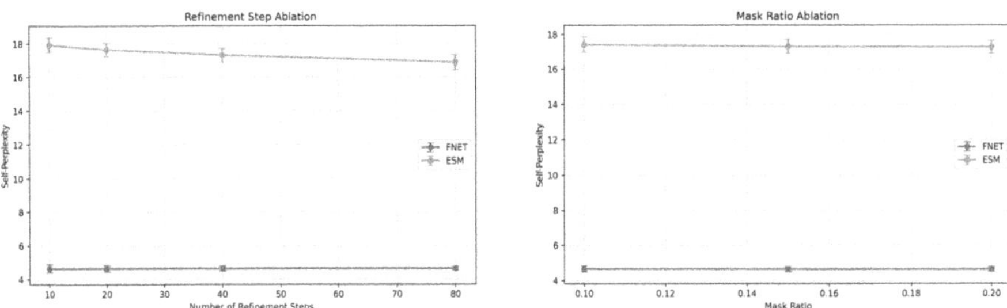

Fig. 4. Ablation of iterative refinement parameters. *Left:* Effect of refinement depth on self-perplexity, showing saturation beyond approximately 40 steps. *Right:* Effect of per-iteration mask ratio on self-perplexity at a fixed refinement depth of 40 steps.

Overall, these results indicate that masked generation is robust to refinement hyperparameters, motivating the use of a fixed schedule of 40 steps and 15% masking in subsequent experiments.

4.3.2   Sensitivity of Efficiency Metrics to Decoding Strategy

To disentangle architectural efficiency from decoding heuristics, we evaluated runtime and peak GPU memory usage across multiple sampling strategies while holding model architecture, sequence length, and hardware constant. Mean runtime, standard deviation, and peak memory usage for each configuration are reported in Table 5.

Across all strategies, runtime behavior is primarily dictated by architectural paradigm rather than sampling choice. FNet exhibits stable generation times with low variance, indicating that Fourier mixing dominates computational cost. ESM2 consistently achieves the lowest latency due to optimized attention kernels, while ProGen exhibits substantially higher latency across all strategies due to sequential autoregressive decoding.

Memory usage remains invariant across decoding strategies for masked models, whereas ProGen shows strong sensitivity to sampling complexity, with peak GPU memory increasing by more than fourfold under Top-k decoding. These results confirm that the efficiency trends reported earlier are robust to decoding strategy and reflect intrinsic architectural differences.

Table 5. Performance metrics for different models and sampling strategies

Model	Strategy	Mean Runtime (s)	Std Dev (s)	Peak Memory (MB)
FNet	Greedy	0.48	±0.011	456.73
	Pure Sampling	0.42	±0.004	456.73
	Top-k (k = 10)	0.59	±0.003	456.73
ESM2	Greedy	0.27	±0.003	51.39
	Pure Sampling	0.31	±0.003	51.39
	Top-k (k = 10)	0.36	±0.003	51.39
ProGen	Greedy	3.21	±0.018	697.30
	Pure Sampling	3.22	±0.008	1286.17
	Top-k (k = 10)	3.23	±0.007	3051.92

4.4   Summary Comparison

Taken together, these results highlight **complementary strengths** across
ESM2, **FNet**, and **ProGen**, rather than identifying a single uniformly superior
model. The main observations are:

- **ESM2** performs strongly across most criteria, combining low latency and high
  throughput with reference-like amino acid statistics, reduced repetition, and
  high embedding-based confidence scores, making it a robust general-purpose
  model for protein sequence generation.
- **FNet** achieves sequence-level and structural proxy metrics largely comparable to ESM2, while offering improved speed and memory efficiency relative
  to ProGen. Despite practical FFT overheads partially offsetting its theoretical $O(n \log n)$ complexity, FNet represents a promising direction for efficient
  long-context protein modeling.
- **ProGen** exhibits distinct generative behavior, including stronger local k-mer
  alignment, alongside higher latency and memory usage, greater divergence in
  global statistics, and lower predicted foldability. This suggests that ProGen
  captures complementary aspects of sequence realism not fully reflected by
  global or structure-oriented metrics.

Overall, compositional, informational, and structural analyses indicate that
ESM2 and **FNet** tend to generate sequences closer to naturally occurring,
structurally plausible proteins, while **ProGen** emphasizes alternative generative properties. To substantiate these conclusions, we conducted targeted ablation studies on refinement hyperparameters and decoding strategies, showing
that sequence quality is largely robust to masking schedules and that observed
efficiency trends are dominated by architectural choices rather than decoding
heuristics.

5   Conclusion

This work highlights complementary strengths across three architectural paradigms for protein sequence generation. **ESM2** achieves the strongest overall balance between generation speed, calibration, and biological signal, reflecting the maturity of attention-based modeling. **FNet**, evaluated in this domain for the first time, produces sequence statistics and structural confidence comparable to ESM2 while offering substantial efficiency gains over autoregressive decoding, although practical FFT overheads limit realization of its full theoretical advantage. **ProGen** preserves short-range motifs effectively but incurs high computational cost due to sequential decoding.

Overall, attention-based and Fourier-based parallel encoders should be viewed as complementary rather than competing approaches, occupying different points in the quality–efficiency trade-off. Fourier mixing emerges as a promising component for scalable protein sequence modeling, particularly when combined with optimized or hybrid architectures.

6   Limitations and Future Work

This study focuses on a single protein family (kinases), and results may not generalize to structurally diverse families such as multi-domain or intrinsically disordered proteins. To enable controlled comparison, sequences were standardized to a fixed length, potentially masking behaviors at longer contexts. The evaluation was limited to three architectures, and larger model checkpoints, efficient attention variants, state-space models, and hybrid designs were not explored due to computational constraints.

Biological assessment was restricted to sequence-level statistics and embedding-based confidence measures, as more computationally intensive structure-level analyses and larger-scale evaluations were not feasible under the available compute budget. Future work includes improving FFT kernel efficiency, extending benchmarks to additional protein families and longer sequences, incorporating structure-aware evaluations, and exploring hybrid or conditional generation strategies to better characterize the roles of Fourier-based, attention-based, and autoregressive models in protein design.

Acknowledgment. The authors thank **ArcLabs, Department of CSE-AIML, Pes University** for providing GPU support throughout the course of this research.

A   Appendix

This appendix describes the evaluation metrics used during fine-tuning and generation for all models.

A.1   Validation Perplexity

For ESM2 and FNet, validation perplexity was derived directly from the masked language modeling (MLM) loss. Given the loss L_{MLM}, perplexity was computed as

$$\mathrm{PPL}_{\mathrm{val}} = \exp(L_{\mathrm{MLM}}). \tag{4}$$

ProGen, being autoregressive, uses the standard left-to-right formulation. The model conditions each token on all previous tokens, and perplexity is

$$\mathrm{PPL}_{\mathrm{val}}^{\mathrm{AR}} = \exp\left(-\frac{1}{T}\sum_{t=1}^{T}\log P_{\mathrm{ProGen}}(x_t \mid x_{<t})\right). \tag{5}$$

References

1. Çelik, M.H., Xie, X.: Efficient inference, training, and fine-tuning of protein language models. iScience **28**(10) (2025)
2. Chen, Y., et al.: Deep generative model for drug design from protein target sequence. J. Cheminf. **15**(1), 38 (2023)
3. Cock, P.J., et al.: Biopython: freely available python tools for computational molecular biology and bioinformatics. Bioinformatics **25**(11), 1422 (2009)
4. Dauparas, J., et al.: Robust deep learning-based protein sequence design using proteinmpnn. Science **378**(6615), 49–56 (2022)
5. Elnaggar, A., et al.: Prottrans: towards cracking the language of life's code through self-supervised deep learning and high performance computing **10**(2007). arxiv 2020. arXiv preprint arXiv:2007.06225
6. Ferruz, N., Höcker, B.: Controllable protein design with language models. Nat. Mach. Intell. **4**(6), 521–532 (2022)
7. Gasteiger, E., et al.: Protein identification and analysis tools on the expasy server. In: The Proteomics Protocols Handbook, pp. 571–607. Springer, Totowa, NJ (2005)
8. Hallee, L., Rafailidis, N., Bichara, D.B., Gleghorn, J.P.: Diffusion sequence models for enhanced protein representation and generation. arXiv preprint arXiv:2506.08293 (2025)
9. Heinzinger, M., et al.: Bilingual language model for protein sequence and structure. NAR Genomics Bioinf. **6**(4), lqae150 (2024)
10. Ingraham, J.B., et al.: Illuminating protein space with a programmable generative model. Nature **623**(7989), 1070–1078 (2023)
11. Koh, H.Y., et al.: AI-driven protein design. Nat. Rev. Bioeng., 1–23 (2025)
12. Kyte, J., Doolittle, R.F.: A simple method for displaying the hydropathic character of a protein. J. Mol. Biol. **157**(1), 105–132 (1982)

13. Lee-Thorp, J., Ainslie, J., Eckstein, I., Ontanon, S.: Fnet: mixing tokens with fourier transforms. In: Proceedings of the 2022 Conference of the north American Chapter of the Association for Computational Linguistics: Human Language Technologies, pp. 4296–4313 (2022)
14. Lin, Z., et al.: Evolutionary-scale prediction of atomic-level protein structure with a language model. Science **379**(6637), 1123–1130 (2023)
15. Madani, A., et al.: Progen: language modeling for protein generation. arXiv preprint arXiv:2004.03497 (2020)
16. Marquet, C., et al.: Embeddings from protein language models predict conservation and variant effects. Hum. Genet. **141**(10), 1629–1647 (2022)
17. Meier, J., Rao, R., Verkuil, R., Liu, J., Sercu, T., Rives, A.: Language models enable zero-shot prediction of the effects of mutations on protein function. Adv. Neural. Inf. Process. Syst. **34**, 29287–29303 (2021)
18. Meshchaninov, V., et al.: Diffusion on language model encodings for protein sequence generation. arXiv preprint arXiv:2403.03726 (2024)
19. Mirdita, M., Von Den Driesch, L., Galiez, C., Martin, M.J., Söding, J., Steinegger, M.: Uniclust databases of clustered and deeply annotated protein sequences and alignments. Nucleic Acids Res. **45**(D1), D170–D176 (2017)
20. Paszke, A., et al.: Pytorch: an imperative style, high-performance deep learning library. Adv. Neural Inf. Process. Syst. **32** (2019)
21. Rives, A., et al.: Biological structure and function emerge from scaling unsupervised learning to 250 million protein sequences. Proc. Natl. Acad. Sci. **118**(15), e2016239118 (2021)
22. Steinegger, M., Söding, J.: Mmseqs2 enables sensitive protein sequence searching for the analysis of massive data sets. Nat. Biotechnol. **35**(11), 1026–1028 (2017)
23. The UniProt Consortium: Uniprot: the universal protein knowledgebase in 2023. Nucleic Acids Res. **51**(D1), D523–D531 (2023). https://doi.org/10.1093/nar/gkac1052
24. Trinquier, J.: Data-driven generative modeling of protein sequence landscapes and beyond. Ph.D. thesis, Sorbonne Université (2023)
25. Wang, L., Zhang, H., Xu, W., Xue, Z., Wang, Y.: Deciphering the protein landscape with protflash, a lightweight language model. Cell Reports Phys. Sci. **4**(10) (2023)
26. Watson, J.L., et al.: De novo design of protein structure and function with rfdiffusion. Nature **620**(7976), 1089–1100 (2023)
27. Wolf, T., et al.: Transformers: state-of-the-art natural language processing. In: Proceedings of the 2020 Conference on Empirical Methods in Natural Language Processing: System Demonstrations, pp. 38–45 (2020)
28. Wootton, J.C., Federhen, S.: Analysis of compositionally biased regions in sequence databases. In: Methods in Enzymology, vol. 266, pp. 554–571. Elsevier, San Diego, CA (1996)
29. Zhang, Z., Wayment-Steele, H.K., Brixi, G., Wang, H., Kern, D., Ovchinnikov, S.: Protein language models learn evolutionary statistics of interacting sequence motifs. Proc. Natl. Acad. Sci. **121**(45), e2406285121 (2024)
30. Zhao, L., et al.: Protein a-like peptide design based on diffusion and esm2 models. Molecules **29**(20), 4965 (2024)

A Structure-Aware Attention Mechanism for Protein Function Prediction

Md Tahmid Islam[✉][iD] and Changhui Yan[iD]

North Dakota State University, Fargo, ND, USA
`{mdtahmid.islam,changhui.yan}@ndsu.edu`

Abstract. Protein structure provides the foundation for a protein's function. Therefore, incorporating structural information is essential in methods for protein function prediction. The attention layer in transformer models has proven to be a powerful tool for encoding rich sequential information in a wide range of machine learning tasks. Although originally developed for sequence inputs, the attention mechanism can be extended to integrate structural information. In this paper, we present a structure-aware attention mechanism that incorporates protein structural information for protein function prediction. We demonstrate that integrating structure-derived features significantly improves the performance of predicting protein functions. Our method outperforms the current state-of-the-art in comprehensive benchmark evaluations, with gains up to 29.6% in F1 score and 31.1% in area under the precision-recall curve (AUPRC). This work introduces a new approach to leveraging the transformer architecture in contexts where structural information is critical, and the framework can be further extended to integrate other types of non-sequential information.

Keywords: Structure-aware · Attention Layer · Transformer · Protein Function Prediction

1 Introduction

Proteins perform various functions. Many systems have been developed to systematically categorize protein functions. Among them the most widely used are Gene Ontology (GO) annotations [6] and Enzyme Commission (EC) numbers [17]. GO categorize gene products hierarchically using three independent ontologies, each corresponding to a different aspect: molecular function (MF), which describes the activity performed by the protein; cellular component (CC), which specifies its location in the cell; and biological process (BP), which defines the broader pathways and networks in which the protein is involved. Each node in these ontologies is associated with a unique GO term and may be subdivided into subclasses that are represented as child nodes in the hierarchy [6].

Similarly, EC numbers classify enzymes based on the chemical reactions they catalyze, using a four-digit code separated by periods. The first three digits

K. L. Kabir et al. (Eds.): BICOB 2026, CCIS 2977, pp. 28–40, 2026.
https://doi.org/10.1007/978-3-032-26028-4_3

indicate the class, subclass, and sub-subclass of the reaction, while the fourth digit uniquely identifies the specific enzyme within the sub-subclass [17].

Using these classification systems, proteins can be annotated using GO terms or EC numbers either manually based on expert knowledge from laboratory experiments or computationally through prediction methods. While manual annotations are more reliable, the process is both costly and time-consuming, making it impractical for large-scale annotations [25]. As a result, despite the UniProt database containing over 250 million protein sequences, only about 0.2% of them have been manually annotated to date [1]. This huge gap between known protein sequences and experimentally validated functions highlights the urgent need for accurate and efficient computational methods for protein function prediction.

Over the past decade, many sequence-based and structure-based methods have been developed. Sequence-based methods aim to identify patterns and subtle features embedded in protein sequences [16]. A simple sequence-based method is to use tools like BLAST [2] to search for similar sequences and transfer annotations from a protein with known functions to another one that shares high sequence similarity under the assumption that similar sequences often lead to similar functions. Recent models incorporate deep learning to capture more complex sequence relationships. For example, DeepPred [20] uses neural networks trained on various sequence-derived features, while DeepGOPlus [14] combines convolutional neural networks (CNNs) with sequence similarity search to improve accuracy. HiFun [27] treats protein sequences as a form of language, embedding them using the BLOSUM62 matrix and FastText, then passes them through convolutional neural network (CNN) and bidirectional long-and short-term memory (BiLSTM) [29] layers with self-attention to extract richer contextual features. Although these studies demonstrate strong performance, the performance of sequence-based methods often degrade when proteins share low sequence similarity, as many functionally related proteins may not share high sequence similarity.

In contrast, structure-based methods utilize the three-dimensional (3D) structure of proteins, often represented as contact maps that capture the spatial relationships between residues [16]. Among these, DeepFRI [9] is a notable method that combines sequence and structure information for function prediction. It pre-trains a language model on the Pfam database [8] using a recurrent neural network with long short-term memory [10] to extract residue-level sequence features. These features are then combined with contact maps in a Graph Convolutional Network (GCN) [13], achieving significant improvements in protein function prediction.

GAT-GO [15] advances this idea by replacing the GCN with a Graph Attention Network (GAT) [24] and using contact maps predicted by RaptorX [26] instead of experimentally determined structures. Other models such as HEAL [11], and Struct2GO [12] further extend this approach using graph attention mechanisms or hierarchical pooling, often incorporating predicted structures from AlphaFold2 [22].

While these methods employ neural networks or graph-based approaches, our work takes a different approach by utilizing a transformer architecture. In recent years, transformer-based language models have gained increasing popularity in genomics and protein sequence analysis [30]. Models such as TALE [5], the large-scale model by Rives et al. [19], and ProteinBERT [4] have demonstrated the effectiveness of transformers in predicting protein function from sequence. Building on this foundation, we employ the pre-trained ProtBERT model [7], a BERT model pre-trained on the UniRef100 [21] dataset, as the sequence encoder. By directly integrating contact maps into the attention layer of the transformer, we introduce ProtCmapBERT, a model designed to improve the accuracy of protein function prediction.

2   Methodology

2.1   Training Datasets

We used the same training datasets curated and preprocessed by Gligorijević et al. [9]. To ensure non-redundancy, Gligorijević et al. clustered the protein chains at 95% sequence identity, and selected one high-quality representative from each cluster. The functions of proteins were annotated with Gene Ontology (GO) terms in one dataset and Enzyme Commission (EC) numbers in the other. The GO-annotated dataset includes approximately 30,000 protein samples, while the EC-annotated dataset includes around 15,000 samples. We trained three separate models on the GO-annotated dataset, one for each GO ontology: Molecular Function (MF), Biological Process (BP), and Cellular Component (CC). An additional model was trained using the EC-annotated dataset.

2.2   Contact Map Generation

We constructed contact maps using the same criteria as DeepFRI [9]. Specifically, two residues were defined as being in contact if the distance between their $C\alpha$ atoms was less than 10 Å. In such cases, the corresponding cell in the contact map was assigned a value of 1; otherwise, it was assigned a value of 0.

2.3   Model Architecture

The model training process consisted of four primary components: a tokenizer, the ProtBERT pre-trained layers, a custom structure-aware self-attention layer, and a classification head (Fig. 1). Protein sequences were tokenized and processed together with their corresponding contact maps as input to the model.

Our architecture was built on the pre-trained ProtBERT model to leverage its capacity in protein sequence embedding. We retained ProtBERT's embedding and encoder layers, but replaced the self-attention module in each transformer layer with our custom structure-aware attention module. The parameters of these layers and the query, key, value weights and biases of the attention module were initialized with the learned values from the pre-trained ProtBERT model.

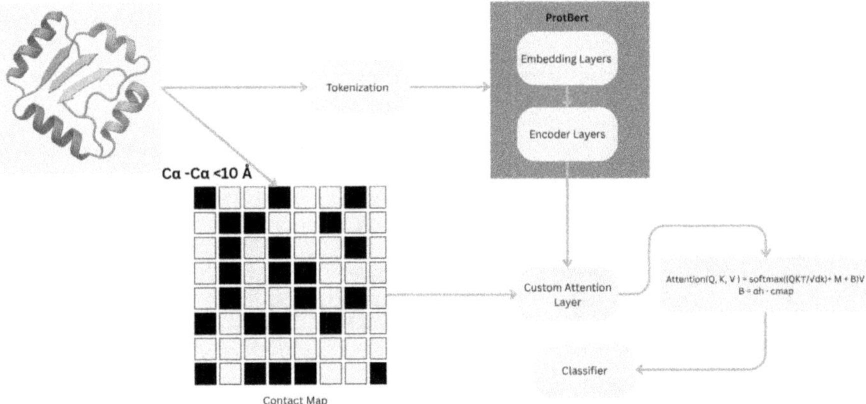

Fig. 1. The Structure-Aware Attention Mechanism. A contact map constructed from the Cα matrix is multiplied by a learnable vector, and the resulting product is added as a bias to the attention scores $(QK^{\top})$. This incorporation adjusts the attention weights to reflect spatial interactions between residues in the 3D structure.

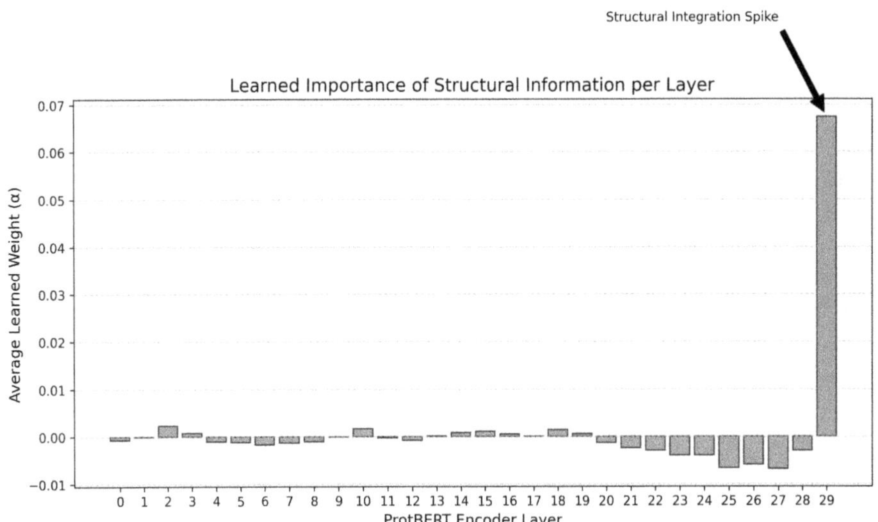

Fig. 2. Distribution of structural importance weights (α) across the 30-layer encoder. The model was initialized with structural bias across all layers, but optimization converged toward a "late-integration" strategy. Early layers (0–28) maintain near-zero weights, indicating a reliance on pre-trained sequence representations. A significant spike in Layer 29 (mean $\alpha \approx 0.07$) demonstrates that the network learned to utilize the 3D contact map as a critical final-stage feature to resolve functional site geometry before classification.

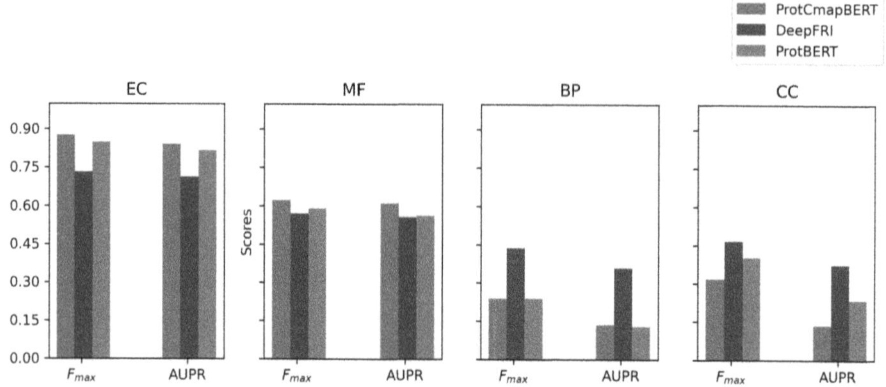

Fig. 3. A comparison of our ProtCmapBERT model (green) with DeepFRI (blue) and ProtBERT (red) in protein function prediction. The figure presents four separate subplots. The first subplot corresponds to protein functions annotated by Enzyme Commission (EC) numbers, while the remaining three correspond to protein classes annotated by GO terms in Molecular Function (MF), Biological Process (BP), and Cellular Component (CC), respectively. Within each subplot, two key evaluation metrics are shown: the protein-centric maximum F1 score (F_{max}) and the area under the precision-recall curve (AUPR). The results highlight the improved performance of the ProtCmapBERT method across the MF and EC categories. (Color figure online)

We modified the attention module to incorporate a learnable bias derived from the contact map. Specifically, we introduced a single learnable vector whose scalar elements correspond to individual attention heads. This vector was then multiplied by the contact map and added to the raw attention scores, providing a soft, head-specific bias that prioritizes attention toward contacting residues. Our approach was inspired by Xia et al. [28], who injected external semantic similarity scores between sentence pairs into the attention mechanism to guide BERT to focus more on semantically aligned token pairs.

The standard attention formulation proposed by Vaswani et al. [23] is:

$$\mathrm{Attention}(Q, K, V) = \mathrm{softmax}\left(\frac{QK^\top}{\sqrt{d_k}} + M\right)V \tag{1}$$

We modified it to:

$$\mathrm{Attention}(Q, K, V) = \mathrm{softmax}\left(\frac{QK^\top}{\sqrt{d_k}} + M + B\right)V \tag{2}$$

where:

- $B = \alpha_h \cdot \mathrm{cmap}$
- $\alpha_h \in \mathbb{R}^H$ is a learnable vector controlling the influence of the contact map for each of the H attention heads.
- $\mathrm{cmap} \in \{0, 1\}^{b \times L \times L}$ is the binary contact map for each protein, with b as the batch size and L as the sequence length.

The structural bias B was added before the softmax computation, thereby allowing the model to integrate both learned sequence dependencies and structurally-informed attention patterns. Although this mechanism was implemented across all encoder layers, the network was allowed to learn the relative importance of structural information at each stage. As shown in (Fig. 2) the model specifically optimized these weights to prioritize structural refinement in the final encoder layer.

A classification layer was added to map the outputs of the attention layer to classes of protein functions. The entire model was then fine-tuned for protein function prediction, using both protein sequence and structure as input, with functions labeled by GO terms or EC numbers.

To establish a baseline and conduct a systematic ablation study, we fine-tuned the original pre-trained ProtBERT model using the same training set and hyperparameters. Since ProtCmapBERT is initialized with ProtBERT weights and maintains an identical architecture, differing only by the inclusion of the contact map bias, the ProtBERT baseline serves as a direct ablation of the structural information component. This allows for the precise isolation of the performance gains attributable to 3D structural constraints.

2.4   Loss Function and Hyperparameter Tuning

We trained one model for each of the three Gene Ontologies and one for EC numbers. For each ontology, proteins were labeled with multiple GO terms (or EC numbers). We formulated this multi-label classification task as a set of binary classification problems, one per label, and used `BCEWithLogitsLoss` as the loss function. To address class imbalance, we assigned positive weights to each class (w_c), inversely proportional to their sample counts and clipped them between 1.0 and 10.0 to maintain training stability. This weighting scheme is defined as the following equation:

$$w_c = \max\left(1.0, \min\left(\frac{\overline{count}}{count_c + \epsilon}, 10.0\right)\right) \tag{3}$$

where $\overline{count}$ is the mean number of samples per class, $count_c$ is the sample count for a given class c, and ϵ is a small constant to avoid division by zero. This strategy reduces bias toward frequent labels and improves robustness in learning from underrepresented classes.

Hyperparameters were selected through iterative refinement based on observed model performance. We used the same validation sets as Gligorijević *et al.* [9], consisting of 10% the training data. To prevent overfitting, we applied early stopping with a patience of 10 epochs— i.e., training was terminated if the validation loss failed to improve for 10 consecutive epochs.

For optimization, we used the Adam optimizer with a learning rate of 7×10^{-6}, a batch size of 16, and a dropout rate of 0.01. The model was trained for up to 200 epochs. Input sequences were padded to a maximum length of 1000, corresponding to the longest sequence in the training set.

2.5   Availability

The source code for training the ProtCmapBERT model, together with pre-trained weights and the necessary evaluation dataset, is available for research and non-commercial use at https://github.com/tahmid1234/ProtCmapBERT.

3   Performance Evaluation

We evaluated separate models for each of the three Gene Ontologies (Molecular Function, Biological Process, and Cellular Component) as well as for EC Numbers, and compared their performance against two baseline models: a fine-tuned ProtBERT model and the DeepFRI model.

3.1   Test Datasets

For evaluation, we used two test datasets provided by DeepFRI [9]: one for GO term prediction and one for EC number prediction. These datasets contains PDB chain identifiers along with their sequence identity percentages relative to the training set. Because the original test sets included only chain names and similarity scores, we retrieved the corresponding protein sequences and 3D structures from the RCSB Protein Data Bank (https://www.rcsb.org) [3]. From these structures, we extracted the Cα (alpha carbon) coordinates and generated contact maps for input to our models. The ground-truth labels were obtained from the annotation files provided in the DeepFRI GitHub repository [9].

   To evaluate model robustness under varying levels of sequence redundancy, we derived five evaluation sets from each test dataset, defined by maximum sequence identity thresholds relative to the training data. The thresholds were set to 95%, 70%, 50%, 40%, and 30%. The 95% set includes all samples from the original test dataset, while each subsequent set contains only those samples below its respective threshold.

3.2   Evaluation Metrics

We adopted two evaluation metrics commonly used in protein function prediction literature: (1) protein-centric maximum F1 score ($\mathbf{F_{max}}$), the official evaluation metric proposed by CAFA [18], and (2) area under the precision-recall curve ($\mathbf{AUPRC}$), following the evaluation protocols used in TALE [5].

1) **Protein-centric $\mathbf{F_{max}}$**: For a given evaluation set of N proteins, let P_i denote the i-th protein, where $i \in \{1, \ldots, N\}$. For protein P_i, let $\mathrm{TP}_i(t)$ be the number of correctly predicted terms (true positives) at a decision threshold $t \in [0, 1]$. Let $P_i(t)$ be the set of predicted terms with score $\geq t$, and T_i be the set of ground-truth (experimentally validated) terms. Then, the per-protein precision and recall are defined as:

$$\mathrm{Precision}_i(t) = \frac{\mathrm{TP}_i(t)}{|P_i(t)|}, \quad \mathrm{Recall}_i(t) = \frac{\mathrm{TP}_i(t)}{|T_i|}$$

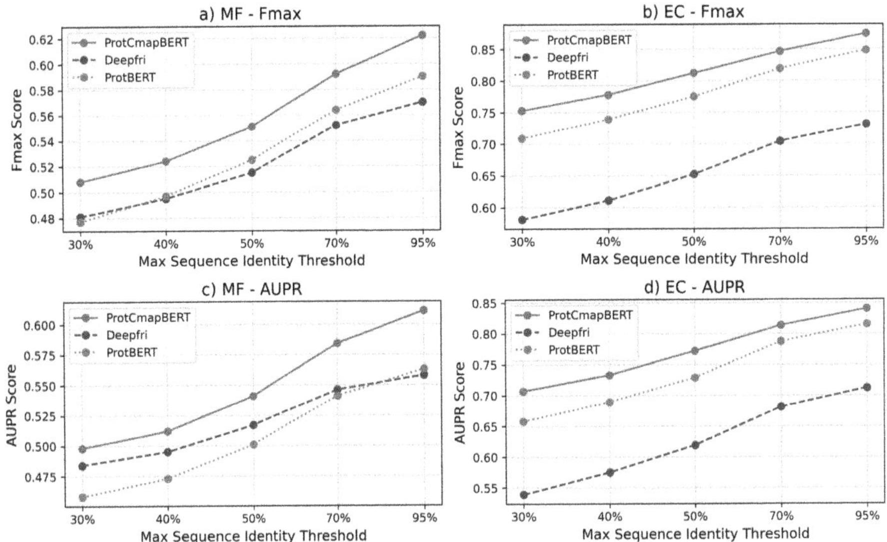

Fig. 4. Performance comparison of the models on test sets across varying sequence identity thresholds relative to the training dataset. This figure shows the $F_{\max}$ and AUPR scores for predicting MF terms and EC numbers using ProtCmapBERT (green), DeepFRI (blue), and ProtBERT (red). (Color figure online)

To obtain an overall measure for the entire set of proteins, we average the per-protein precision and recall values across all proteins:

$$\text{Precision}_{\text{avg}}(t) = \frac{1}{N} \sum_{i=1}^{N} \text{Precision}_i(t) \tag{4}$$

$$\text{Recall}_{\text{avg}}(t) = \frac{1}{N} \sum_{i=1}^{N} \text{Recall}_i(t) \tag{5}$$

The F_1 score for a given threshold t is then computed as the harmonic mean of the overall average precision and recall:

$$F(t) = \frac{2 \cdot \text{Precision}_{\text{avg}}(t) \cdot \text{Recall}_{\text{avg}}(t)}{\text{Precision}_{\text{avg}}(t) + \text{Recall}_{\text{avg}}(t)} \tag{6}$$

Finally, we identify the maximum F_1 score, denoted as $F_{\max}$, from the list of F_1 scores computed across all thresholds $t \in [0, 1]$ with an increment of 0.01.

2) Term-centric AUPRC: We treated the multi-label prediction task as a binary classification problem across all protein-function pairs. To compute the precision-recall (PR) curve, we first vectorized the predicted scores and ground-truth annotations. Precision and recall were then calculated over a sliding decision $t \in [0, 1]$threshold , and the area under the curve was determined using the trapezoidal rule. This approach is particularly appropriate for imbalanced datasets, as it emphasizes the model's ability to prioritize true functional annotations ahead of negatives.

Table 1. Prediction of EC numbers (TEST: $\leq$ 30% SEQ IDENTITY TO TRAIN)

Model	F_{max}	AUPR
ProtCmapBERT	0.753	0.707
ProtBERT	0.709	0.658
DeepFRI	0.581	0.539

Table 2. Prediction of MF terms (TEST: $\leq$ 30% SEQ IDENTITY TO TRAIN)

Model	F_{max}	AUPR
ProtCmapBERT	0.508	0.498
ProtBERT	0.477	0.458
DeepFRI	0.481	0.484

Attention Head Visualization

To analyze the impact of structural integration, we visualized the final-layer attention heads (Fig. 5) for protein *5W8S-A*. The sequence-only ProtBERT exhibits diffuse attention patterns that fall below visualization thresholds, which indicates a lack of specific residue-residue focus. In contrast, our ProtCmapBERT model demonstrates concentrated attention. By constraining the search space to a 10 Åspatial distance, the model successfully sharpens its focus. The learnable parameter α allows the model to prioritize specific functional neighbors within this 3D radius and effectively recovers the physical interaction network of the protein.

3.3   Results

For a fair comparison, all models were fine-tuned on the same training, validation datasets and evaluated on the same test datasets using the same evaluation metrics.

We started with the test datasets with a maximum sequence identity of 95% relative to the training data. As shown in Fig. 3), our model significantly outperforms both ProtBERT and DeepFRI in protein function prediction. When protein functions are labeled using MF ontology terms (Fig. 3, Subplot MF) or EC numbers (Fig. 3, Subplot EC) our model consistently outperforms both ProtBERT and DeepFRI in terms of F_{max} and AUPR. Specifically, for MF terms

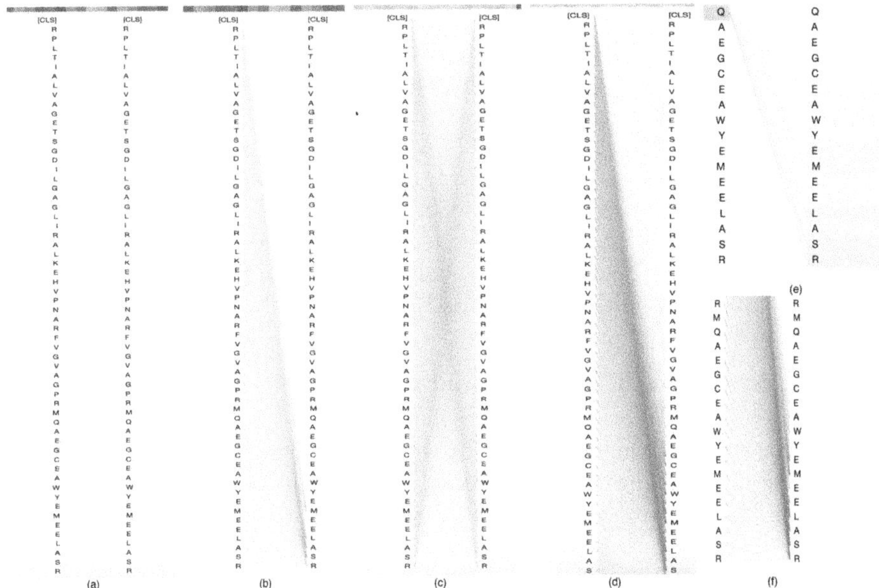

Fig. 5. Attention Map Comparison between ProtCmapBERT and ProtBERT. (a) Prot-BERT with all heads active, showing diffuse, low-intensity attention. (b) ProtCmap-BERT with all heads active, demonstrating "structured sparsity." (c) Single-head view (Head 16) for ProtBERT. (d) Single-head view (Head 16) for ProtCmapBERT showing high-intensity focal points. (e) & (f) reveals a high-intensity structural interaction hub. In this region, residues 48 through 60 receive concentrated attention from multiple distal residues, a direct result of the 3D spatial constraints integrated into the model.

prediction, Our model achieves 0.62 F_{max} and 0.61 AUPR, compared with 0.57 and 0.56 for DeepFRI and 0.59 and 0.56 for ProtBERT. These results represent improvements of 9.1% and 5.4% in F_{max} and 9.4% and 8.5% in AUPR. For EC numbers prediction, the improvements are 19. 6% and 3% in F_{max} and 18% and 3% in AUPR.

It is worth noting that when BP and CC terms are used as function labels, our model doesn't exhibit the same level of improvement over DeepFRI and ProtBERT (Fig. 3, Subplots BP and CC) . This outcome is expected, since the cellular component a protein is located and the biological process it participates in are not as directly dependent on protein structure as its molecular function. Proteins with very different structures may localize to the same cellular component, and a single biological process may involve structurally diverse proteins. Consequently, incorporating structural information provides limited benefit for predicting BP and CC terms and may even reduce performance by introducing noise.

To further evaluate the robustness of our model, we filtered the test datasets to remove proteins with high sequence similarity to the training data using different thresholds. As shown in Fig. 4, when MF terms and EC numbers are used as functional labels, our method consistently outperforms DeepFRI and ProtBERT across all sequence identity thresholds, underscoring the robustness of the proposed structure-aware attention mechanism in leveraging structural information for protein function prediction.

Notably, at the most challenging 30% sequence identity threshold (Tables 1 & 2), our model achieves an F_{max} of 0.753 for EC prediction, representing improvements of 29.6% over DeepFRI ($F_{max} = 0.581$) and 6.2% over ProtBERT ($F_{max} = 0.709$). When AUPR is considered, the gains increase to 31.1% and 7.4%, respectively. For MF prediction, our model achieves improvements of 5.6% and 6.4% in F_{max} over DeepFRI and ProtBERT, respectively, and 2.9% and 8.7% in AUPR.

Ablation Study and Computational Efficiency

Ablation Study. To isolate the contribution of the structure-aware attention mechanism, we performed an ablation study by comparing ProtCmapBERT against the original ProtBERT baseline. Since both models utilize the same Transformer architecture and were initialized with the same pre-trained weights, ProtBERT serves as an ablated version of our model with the structural bias $\alpha.cmap$ removed.

The performance gap between ProtCmapBERT and ProtBERT serves as a quantitative measure of the structural information's contribution. By isolating this variable, we observe that the significant gains in F_{max} and AUPRC are directly localized to the structure-aware attention mechanism. This is further supported by our attention visualizations (Fig. 5), which show that the model switches from the diffuse attention seen in the ablated version (ProtBERT) to highly concentrated structural hubs in ProtCmapBERT.

Computational Overhead and Memory Footprint. We evaluated the computational costs associated with integrating 3D structural information. Using an NVIDIA H100 GPU, ProtCmapBERT required an average inference time of 13.07 ms per sample, compared to 5.19 ms for the sequence-only baseline. While the addition of contact map processing and the structural bias term increases latency, the model remains capable of high-throughput annotation. Furthermore, the memory footprint remains efficient; peak VRAM usage for ProtCmapBERT was 4.39 GB. This low memory requirement is significant as it demonstrates that the model, despite being benchmarked on enterprise-grade hardware, is fully compatible with widely available consumer-grade GPUs.

3.4   Discussion

Computational methods that can accurately predict protein function at scale have profound implications across many areas of research. Most existing approaches follow the sequence–structure–function paradigm. Encoding protein sequences for computational models is relatively straightforward, since machine learning methods are designed to operate on vectorized inputs. In contrast, efficiently leveraging protein structure for function prediction remains an open challenge.

In this paper, we investigate the feasibility of using protein structural information to guide the self-attention mechanism in transformer models for protein function prediction. Our results demonstrate that the proposed structure-aware attention mechanism significantly improves predictive performance. This mechanism offers a novel approach to integrating 3D structural data into transformer architectures.

Transformer-based language models for protein sequences have been widely developed and successfully applied to numerous problems. It is therefore desirable to harness the strengths of these models when developing new methods that combine sequence features with structure-derived information. The proposed method provides an efficient framework for integrating structural and sequence information, while leveraging the proven power of existing transformer-based models. Furthermore, this approach has the potential to be extended to incorporate other types of non-sequence features.

Disclosure of Interests. The authors declare that they have no competing interests.

References

1. Uniprot: the universal protein knowledgebase in 2025. Nucleic Acids Res. **53**(D1), D609–D617 (2025)
2. Altschul, S.F., Gish, W., Miller, W., Myers, E.W., Lipman, D.J.: Basic local alignment search tool. J. Mol. Biol. **215**(3), 403–410 (1990)
3. Berman, H.M., et al.: The protein data bank. Nucleic Acids Res. **28**(1), 235–242 (2000). https://doi.org/10.1093/nar/28.1.235
4. Brandes, N., Ofer, D., Peleg, Y., Rappoport, N., Linial, M.: Proteinbert: a universal deep-learning model of protein sequence and function. Bioinformatics **38**(8), 2102–2110 (2022)
5. Cao, Y., Shen, Y.: Tale: Transformer-based protein function annotation with joint sequence-label embedding. Bioinformatics **37**(18), 2825–2833 (2021)
6. Dessimoz, C., Škunca, N.: The gene ontology handbook. Springer Nature (2017)
7. Elnaggar, A., et al.: Prottrans: towards cracking the language of life's code through self-supervised learning. IEEE Trans. Patt. Anal. Mach. Intell. **44**, 7112–7127 (2021)
8. Finn, R.D., et al.: Pfam: the protein families database. Nucleic Acids Res. **42**(D1), D222–D230 (2014)
9. Gligorijević, V., et al.: Structure-based protein function prediction using graph convolutional networks. Nat. Commun. **12**(1), 3168 (2021). https://github.com/flatironinstitute/DeepFRI
10. Graves, A.: Generating sequences with recurrent neural networks. arXiv preprint arXiv:1308.0850 (2013)
11. Gu, Z., Luo, X., Chen, J., Deng, M., Lai, L.: Hierarchical graph transformer with contrastive learning for protein function prediction. Bioinformatics **39**(7), btad410 (2023)

12. Jiao, P., Wang, B., Wang, X., Liu, B., Wang, Y., Li, J.: Struct2go: protein function prediction based on graph pooling algorithm and alphafold2 structure information. Bioinformatics **39**(10), btad637 (2023)
13. Kipf, T.: Semi-supervised classification with graph convolutional networks. arXiv preprint arXiv:1609.02907 (2016)
14. Kulmanov, M., Hoehndorf, R.: Deepgoplus: improved protein function prediction from sequence. Bioinformatics **36**(2), 422–429 (2020)
15. Lai, B., Xu, J.: Accurate protein function prediction via graph attention networks with predicted structure information. Briefings Bioinf. **23**(1), bbab502 (2022)
16. Lin, B., Luo, X., Liu, Y., Jin, X.: A comprehensive review and comparison of existing computational methods for protein function prediction. Briefings Bioinf. **25**(4), bbae289 (2024)
17. McDonald, A.G., Boyce, S., Moss, G.P., Dixon, H.B., Tipton, K.F.: Explorenz: a mysql database of the iubmb enzyme nomenclature. BMC Biochem. **8**(1), 14 (2007)
18. Radivojac, P., et al.: A large-scale evaluation of computational protein function prediction. Nat. Methods **10**(3), 221–227 (2013)
19. Rives, A., et al.: Biological structure and function emerge from scaling unsupervised learning to 250 million protein sequences. Proc. Natl. Acad. Sci. **118**(15), e2016239118 (2021)
20. Sureyya Rifaioglu, A., Doğan, T., Jesus Martin, M., Cetin-Atalay, R., Atalay, V.: Deepred: automated protein function prediction with multi-task feed-forward deep neural networks. Sci. Rep. **9**(1), 7344 (2019)
21. Suzek, B.E., Wang, Y., Huang, H., McGarvey, P.B., Wu, C.H., Consortium, U.: Uniref clusters: a comprehensive and scalable alternative for improving sequence similarity searches. Bioinformatics **31**(6), 926–932 (2015)
22. Varadi, M., et al.: Alphafold protein structure database: massively expanding the structural coverage of protein-sequence space with high-accuracy models. Nucleic Acids Res. **50**(D1), D439–D444 (2022)
23. Vaswani, A., et al.: Attention is all you need. Adv. Neural Inf. Process. Syst. **30** (2017)
24. Velickovic, P., Cucurull, G., Casanova, A., Romero, A., Lio, P., Bengio, Y., et al.: Graph attention networks. stat **1050**(20), 10–48550 (2017)
25. Vu, T.T.D., Jung, J.: Protein function prediction with gene ontology: from traditional to deep learning models. PeerJ **9**, e12019 (2021)
26. Wang, S., Sun, S., Li, Z., Zhang, R., Xu, J.: Accurate de novo prediction of protein contact map by ultra-deep learning model. PLoS Comput. Biol. **13**(1), e1005324 (2017)
27. Wu, J., et al.: Hifun: homology independent protein function prediction by a novel protein-language self-attention model. Briefings Bioinf. **24**(5) (2023)
28. Xia, T., Wang, Y., Tian, Y., Chang, Y.: Using prior knowledge to guide Bert's attention in semantic textual matching tasks. In: Proceedings of the Web Conference 2021, pp. 2466–2475 (2021)
29. Xu, G., Meng, Y., Qiu, X., Yu, Z., Wu, X.: Sentiment analysis of comment texts based on bilstm. IEEE Access **7**, 51522–51532 (2019)
30. Zhang, S., Fan, R., Liu, Y., Chen, S., Liu, Q., Zeng, W.: Applications of transformer-based language models in bioinformatics: a survey. Bioinf. Adv. **3**(1), vbad001 (2023)

Deep Learning Based Wound Segmentation and Characterization with LLM-Assisted Clinical Interpretability

Fisha Mehabaw Alemayoh[1], Hailemicael Lulseged Yimer[2], Xiaohong Yuan[2], and Letu Qingge[2(✉)]

[1] Department of Data Science, University of Verona, Strada Le Grazie 15, 37134 Verona, Italy
`fishamehabaw.alemayoh@studenti.univr.it`
[2] Department of Computer Science, North Carolina A&T State University, Greensboro, NC 27411, USA
`hlyimer@aggies.ncat.edu, {xhyuan,lqingge}@ncat.edu`

Abstract. Chronic wound assessment requires accurate segmentation and interpretable analysis for effective treatment planning. Although deep learning models show promise in automated wound segmentation, clinical adoption is hampered by limited interpretability. In this paper, we present a framework that integrates advanced wound segmentation with conversational interfaces powered by a large language model (LLM). Our EfficientNet-Attention U-Net trained on 2,760 wound images achieves state-of-the-art performance with an accuracy of 99.76%, Dice score of 0.9065, and intersection over Union (IoU) of 0.8324. Beyond segmentation, we extract morphological features, color distributions, texture properties, and edge characteristics. We integrate GPT 2.0 through an interface that translates quantitative measurements into clinical narratives, allowing healthcare providers to naturally interact with AI assessments, by We integrate GPT 2.0 to enable a conversational AI system that provides healthcare providers with clear, clinically meaningful explanations of extracted wound properties. The LLM-generated explanations were evaluated by healthcare professionals and received high scores across multiple dimensions. Interpretability and clarity received high scores, with a median of 4.33 (IQR 4.0–5.0), indicating that most respondents found the explanations easy to understand and well-presented and 85.2% of participants agreed this assessment. Consistency, reflecting the logical coherence and reliability of information across explanations, achieved a median rating of 4.00 (IQR 3.7–4.7), showing moderate variability and 72.8% agreement among evaluators. Usefulness, capturing how informative and actionable the explanations were also received a median of 4.00 (IQR 4.0–4.7), with 77.8% agreement, indicating generally favorable evaluations. Finally, the absence of unsupported or hallucinated information was rated positively, with a median of 4.00 (IQR 4.0–5.0) and 91.4% agreement, indicating that most participants found the explanations reliable and factually accurate. The results demonstrate superior segmentation accuracy alongside clinically meaningful interpretations, addressing the need for explainable AI in healthcare. Our system offers a practical solution for remote and clinical wound care settings.

Keywords: Wound Image Segmentation · EfficientNet · Large Language Model · Conversational AI

1   Introduction

Chronic wounds pose a major global healthcare challenge, affecting millions of patients each year and placing a substantial economic burden on healthcare systems worldwide [1,2]. Accurate assessment and monitoring of wound healing progression are critical to determining appropriate treatment strategies and improving patient outcomes [3]. Traditional wound assessment methods are highly dependent on subjective visual inspection by healthcare professionals, which can lead to variability between observers and inconsistent documentation practices [4]. Furthermore, the increasing shortage of specialized wound care professionals, particularly in rural and underserved areas, necessitates the development of automated, objective, and accessible wound assessment tools [5,6].

Recent advances in deep learning and computer vision have demonstrated remarkable success in medical image analysis tasks [7–9], including lesion detection, tumor segmentation, and disease classification. Among these techniques, convolutional neural networks (CNNs), particularly the U-Net architecture, have shown exceptional performance in biomedical image segmentation tasks due to their ability to capture both local and global contextual information through encoder-decoder structures with skip connections [10]. Despite these technological advances, the adoption of AI-based wound assessment tools in clinical practice remains limited, primarily due to concerns regarding interpretability, trust, and the lack of clinically meaningful explanations accompanying automated predictions.

The integration of large language models (LLMs) with computer vision systems presents a promising approach to bridge this interpretability gap [11]. LLMs have demonstrated remarkable capabilities in understanding context, generating human-like explanations, and engaging in meaningful dialogue across various domains, including healthcare. By combining automated wound segmentation with LLM-powered conversational interfaces, we can create systems that not only perform accurate quantitative analysis but also provide clinically relevant interpretations and respond to healthcare providers' queries in natural language [12,13].

In this work, we present a comprehensive framework for an automated wound segmentation, characterization, and interpretation that addresses these challenges. Our system leverages an EfficientNet-attention U-Net-based for precise wound boundary delineation, extracts clinically relevant morphological, color, texture, and edge-based features, and integrates an LLM-powered conversational interface to provide interpretable wound assessments. Specifically, our contributions are as follows.

- Integrate a large language model (GPT 2.0) that interprets the extracted features and provides natural language explanations, making the AI's assessment understandable and interactive for healthcare providers by making the model's predictions more interpretable and clinically understandable for healthcare providers. This capability is deployed through a Streamlit application, enabling real-time, conversational interaction suitable for both clinical environments and remote healthcare settings.

The remainder of this paper is organized as follows. Section 2 reviews related work in wound segmentation, feature extraction, and AI interpretability in healthcare. Section 3 describes our methodology, including dataset preparation, segmentation architecture, feature extraction techniques, and LLM integration. Section 4 presents experimental results and performance evaluation. Section 5 compares our results with existing works, and Sect. 6 concludes with future research directions.

2   Literature Review

This section reviews existing works on wound image segmentation, model interpretability, and AI-assisted decision support. Deep learning models, particularly CNN and U-Net-based architectures are used to automate wound detection and segmentation.

Recent research in wound image analysis has explored a wide range of deep learning architectures for improving segmentation quality. The encoder–decoder models such as U-Net, Attention U-Net (AttUNet), and Efficient Attention U-Net (EffUNet) have been widely evaluated. For example, Kang et al. reported that Attention U-Net achieved the strongest performance in their study, obtaining the highest sensitivity (0.622) and a Dice score of 0.608 [14]. In work focused on pressure injury segmentation, Liu et al. compared U-Net and Mask R-CNN, achieving Dice scores of 0.8448 and 0.5006, respectively [15].

Another direction has been detection-led segmentation pipelines. Scebba et al. showed that combining object detection with pixel-level segmentation substantially improved performance, increasing the Matthews' correlation coefficient (MCC) from 0.15 to 0.85 compared with manual annotation–based pipelines [16]. Beyond RGB images, ultrasound-based wound assessment has also been explored. Schlereth et al. evaluated U-Net, FCN, and ResNet–U-Net on noninvasive ultrasound images, reporting Dice scores between 0.27 and 0.34, highlighting both the difficulty and the potential of this imaging modality [17].

A growing number of studies have focused on improving U-Net itself. Wu et al. incorporated attention mechanisms for tumor segmentation and reported improvements over the baseline U-Net, achieving an IoU of 0.48 and an accuracy of 0.99 [18]. Similarly, Alhuda et al. introduced Triplet Attention into several U-Net variants including DenseNet-U-Net, ResNet-U-Net, Efficient-U-Net, and MobileNet-U-Net to enhance skin lesion segmentation. Their experiments showed improvements across metrics such as mIoU, recall, precision, and F1-score, demonstrating the value of attention for focusing on relevant lesion structures [19].

Other architectural modifications have targeted multi-scale feature extraction. A multi-scale U-Net design proposed for histopathological segmentation added explicit multi-resolution feature maps at each encoder stage, leading to the best performance among tested models, with mIoU scores of 0.7842 and 0.6047 for germinal center and sinus segmentation, respectively [20]. Additional work has explored integrating spatial and channel attention into U-Net for semantic segmentation, showing consistent improvements in capturing important features across complex scenes, while also examining the computational trade-offs [21].

Several application-specific systems have also been developed. An RGB-D wound-measurement framework using HarDNet-FSEG demonstrated strong segmentation performance (Dice 0.86 and accuracy 0.95) by combining color and depth information [22]. Encoder–decoder architectures such as FCN, PSPNet, U-Net, SegNet, and DeepLabV3 have been evaluated for pressure wound segmentation, with MobileNet-U-Net achieving the highest accuracy (99.67%) [23]. Additionally, the AutoTrace model, designed for mobile devices and incorporating attention in skip connections, achieved mean IoU values of 0.8644 and 0.7192 in a clinical cohort study focused on tissue-type segmentation and inter-rater consistency [24].

Recent work has also introduced new datasets and benchmarks addressing the scarcity of annotated wound data. The Foot Ulcer Segmentation Challenge (FUSeg) provided 1210 pixel-wise annotated images from 889 patients and evaluated multiple segmentation architectures. Ensembled U-Net and LinkNet achieved the strongest performance with a Dice score of 0.888, outperforming HarDNet-MSEG and double encoder–decoder models. The authors also highlighted limitations in previous datasets like Medetec and WoundDB, emphasizing the need for large fully annotated datasets for fair benchmarking [27].

Another major contribution is WSNet, which introduced the WoundSeg1 dataset containing diverse wound images and proposed a global–local hybrid segmentation model leveraging wound-domain pretraining. Experiments showed that DenseNet121–LinkNet achieved the best performance, with an IoU of 0.713 and a Dice score of 0.847. WSNet also established new state-of-the-art results on AZH Woundcare and Medetec datasets and demonstrated how segmentation improves downstream tasks such as wound area and volume estimation [28].

A new line of research has evaluated tissue-specific segmentation. Morgado et al. conducted agreement analyses across multiple expert annotators and evaluated DeepLabV3-R50 and SegFormer-B0 for chronic wound tissue segmentation. Their best model achieved a mean IoU of 62.95% and a Dice score of 76.82%, highlighting the need for reliable tissue-level annotations [26]. Complementary to this, Kabir et al. performed a comprehensive evaluation of deep learning models for wound tissue segmentation using a novel dataset, reporting that FPN+VGG16 achieved the highest Dice score (82.25%), demonstrating strong performance in multi-class tissue segmentation tasks [29].

Additional work has explored computational efficiency. Ribeiro da Costa et al. compared U-Net, DeepLabV3, SegNet, and FCN-VGG19 for malignant wound segmentation and applied network pruning, finding that FCN achieved the best performance with an average Dice score of 0.8625 while maintaining computational efficiency [30]. A dual-attention U-Net with VGG16 backbone and transfer learning, reached a Dice score of 94.1% and an IoU of 89.3%, demonstrating how attention mechanisms significantly enhance wound boundary localization [31].

Earlier patch-based and post-processing approaches also showed competitive performance. Cui et al. reported that their post-processing U-Net achieved a Dice score of 0.845 on diabetic wound images [32], while Wang et al. proposed a MobileNetV2-based fully automatic framework achieving a Dice of 0.9047, outperforming several recent models through effective transfer learning and false-positive reduction techniques [33].

While these studies collectively showed steady progress in segmentation accuracy across different wound types and imaging modalities, most of them focus purely on pixel-level performance and do not address the broader clinical usability gap, namely, how to interpret model outputs in a way that supports clinicians and non-experts. Recent advances in large language models (LLMs) offer a promising complement by enabling natural-language explanations, interactive feedback, and context-aware interpretation of model predictions [11]. Integrating LLMs with deep learning pipelines may therefore help bridge the gap between complex segmentation outputs and practical, user-friendly wound assessment.

3   Methodology

Figure 1 illustrates the overall workflow of the proposed framework, which consists of three main stages: wound segmentation, wound property characterization, and LLM-based interpretability. In the first stage, an RGB wound image is provided as input to an EfficientNet-based Attention U-Net model, which produces a binary segmentation mask that delineates the wound region from the surrounding skin.

The proposed model is an EfficientNet-Attention U-Net. EfficientNet-B7, pretrained on ImageNet, is used to extract hierarchical feature representations from input images [35]. Intermediate encoder feature maps are forwarded to the decoder through attention gates, which selectively emphasize relevant spatial regions while suppressing irrelevant background information. The decoder progressively upsamples the feature maps, concatenates them with the attention-filtered skip connections, and refines them using convolutional blocks [36]. A final 1×1 convolution followed by a sigmoid activation produces a pixel-wise binary segmentation mask at the original image resolution.

In the second stage, the segmented wound region is isolated using the predicted mask and processed to extract quantitative wound properties. These include geometric features such as wound area, as well as image-derived texture features computed from the segmented region using OpenCV-based implementations. In the final stage, the extracted wound properties are formatted as

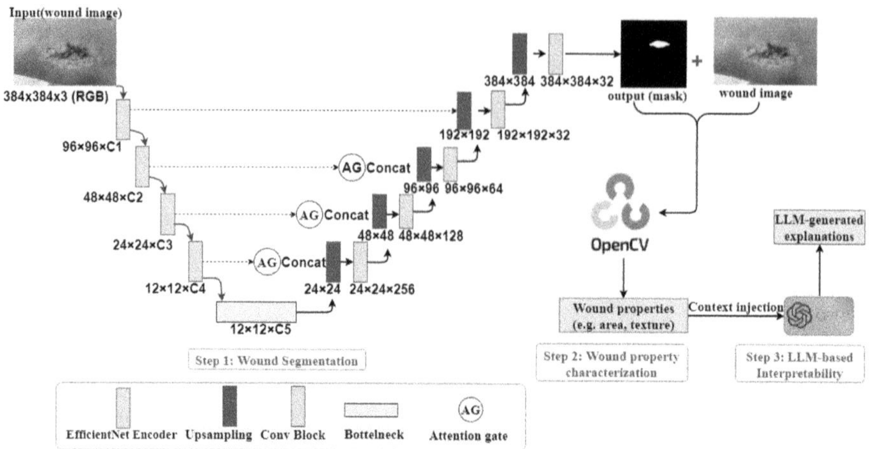

Fig. 1. Overview of the proposed wound analysis framework, where an RGB wound image is segmented using an EfficientNet-based Attention U-Net, features are extracted from the wound region, and fed to an LLM for natural language explanations.

structured inputs and supplied to an LLM (i.e. GPT-2), which generates natural language explanations describing wound characteristics.

3.1   Dataset

2760 wound images were used. The images were collected and merged from three different sources; 374 samples from Medetec Wound Database [25, 26], 1210 samples from FUSeg [27], and 1176 samples from WSNet [28]. The dataset contains wound images and their corresponding masks. Masks are binary images which are represented with pixels in the foreground (white), which allows operations to be performed in the same area on the original image and background (black) that hide the image and prevent processing [34].

3.2   Preprocessing Pipeline

Originally, the images were annotated and resized to 512×512 maintaining aspect ratio by data owners. Further preprocessing was performed like images were resized to 384×384, normalized their pixel values from the [0, 255] to [0, 1] range, and masks were binarized to ensure clean segmentation target, and reshaped to a single-channel form by adding channel dimension to smooth them and help segmentation algorithms train faster and stably. Data were then organized into mini-batches for efficient GPU training, with shuffling applied at the end of each epoch assisting enhanced generalization and preventing the model from learning data order. The dataset was initially partitioned into training and testing subsets using an 80:20 ratio. Then, the training subset was further divided into training and validation sets in an 80:20 ratio respectively. To check the distribution between wound and non-wound (background) pixels, class imbalance analysis was performed.

The dataset shown an extreme imbalance with wound region occupying a much smaller proportion of the image compared to the background (non-wound region) and that can negatively affect model learning, introducing bias towards the dominant non-wound class. To address the class imbalance problem, we applied a positive class weighting [15,16] with a value of $\log\left(\frac{1-\text{train\_wound\_ratio}}{\text{test\_wound\_ratio}}\right)$. In addition, we implemented augmentation that involves random horizontal flipping, vertical flipping, and small-angle rotations to increase data variability and improve model generalization.

3.3 Wound Segmentation Models

We employed four U-Net-based architectures for automatic wound segmentation: U-Net, Attention U-Net, U-Net++, and EfficientNet-Attention U-Net.

U-Net applies an encoder-decoder architecture with skip connections that transfer spatial information from the contracting path to the expanding path, preserving fine details needed for accurate segmentation. Attention U-Net enhances U-Net by adding attention gates between encoder and decoder layers, which help the model focus on wound regions while suppressing irrelevant background features.

U-Net++ introduces nested skip connections with intermediate convolutional layers, enabling better multi-scale feature integration between encoder and decoder paths. EfficientNet-Attention U-Net model combines an EfficientNet encoder (pretrained on ImageNet) with attention-gated skip connections, leveraging both transfer learning and spatial attention for improved segmentation performance. EfficientNet encoders provide strong feature representation with fewer parameters, making the architecture suitable for heterogeneous wound images with varying illumination, texture, and shape.

All models were trained using the Adam optimizer (learning rate $= 1 \times 10^{-5}$) with gradient clipping to improve stability. Weighted binary cross-entropy and Dice score were used as the loss function. To prevent overfitting and ensure convergence, Early Stopping was applied by monitoring the validation Dice score with a patience of 20 epochs and automatic restoration of the best weights. A *ReduceLROnPlateau* scheduler further adjusted the learning rate by a factor of 0.5 when validation performance plateaued. *ModelCheckpoint* was employed to save the best-performing model based on the highest validation Dice score. The network was trained for up to 300 epochs using separate training and validation datasets.

Both overall and dataset-specific performance evaluations were conducted using accuracy, Dice Similarity Coefficient (DSC), and Intersection-over-Union (IoU). Accuracy measures the proportion of correctly classified pixels across the entire image. The DSC assesses the overlap between the predicted wound mask and the ground-truth annotation (masks), providing a balanced measure that penalizes both false positives and false negatives. A higher DSC shows a stronger agreement between the model output and the true wound region. The IoU is the ratio between the intersection and the union of the predicted and actual wound areas.

$$\text{Accuracy} = \frac{TP + TN}{TP + TN + FP + FN} \tag{1}$$

$$\text{DSC} = \frac{2TP}{2TP + FP + FN} \tag{2}$$

$$\text{IoU} = \frac{TP}{TP + FP + FN} \tag{3}$$

3.4   Wound Feature Extraction

After segmentation, wound properties such as morphological, color, texture, and edge-based features were extracted from the wound region to quantify wound characteristics. The Wound area was isolated using the predicted binary mask in combination with its corresponding wound image and the following properties were computed using OpenCV [37]. These wound properties correspond to clinically relevant visual cues commonly used in wound assessment, such as size, shape, tissue color, and surface characteristics. They support transparent and interpretable explanations for wounds analysis.

Morphological Features: Area - the total number of pixels in the segmented region. Perimeter - measures the arc length of the wound's boundary contour. Circularity - measures shape regularity with values closer to 1 indicate a circular wound where as lower values refer to irregular wound shapes.

$$C = \frac{4\pi A}{P^2} \tag{4}$$

where:

- A is the area of the shape
- P is the perimeter of the shape.

Color Features: Average red, green and blue (RGB) color intensities of the wound pixels. RGB color features have been reported in prior studies to correlate with visual characteristics of wound tissues, such as granulation and necrosis [38,39]. There is no one-to-one deterministic mapping between specific RGB values and wound tissue types. For example, necrotic tissue often appears darker, while granulation tissue may exhibit higher red channel intensity; however, these visual characteristics are influenced by multiple factors and should be interpreted probabilistically rather than conclusively.

Texture Features: Contrast, homogeneity and dissimilarity values were calculated using the gray-level co-occurrence matrix of the segmented wound region to capture local intensity variations, uniformity and dissimilarity of the wound texture to show how consistent is the wound [40,41].

Edge Features: presented with edge sharpness, indicating boundary definition and lesion severity by calculating the mean values of edges detected.

Wound Complexity Index: We compute the mean of contrast, edge sharpness, and shape irregularity to derive an image-derived, relative measure of wound visual complexity. The resulting Wound Complexity Index is designed to capture visual complexity patterns across wound images and has not been aligned or validated against established clinical wound severity scales. Accordingly, it is used exclusively for comparative and exploratory analysis within this study.

3.5 Conversational Wound Analysis Application

To enable real-time wound analysis, a diagnostic application was developed using Streamlit. Users upload wound image, and the trained U-Net based segmentation model performs preprocessing, segmentation and wound features extraction. The application displays the original wound image, segmented wound with bound-boxing and color overlay, and wound features. The system incorporates an AI-powered medical assistant in which the extracted wound features are provided as contextual input to a large language model (GPT 2.0) accessed via OpenRouter API. The LLM is guided by a system prompt that positions it as a medical assistant specialized in interpreting wound analysis results based on detected image properties. The conservational medical assistant interprets the wound properties and answers user queries through a chat-based interaction.

The LLM-enabled medical assistant was evaluated by eight healthcare professionals voluntary, including physicians and nurses with varying levels of experience in wound care and diverse clinical practice settings. The assessors were recruited from Ayder Comprehensive Specialized Hospital and St. Peter's Hospital, and Redat Medical Plaza in Ethiopia. The evaluations were provided using exported explanations and LLM-generated explanations reviewed across multiple wound images. A structured questionnaire was specifically designed for this study to systematically capture both evaluator background and perceptions of the generated explanations.

The questionnaire consisted of three sections. The first section collected evaluator characteristics, including professional role, years of experience in wound care, and primary clinical setting. The second section focused on explanation-level assessment and was completed separately for each wound image, identified using a unique wound ID. This section comprised a set of 5-point Likert scale (1 = strongly disagree, 5 = strongly agree) items evaluating multiple aspects of the LLM-generated explanations, including clarity and ease of understanding, appropriateness of clinical language, understandability without prior technical or AI-related knowledge, consistency with the visual appearance of the wound, alignment with extracted quantitative wound features, and the absence of unsupported or hallucinated information. Additional items assessed perceived usefulness, such as the ability of the explanation to summarize wound characteristics, add value beyond raw numerical outputs, and support clinical documentation or reporting (without implying diagnostic or treatment decision-making). An overall quality rating and a binary usefulness judgment were also included. The final section provided an optional free-text field for evaluators to report unclear, misleading, or hallucinated statements, or to offer additional comments on interpretability and usefulness.

Evaluation results were analyzed using both quantitative and qualitative methods. Likert-scale responses were grouped into four interpretability-related dimensions: interpretability and clarity, consistency with visual and quantitative information, usefulness, and hallucination awareness. Descriptive statistics (median, IQR, agreement percentage) were computed to summarize evaluator responses across wound cases and questionnaire items. The percentage of agreement was calculated as the proportion of evaluators assigning ratings of 4 or 5 on the 5-point Likert scale, divided by the total number of evaluator responses. Free-text comments were analyzed qualitatively to identify recurring themes related to ambiguity, misalignment with wound appearance or features, and perceived limitations of the explanations. This evaluation framework was designed to assess the interpretability and perceived clinical usefulness of LLM-generated explanations for decision-support, rather than to validate diagnostic accuracy or clinical outcomes.

4   Result

4.1   Dataset Properties

A total of 2,760 wound images were used in this study. The dataset was divided into 80% for training and 20% for testing. Within the training portion, an additional 80/20 split was applied to create the final training and validation sets. Consistent with real-world wound-care data, the dataset showed a strong class imbalance, with a wound-to-non-wound pixel ratio of approximately 1:21.

During data inspection, we identified a subset of images whose masks contained zero wound pixels, even though the corresponding images clearly displayed visible wounds. Specifically, 11 such cases were found in the training set and 13 in the test set. Because these masks were incorrectly annotated, they were removed from the dataset to avoid introducing label noise and misleading the model during training and evaluation.

4.2   Segmentation Performance

The baseline U-Net achieved an accuracy of 0.988, with a Dice score of 0.674 and an IoU of 0.512 (Table 1). Introducing attention gates improved overlap measures: Attention U-Net reached a Dice score of 0.724 and an IoU of 0.577 (Table 1), although with a slightly lower accuracy, likely due to its stronger focus on foreground regions. U-Net++ showed a further increase in segmentation quality, achieving a Dice score of 0.748 and an IoU of 0.606 (Table 1), reflecting the benefit of its nested skip-connection design. The best performance was obtained with the EfficientNet-based Attention U-Net, which reached an accuracy of 0.997, a Dice score of 0.9065, and an IoU of 0.8324 (Table 1). This model showed the greatest reduction in loss and the strongest ability to capture wound regions accurately, suggesting that combining EfficientNet's powerful encoder with attention mechanisms provides the most robust segmentation results for this dataset. In addition, our model can segment multiple wound regions from a single wound image input and compute the total wound area and perimeter, while reporting the mean values for all other wound properties (Fig. 2).

The segmentation performance across the three wound datasets by EfficientNet-Attention U-Net model is summarized in Table 2. The model achieved consistently high Dice and IoU scores on all sources, indicating robust generalization across heterogeneous data. The best performance was observed on the Medetec dataset (Dice = 0.938, IoU = 0.884), while comparable results were obtained on the FUSC and WSNet datasets, demonstrating stable performance across different imaging conditions. Training curves for loss, Dice, IoU, and accuracy are shown in Figure 3. The curves demonstrate steady convergence of the model, with closely aligned training and validation metrics, indicating stable training and no evidence of overfitting. The smooth progression of the curves across all metrics provides further evidence of the robustness and reliability of the model's segmentation capabilities.

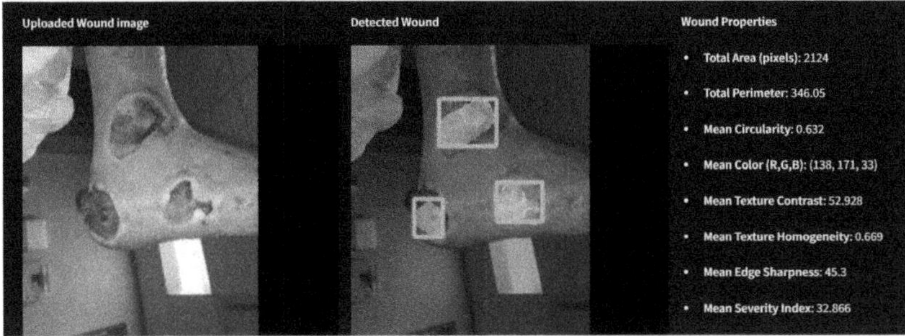

Fig. 2. Multi-region wound segmentation and properties extraction.

Table 1. Segmentation performance of our four models on test dataset

Model	Accuracy	Dice	IoU	Loss
U-Net	0.9879	0.6735	0.5125	0.4509
Attention U-Net	0.9720	0.7236	0.5771	0.4798
U-Net++	0.9920	0.7479	0.6063	0.3505
EfficientNet-Attention U-Net	**0.9976**	**0.9065**	**0.8324**	**0.1508**

Table 2. EfficientNet-Attention U-Net model segmentation performance across different wound datasets

Dataset	Accuracy	Dice	IoU	Loss
FUSeg	0.9980	0.9038	0.8279	0.1743
Medetec	0.9897	0.9379	0.8840	0.1157
WSNet	0.9911	0.9022	0.8343	0.1629

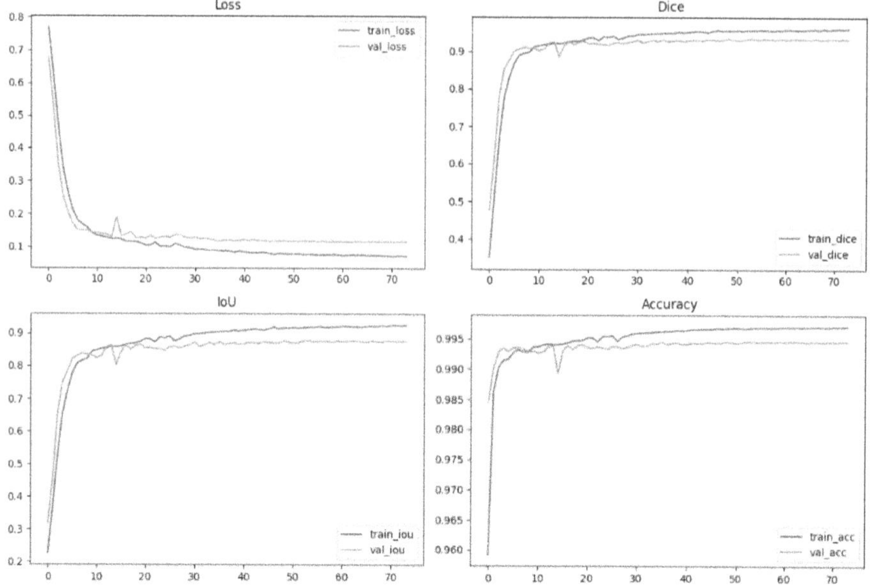

Fig. 3. EfficientNet based attention U-Net model training curve.

4.3 Conversational AI Assistant

To complement the segmentation model with interpretability, we implemented an AI-powered conversational assistant that interprets the extracted wound properties and provides human-readable clinical insights. Figure 4 demonstrates an example interaction between the user and the AI assistant. After segmentation, we successfully computed wound characteristics such as area (2956 pixels), circularity (0.486), texture contrast (30.091), and mean color values. The conversational assistant transformed these quantitative values into an easy-to-understand clinical narrative, describing possible wound severity, tissue characteristics, and potential concerns (e.g., necrotic tissue, tissue integrity variations, healing discrepancies).

Evaluators' Characteristics: A total of eight healthcare professionals participated in the evaluation, including six physicians and two nurses. Evaluators had varying levels of wound care experience: among physicians, four had 2–5 years of experience, one had 6–10 years, and one had less than 2 years; among nurses, one had less than 2 years and one had more than 10 years of experience. The clinical settings of evaluators were diverse: five physicians worked in hospitals and one in home care, whereas the two nurses were distributed across hospital and outpatient clinic settings. Each evaluator assessed a total of ten wound images, each accompanied by three LLM-generated explanations, reflecting responses to different clinical questions in order to capture diverse explanatory perspectives for the same wound.

Item-Level Evaluation of LLM-Generated Explanations: Item-level results based on the 5-point Likert-scale questionnaire (1 = strongly disagree, 5 = strongly agree) are summarized in Table 3. Overall, LLM-generated explanations received consistently high ratings across most evaluation items. Measures of interpretability and clarity were particularly strong, with the explanation being rated as clear and easy to understand (median = 4.0, IQR = 4.0–5.0), using appropriate clinical language (median = 4.0, IQR = 4.0–5.0), and being understandable without technical or AI-related knowledge (median = 4.0, IQR = 4.0–5.0). Agreement levels for these items were high, ranging from 88.9% to 97.5%.

Items related to consistency also demonstrated favorable evaluations. Consistency with the visual appearance of the wound achieved a median score of 4.0 (IQR = 4.0–5.0), with 75.3% of responses indicating agreement or strong agreement. Alignment with the provided quantitative wound features showed a median score of 4.0 with a narrower interquartile range (IQR = 4.0–4.0), and 79.0% agreement. The absence of unsupported or hallucinated information was rated positively (median = 4.0, IQR = 4.0–5.0), with 91.4% agreement.

Usefulness-related items were also rated highly. Evaluators agreed that the explanations effectively summarized wound characteristics (median = 4.0, IQR = 4.0–5.0; 84.0% agreement), added value beyond raw numerical outputs (median = 4.0, IQR = 4.0–5.0; 85.2% agreement), and could support documentation or reporting without implying clinical decision-making (median = 4.0, IQR = 4.0–5.0; 84.0% agreement).

The overall quality of the LLM-generated explanations was rated with a median score of 4.0 (IQR = 3.0–4.0). While the majority of evaluators provided positive ratings, the percentage of agreement for overall quality (69.1%) was lower than that observed for individual interpretability and usefulness items, indicating greater variability in holistic assessments across wound cases.

Dimension-level Evaluation: Dimension-level analysis was conducted by aggregating item-level Likert scores into three predefined dimensions: interpretability and clarity, consistency, and usefulness. As shown in Table 4, all dimensions achieved median scores of 4.0 or higher, indicating overall positive evaluator perceptions.

Interpretability and clarity received the highest ratings, with a median score of 4.33 (IQR: 4.0–5.0) and 85.2% agreement, suggesting that the LLM-generated

Table 3. Item-level evaluation of LLM-generated explanations (Likert scale: 1 = strongly disagree, 5 = strongly agree). n = 80 item-level evaluations aggregated across 10 wound images and 8 evaluators.

Item	Median	IQR	% Agree (4–5)
The explanation is clear and easy to understand.	4.0	4.0–5.0	97.5
The explanation uses appropriate clinical language.	4.0	4.0–5.0	88.9
The explanation is understandable without technical AI knowledge.	4.0	4.0–5.0	95.1
The explanation is consistent with the visual appearance of the wound.	4.0	4.0–5.0	75.3
The explanation aligns with the provided quantitative wound features.	4.0	4.0–4.0	79.0
The explanation does not introduce unsupported or hallucinated information.	4.0	4.0–5.0	91.4
The explanation helps summarize wound characteristics.	4.0	4.0–5.0	84.0
The explanation adds value beyond raw numerical outputs.	4.0	4.0–5.0	85.2
The explanation could support documentation or reporting.	4.0	4.0–5.0	84.0
Overall quality of the LLM-generated explanations for this wound.	4.0	3.0–4.0	69.1

explanations were generally perceived as clear, accessible, and appropriately worded for clinical audiences. The usefulness dimension also demonstrated favorable evaluations, achieving a median score of 4.00 (IQR: 4.0–4.7), with 77.8% of responses indicating agreement that the explanations added value beyond raw numerical outputs and could support documentation or reporting.

The consistency dimension exhibited comparatively lower agreement, with a median score of 4.00 (IQR: 3.7–4.7) and 72.8% agreement. This variability suggests that, while explanations were generally aligned with visual and quantitative wound information, some wound cases exhibited perceived inconsistencies between the generated narratives and the underlying image-based or feature-based representations.

Evaluators generally found the LLM-generated explanations clear and useful for understanding wound characteristics. Minor issues were identified, primarily related to occasional color misinterpretation, and discrepancies in severity classification. Overall, these issues were infrequent, and the explanations were largely accurate and clinically informative.

Table 4. Dimension-level evaluation of LLM-generated explanations (Likert scale: 1 = strongly disagree, 5 = strongly agree). n = 80 item-level evaluations aggregated across 10 wound images and 8 evaluators.

Dimension	Items (n)	Median	IQR	% Agree (4–5)
Interpretability & Clarity	3	4.33	4.0–5.0	85.2
Consistency	3	4.00	3.7–4.7	72.8
Usefulness	3	4.00	4.0–4.7	77.8

5 Comparison

In this study, we developed and evaluated multiple deep learning models for wound segmentation, including U-Net, Attention U-Net, U-Net++, and EfficientNet-based Attention U-Net. Across all models, the EfficientNet-based Attention U-Net demonstrated the strongest performance, achieving the highest Dice and IoU scores. In addition, we integrated a large language model for interpreting the characteristics extracted from the segmented wound.

The EfficientNet encoder enhances multi-level feature extraction through compound scaling, enabling the model to capture hierarchical contextual cues essential for accurate wound boundary localization [42]. Combined with attention gates, which suppress irrelevant background structures and emphasize clinically meaningful lesion regions [18,19], this architecture offers a clear advantage over the baseline U-Net. Traditional U-Net struggles to model complex, multi-scale wound textures and is more susceptible to background noise, consistent with observations from earlier studies [20,21].

Two major recent datasets, FUSeg and WSNet/WoundSeg1, are closely related to our work and provide important benchmarks. The Foot Ulcer Segmentation Challenge (FUSeg) [27], which includes 1210 expertly annotated diabetic foot ulcer images, reported its best performance using an ensemble of U-Net and LinkNet (Dice 0.888) (Table 5). Other approaches on FUSeg include SE AG VGG16 U-Net [31] with Dice 0.896 and Attention U-Net + autoaugment [43] with Dice 0.859. On the same FUSeg dataset, our EfficientNet–Attention U-Net achieves a Dice score of 0.9038, exceeding all reported results while relying on a single, end-to-end architecture without ensembling or heavy encoder backbones. This performance gain is notable given that several competing methods depend on complex architectural components, extensive augmentation, or ensemble strategies. The results suggest that the combination of EfficientNet's multi-scale feature extraction and attention mechanisms enables effective modeling of wound boundaries and tissue variability, even within a dataset designed around a specific wound type.

Similarly, the WSNet study [28] introduced WoundSeg1, a large and diverse dataset with severe intra-class variability. Their best model, DenseNet121–LinkNet, achieved Dice 0.847. On the same dataset, our model outperforms with Dice 0.9022 (Table 5), showing that the hierarchical feature extraction and attention mechanisms of EfficientNet–Attention U-Net can handle high variability and complex wound textures without relying on domain-adaptive pretraining or hybrid global–local designs.

Other contemporary studies also help position our results. Wang et al. [33] reported a Dice score of 0.9405 using MobileNetV2 combined with extensive post-processing, including connected component filtering. While this approach achieves slightly higher accuracy, it relies on multi-stage processing and task-specific tuning tailored to a single wound category. In comparison, our method attains a comparable Dice score of 0.9379 (Table 5) while remaining fully end-to-end and demonstrating stronger generalization across varied wound morphologies without dependence on auxiliary processing stages.

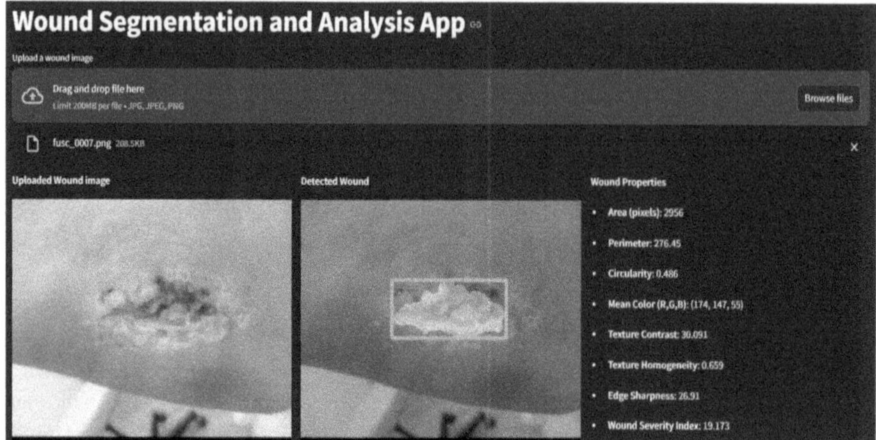

A. Wound segmentation

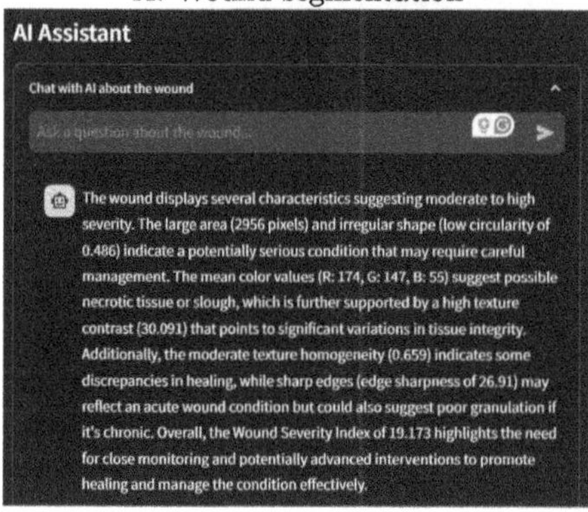

B. Conversational AI

Fig. 4. AI conversational assistant responses about wound properties.

A Dual Attention U-Net [31] reported Dice 0.896 on FUSeg. This model incorporates dual stacked U-Nets, ASPP modules, squeeze-and-excitation blocks, and extensive transfer learning. On the same dataset, our EfficientNet–Attention U-Net achieves a slightly higher Dice score of 0.9038 (Table 5). While the Dual Attention U-Net performs well, it relies on a more complex architecture with heavy encoders (e.g., VGG16) and dual-attention cascades, making it computationally intensive. In contrast, our model maintains a balance between performance, efficiency, and generalizability while achieving comparable or better segmentation performance. This demonstrates that our approach can deliver high accuracy on heterogeneous wound imagery without requiring multi-stage pipelines or excessive computational resources.

Overall, most literature [29,31] confirm that wound segmentation performance varies widely depending on dataset focus (single vs. mixed wound types), model complexity, and the availability of domain-specific pretraining. Within this landscape, our EfficientNet–Attention U-Net consistently outperforms most RGB-based models including WSNet, FPN+VGG16, MobileNet, FCN-VGG19, and several U-Net variants and approaches the performance of state-of-the-art ensemble or dual-stage architectures. This demonstrates that combining EfficientNet with attention gating offers a strong and efficient solution for precise wound boundary segmentation across diverse clinical scenarios.

Although our segmentation model performs on comparable with, or exceeds, many state-of-the-art approaches, the significance of this work extends beyond achieving high pixel-level accuracy. Our contribution lies in coupling a strong segmentation backbone with an LLM-driven interpretability module, enabling clinically meaningful explanations and more transparent model outputs. This integration bridges the gap between automated wound segmentation and understandable, decision-support insights suitable for real-world clinical use. While most existing wound segmentation studies stop at generating masks or quantitative measurements, our system goes further by translating raw wound attributes into human-readable explanations through a conversational AI assistant. This integration adds substantial clinical value, enabling clinicians and patients to interact with the model's outputs in an intuitive and interpretable manner, thus addressing one of the major limitations of conventional "black-box" segmentation models.

Table 5. Comparison of wound segmentation models utilizing same datasets with our study. Reported values are Dice scores

Model/Study	FUseg	Medetec	WSnet
Ensembled U–Net + LinkNet (FUSeg) [27]	0.888	–	–
WSNet (WoundSeg1) [28]	–	–	0.847
SE AG VGG16 U-Net [31]	0.896	–	–
MobileNetV2 CNN [33]	–	0.9405	–
Attention U-Net + autoaugment [43]	0.859	–	–
Attention Dense U-Net [44]	–	0.8655	–
Ours (EfficientNet-attention U-Net)	**0.9038**	**0.9379**	**0.9022**

- indicates no reported result for that dataset.

6   Conclusion

In this study, we developed a wound segmentation system that incorporate deep learning models and an LLM-based clinical assistant. Among the models we tested, the EfficientNet-based Attention U-Net achieved the best segmentation performance, showing strong accuracy and good overlap with the true wound regions. These results demonstrate that combining a powerful encoder with attention mechanisms is effective for wound image segmentation.

Beyond high-quality segmentation, our main contribution is the combined development of an accurate EfficientNet–Attention U-Net model and an LLM-driven interpretability module that together provide both precise wound delineation and clinically meaningful explanations. By extracting meaningful wound properties and passing them to an LLM, the system can explain its outputs in an understandable way and answer user questions. This reduces the "black-box" nature of deep learning and makes the results more accessible for clinical use. Overall, the system shows promise for both automated wound analysis and improved communication between AI tools and end users.

Future work should include exploring transformer-based architectures, incorporating multi-modal data such as depth or thermal images, clinical wound descriptions, and expanding the dataset to develop wound-specific large language model. In addition, although the LLM-generated explanations were evaluated for interpretability, clarity, consistency, hallucination, and usefulness, formal validation for clinical deployment remains future work. Furthermore, because a heterogeneous dataset was utilized, variability in RGB values caused by differences in skin tone, illumination, camera sensors, and acquisition conditions could be a limitation. Such variability may not only affect color consistency across images but also propagate to texture features derived from the images, including contrast, homogeneity, and dissimilarity. These factors can influence both color- and texture-based feature interpretations and limit their generalizability. Addressing these variations through color normalization, standardized imaging protocols, or controlled acquisition settings is an important direction for future work.

Acknowledgement. This work is supported by the National Science Foundation of the United States under Award 2434487. We thank anonymous reviewers for their insightful comments and inputs.

Data and Code Availability. The dataset used in this study is publicly available on Kaggle at: https://www.kaggle.com/datasets/leoscode/wound-segmentation-images/data. All experiments were performed using only the images and annotations provided in this dataset.

The code used for model training, evaluation, and analysis is openly available at: GitHub repository.

References

1. Chitca, D.-D., Popescu, V., Mastalier, B., Busu, C.: The economic burden of chronic wounds on global healthcare systems. In: Proceedings of the International Conference on Business Excellence, vol. 19, no. 1, pp. 1995–2002 (2025). https://doi.org/10.2478/picbe-2025-0155
2. Papadopoulou, V., et al.: Overcoming biological barriers to improve treatment of a *Staphylococcus aureus* wound infection. Cell Chem. Biol. **30**(5), 513–526.e5 (2023). https://doi.org/10.1016/j.chembiol.2023.04.009
3. Dave, P.: The challenges of chronic wound care and management. Asian J. Dental Health Sci. **4**(1), 45–50 (2024). https://doi.org/10.22270/ajdhs.v4i1.70

4. Mohammed, H.T., et al.: An equitable vision for wound assessment: a comparative case study of AI and human wound tissue assessment across skin tones. Limb Preserv. J. **6**(1) (2025). https://doi.org/10.56885/061916taiknw

5. Freeman, S., Sargent, M., Galarza, L., McAloney, R., Rossnagel, E.: Use of artificial intelligence–driven wound care management to enhance access to care in rural and northern communities. Int. Wound J. **22**(10) (2025). https://doi.org/10.1111/iwj.70767

6. Yimer, H.L., Degefa, H.D., Cristani, M., Cunico, F.: IoT-based coma patient monitoring system. In: IEEE International Multi-Conference on Smart Systems & Green Process (2024). https://doi.org/10.1109/SSGREC64765.2024.10807814

7. Eli, A.A., Ali, A.: Deep learning applications in medical image analysis: advancements, challenges, and future directions. arXiv. https://doi.org/10.48550/arxiv.2410.14131 (2024)

8. Elyan, E., et al.: Computer vision and machine learning for medical image analysis: recent advances, challenges, and way forward. Artif. Intell. Surg. **2**, 24–45 (2022). https://doi.org/10.20517/ais.2021.15

9. Qingge, L., Yimer, H.L., Sam, M., Annan, R., Newman, R., Qin, H.: C2BM: causal concept disentanglement for fair multimodal COVID-19 detection. Proc. AAAI Sympos. Ser. **7**(1), 567–575 (2024). https://doi.org/10.1609/aaaiss.v7i1.31734

10. Guo, X., Yang, H.: An improved segmentation network based on an encoder-decoder architecture 122–128 (2025). https://doi.org/10.1109/iccnse66404.2025.11144213

11. Singh, C., Inala, J. P., Galley, M., Caruana, R., Gao, J.: Rethinking interpretability in the era of large language models. arXiv preprint arXiv:2402.01761, https://doi.org/10.48550/arxiv.2402.01761 (2024)

12. Patil, V.R., Golbhavi, S.R.: Advances and applications of natural language processing in healthcare. Int. J. Sci. Technol. Eng. **13**(7), 2835–2842 (2025). https://doi.org/10.22214/ijraset.2025.73380

13. Salve, M.J., Bais, V.R., Brahmane, P.V., Shikalgar, N.F.K.: Exploring the applications of natural language processing in precision health and healthcare research. In: Advances in Medical Technologies and Clinical Practice, pp. 1–16 (2024). https://doi.org/10.4018/979-8-3693-4422-4.ch001

14. Kang, H., Chu, Y., Oh, B.H., Lee, S., Kim, J., Yang, S.: Development and application of automated wound tissue segmentation algorithm. In: IEEE International Conference on Bioinformatics and Biomedicine (BIBM), pp. 7064–7066 (2024). https://doi.org/10.1109/bibm62325.2024.10821778

15. Liu, T.J., Wang, H., Christian, M., Chang, C.-W., Lai, F., Tai, H.-C.: Automatic segmentation and measurement of pressure injuries using deep learning models and a LiDAR camera. Dental Sci. Rep. **13**(1) (2023). https://doi.org/10.1038/s41598-022-26812-9

16. Scebba, G., et al.: Detect-and-segment: a deep learning approach to automate wound image segmentation Comput. Vis. Pattern Recogn. (2021). arXiv: https://arxiv.org/abs/2111.01590

17. Schlereth, M., et al.: Initial investigations towards non-invasive monitoring of chronic wound healing using deep learning and ultrasound imaging, pp. 261–266 (2022). https://doi.org/10.1007/978-3-658-36932-3_56

18. Wu, Z., Liu, W.M., Chang, J.: Improving u-net performance for tumor segmentation using attention mechanisms. J. Comput. Sci. Res. *6*(4), 66–72 (2024). https://doi.org/10.30564/jcsr.v6i4.7271

19. Al-Huda, Z., Al-antari, M.A., Algburi, R.N.A., Al-maqtari, O., Rajeh, T.M., Khan, G.A.: Triplet attention-enhanced UNet architectures for advanced skin lesion segmentation 1–8 (2024). https://doi.org/10.1109/idap64064.2024.10710922
20. Deep multi-scale u-net architecture and label-noise robust training strategies for histopathological image segmentation (2022). https://doi.org/10.1109/bibe55377.2022.00027
21. Pochamreddy, M.R., Reddy, K.A., Sharma, R.: Enhancing u-net for semantic segmentation with integrated attention mechanisms 1–6 (2024). https://doi.org/10.1109/ariia63345.2024.11051872
22. RGB-D camera-based automatic wound-measurement system. IEEE Trans. Instrum. Measure., 1 (2023). https://doi.org/10.1109/tim.2023.3265758
23. Eldem, H., Ülker, E., and Işıklı, O.Y.: Encoder–decoder semantic segmentation models for pressure wound images. Imaging Sci. J. **70**(2), 75–86 (2022). https://doi.org/10.1080/13682199.2022.2163531
24. Ramachandram, D., et al.: Fully automated wound tissue segmentation using deep learning on mobile devices: cohort study. JMIR mHealth and uHealth **10**(4), e36977 (2022). https://doi.org/10.2196/36977
25. Medetec medical images: wound and dressing database. https://www.medetec.co.uk/files/medetec-dressing-image-databases.html. Accessed 21 Nov 2025
26. Morgado, A.C., Carvalho, R., Sampaio, A.F., Vasconcelos, M.J.M.: Enhancing chronic wound assessment through agreement analysis and tissue segmentation. Sci. Rep. **15**(1), 22244 (2025). https://doi.org/10.1038/s41598-025-06703-5. PMID: 40594951; PMCID: PMC12217643
27. Wang, C., et al.: FUSeg: the foot Ulcer segmentation challenge. Information **15**(3), 140 (2024). https://doi.org/10.3390/info15030140
28. Oota, S. R., Rowtula, V., Mohammed, S., Liu, M., Gupta, M.: WSNet: towards an effective method for wound image segmentation. In: Proceedings of the IEEE/CVF Winter Conference on Applications of Computer Vision (WACV), pp. 3233–3242 (2023). https://doi.org/10.1109/WACV56688.2023.00325
29. Kabir, M.A., Roy, N., Hossain, M.E., Featherston, J., Ahmed, S.: Deep learning for wound tissue segmentation: a comprehensive evaluation using a novel dataset. arXiv preprint arXiv:2502.10652 (2025)
30. Ribeiro da Costa, R.E.A., Silva, B., de Souza, R.B., Lins, R.C., Rodrigues, I.R.: Compressed deep neural networks for segmentation of cutaneous malignant wounds in images (2023). https://doi.org/10.21203/rs.3.rs-3266198/v1
31. Niri, R., et al.: Wound segmentation with u-net using a dual attention mechanism and transfer learning (2025). https://doi.org/10.1007/s10278-025-01386-w
32. Cui, C., Thurnhofer-Hemsi, K., Soroushmehr, R., et al.: Diabetic wound segmentation using convolutional neural networks. IEEE EMBC (2019)
33. Wang, C., Anisuzzaman, D. M., Williamson, V., et al.: Fully automatic wound segmentation with deep convolutional neural networks (2020). https://arxiv.org/abs/2010.05855
34. Gonzalez, R.C., Woods, R.E.: Digital Image Processing (4th ed.). Pearson (2018)
35. Pratik, S., Sharma, P., Rana, D., Balabantaray, B.K.: EU-net: efficient u-shaped deep convolutional neural network for colon polyps segmentation 1–5 (2024). https://doi.org/10.1109/icepe63236.2024.10668906
36. Yadav, A.C., Kolekar, M.H., Sonawane, Y., Kadam, G., Tiwarekar, S., Kalbande, D.: EffUNet++: a novel architecture for brain tumor segmentation using FLAIR MRI images. IEEE Access **1** (2024). https://doi.org/10.1109/access.2024.3480271

37. Pavlovčič, U., Diaci, J., Možina, J., Jezeršek, M.: Wound perimeter, area, and volume measurement based on laser 3D and color acquisition. BioMedical Eng. OnLine **14**, 39 (2015). https://doi.org/10.1186/s12938-015-0031-7
38. Dorileo, E.A.G., Frade, M.A.C., Rangayyan, R.M., de Azevedo-Marques, P.M.: Segmentation and analysis of the tissue composition of dermatological ulcers. In: Proceedings of the Canadian Conference on Electrical and Computer Engineering (CCECE), pp. 1–4 (2010). https://doi.org/10.1109/CCECE.2010.5575143
39. Filko, D., Antonić, D., Huljev, D.: WITA — application for wound analysis and management. In: Proceedings of the International Conference on E-Health Networking, Applications and Services (HealthCom), pp. 68–73 (2010). https://doi.org/10.1109/HEALTH.2010.5556533
40. Zubair, A.R., Alo, O.A.: Grey level co-occurrence matrix (GLCM) based second order statistics for image texture analysis (2024). https://doi.org/10.48550/arxiv.2403.04038
41. Gupta, C., Gondhi, N.K., Lehana, P.: Gray level co-occurrence matrix (GLCM) parameters analysis for pyoderma image variants. J. Comput. Theor. Nanosci. **17**(1), 353–358 (2020). https://doi.org/10.1166/JCTN.2020.8674
42. Wu, Y., Yang, L., Xu, T., Meng, X., Liu, H., Kang, T.D.: Multi-scale feature integration and spatial attention for accurate lesion segmentation (2025). https://doi.org/10.20944/preprints202506.1010.v1
43. Hresko, D.J., Drotar, P., Ngo, Q.C., et al.: Enhanced domain adaptation for foot ulcer segmentation through mixing self-trained weak labels. J. Digit. Imaging **38**, 455–466 (2025). https://doi.org/10.1007/s10278-024-01193-9
44. Alhababi, M., Auner, G., Malik, H., Aljasem, M., Aldoulah, Z.: Unified wound diagnostic framework for wound segmentation and classification. Mach. Learn. Appl. **19**, 100616 (2025). https://doi.org/10.1016/j.mlwa.2024.100616

Averaged Dynamics of Eco-epidemic Processes on Graph

Maria Vasilyeva[✉], Aleksei Krasnikov, Hyangim Ji, and Alexey Sadovski

Department of Mathematics, Texas A&M University – Corpus Christi, Corpus Christi, TX, USA
`maria.vasilyeva@tamucc.edu`

Abstract. We investigate eco-epidemic models to analyze the averaged dynamics of spatio-temporal systems. Specifically, we study a coupled Lotka–Volterra–Susceptible–Infected–Susceptible model with diffusion, posed on various graph structures. This eco-epidemic framework, formulated as ordinary differential equations on graphs (Graph ODEs), consists of four equations describing the evolution of susceptible and infected prey and predator populations, capturing ecological interactions, disease transmission, and spatial dispersal on a graph. Our primary goal is to understand how spatial coupling influence the system's averaged temporal dynamics. Numerical experiments demonstrate that diffusion strength has a significant impact on the temporal behavior, underscoring the crucial role of spatial processes in shaping eco-epidemic dynamics.

Keywords: Eco-epidemic model · spatio-temporal systems · graph structure · averaged temporal dynamics

1   Introduction

Understanding the interplay between ecological interactions and epidemic processes is essential for predicting population dynamics and assessing ecosystem resilience [1,9,11,12]. Infectious diseases can alter predator–prey dynamics by reducing prey abundance, changing predator feeding behavior, or even threatening predator survival when both species are susceptible to infection. Classical predator–prey models, such as the Lotka–Volterra system, capture fundamental ecological interactions of growth, mortality, and predation, while epidemic models such as the Susceptible–Infected–Susceptible (SIS) framework describe disease transmission within populations. When combined, these models provide a powerful tool to study eco-epidemiological systems where population interactions and disease spread are tightly coupled [2–6].

Spacial interaction plays a crucial role in eco-epidemiological systems by allowing individuals or populations to move across space, thereby influencing local interactions and disease spread [8,10,13,14,19,21,24]. To capture such spatial and structural complexity, eco-epidemic models, combining classical frameworks can be extended to graph-based domains, where nodes represent local pop-

K. L. Kabir et al. (Eds.): BICOB 2026, CCIS 2977, pp. 62–75, 2026.
https://doi.org/10.1007/978-3-032-26028-4_5

ulations and edges encode spatial or ecological connectivity [17,18]. Understanding the effect of diffusion is essential for capturing realistic ecological and epidemic dynamics, motivating the need to explore how varying diffusion strengths impact average population trajectories and overall system behavior. In this work, we study the averaged dynamics of such coupled eco-epidemic processes defined on various graph topologies. By analyzing the spatially averaged quantities, we aim to understand how diffusion strength, network structure, and coupling mechanisms influence the emergent temporal behavior of predator–prey–disease systems.

The paper is organized as follows. Section 2 introduces the eco-epidebreakmiological modeling framework formulated as a system of ODEs. Section 3 incorporates discrete diffusion via graph-based spatial coupling. In Sect. 4, we present the time-discretization of the full system. Section 5 investigates the influence of spatial diffusion on the averaged dynamics. We construct spatial networks by randomly distributing nodes within the domain and coupling them through a diffusion kernel acting on nearest neighbors, and we analyze how variations in diffusion strength and connectivity patterns shape the temporal evolution of the spatially averaged populations. The paper ends with a conclusion.

2   Temporal Model

We consider a coupled eco-epidemic interaction that extends the classical predator–prey framework by incorporating disease transmission within both prey and predator populations. Specifically, the model distinguishes between susceptible and infected classes in each species: u_s and u_i represent the susceptible and infected prey populations, while v_s and v_i denote the susceptible and infected predator populations. The temporal evolution of these four interacting populations is described by the following system of ordinary differential equations (ODEs) [6]

$$
\begin{aligned}
\frac{du_s}{dt} &= \alpha_s u_s + \alpha_i u_i - \beta_s^s u_s v_s - \beta_i^s u_s v_i - \gamma_s^u u_s^2 - \sigma_i^s u_s u_i - \sigma_{vi}^s u_s v_i + \omega_i u_i, \\
\frac{du_i}{dt} &= -\beta_s^i u_i v_s - \beta_i^i u_i v_i - \gamma_i^u u_i^2 + \sigma_i^s u_s u_i + \sigma_{vi}^s u_s v_i - \omega_i u_i, \\
\frac{dv_s}{dt} &= -\gamma_s^v v_s + \delta_s^s v_s u_s + \delta_i^s v_s u_i - \sigma_i^{vs} v_s u_i - \sigma_{vi}^{vs} v_s v_i + \omega_{vi} v_i, \\
\frac{dv_i}{dt} &= -\gamma_i^v v_i + \delta_s^i v_i u_s + \delta_i^i v_i u_i + \sigma_i^{vs} v_s u_i + \sigma_{vi}^{vs} v_s v_i - \omega_{vi} v_i,
\end{aligned}
\tag{1}
$$

where α_s and α_i are the intrinsic growth rates of susceptible and infected prey; $\beta_s^s, \beta_i^s, \beta_s^i$, and β_i^i denote the predation rates of susceptible and infected predators on susceptible and infected prey; γ_s^u and γ_i^u are the density-dependent mortality rates of susceptible and infected prey, while γ_s^v and γ_i^v are the intrinsic mortality rates of susceptible and infected predators; $\delta_s^s, \delta_i^s, \delta_s^i$, and δ_i^i represent the reproduction rates of susceptible and infected predators arising from consumption of susceptible and infected prey; σ_i^s and σ_{vi}^s are the transmission rates from infected prey and infected predators to susceptible prey, and σ_i^{vs} and σ_{vi}^{vs} are the corresponding transmission rates to susceptible predators; ω_i and ω_{vi} are the recovery rates of infected prey and infected predators returning to the susceptible state.

The system (1) is supplemented with the following initial conditions

$$u_s(0) = u_s^0, \quad u_i(0) = u_i^0, \quad v_s(0) = v_s^0, \quad v_i(0) = v_i^0, \quad t = 0.$$

Let $y = (u_s, u_i, v_s, v_i)$ be a state vector of susceptible/infected prey and predator populations. Then, we can write (1) as follows

$$\frac{dy}{dt} = f(t, y), \tag{2}$$

with initial condition

$$y(0) = y^0$$

and

$$f(t, y) = \begin{pmatrix} \alpha_s u_s + \alpha_i u_i - \beta_s^s u_s v_s - \beta_i^s u_s v_i - \gamma_s^u u_s^2 - \sigma_i^s u_s u_i - \sigma_{vi}^s u_s v_i + \omega_i u_i \\ -\beta_s^i u_i v_s - \beta_i^i u_i v_i - \gamma_i^u u_i^2 + \sigma_i^s u_s u_i + \sigma_{vi}^s u_s v_i - \omega_i u_i \\ -\gamma_s^v v_s + \delta_s^s v_s u_s + \delta_i^s v_s u_i - \sigma_i^{vs} v_s u_i - \sigma_{vi}^{vs} v_s v_i + \omega_{vi} v_i \\ -\gamma_i^v v_i + \delta_s^i v_i u_s + \delta_i^i v_i u_i + \sigma_i^{vs} v_s u_i + \sigma_{vi}^{vs} v_s v_i - \omega_{vi} v_i \end{pmatrix},$$

where $y(0) = (u_s^0, u_i^0, v_s^0, v_i^0)$.

3   Spatio-temporal Model on Graph

We consider a spatial domain where a set of points (point cloud) is given as follows

$$\mathcal{X} = \{x_1, x_2, \ldots, x_N\}, \quad x_i \in \Omega \subset \mathbb{R}^d.$$

and let $y(x, t) = (y_1(x, t), \ldots, y_N(x, t))$ be a vector variable and N denotes number of points.

To model diffusion on this discrete spatial domain, we construct a weighted graph over the points, where edges encode the strength of dispersal or movement between locations. The edge weights are assigned according to the Gaussian kernel [7]

$$W_{ij}^{GK} = \exp\left(-d_{ij}^2/\sigma^2\right),$$

where d_{ij} is the pairwise distance matrix

$$d_{ij} = \|x_i - x_j\|, \quad i, j = 1, \ldots, N.$$

and parameter $\sigma > 0$ controlling the diffusion range. The diagonal entries are set to zero, $W_{ii} = 0$, to exclude self-interaction. The Gaussian kernel is widely used in graph-based diffusion models because it assigns large weights to nearby points while rapidly suppressing long-range connections, ensuring locality of interactions. Its smooth decay also makes the resulting graph Laplacian a consistent discretization of continuous diffusion operators as the point cloud becomes dense.

In many applications, it is desirable to retain only the k nearest neighbors for each node, which effectively sparsifies the connectivity of the graph. To implement this, for each node i, we select the k smallest distances and keep only the corresponding edges. Let m_{ij} denote a boolean indicator of whether node j is among the k nearest neighbors of node i

$$m_{ij} = \begin{cases} 1, & \text{if } j \text{ is among the } k \text{ nearest neighbors of } i, \\ 0, & \text{otherwise.} \end{cases}$$

The adjacency matrix is then updated by setting

$$W_{ij} = W_{ij}^{GK} \, m_{ij},$$

and setting $W \leftarrow \frac{1}{2}\left(W + W^\top\right)$ to symmetrize.

From the weighted adjacency matrix W, the graph Laplacian is constructed as

$$L = D - W, \quad D = \operatorname{diag}\left(\sum_j W_{1j}, \ldots, \sum_j W_{Nj}\right), \tag{3}$$

which provides a discrete operator for modeling diffusion on the network. In particular, the Laplacian encodes nonlocal interactions: large weights correspond to strong coupling between distant patches, while small weights reduce influence.

By incorporating the graph Laplacian into the diffusion term of the spatio-temporal system, we obtain the following system (Graph ODEs)

$$\frac{dy}{dt} = f(t, u) - Ay, \quad A = \operatorname{diag}(L_1, \ldots, L_K), \quad L_k = \kappa_k L, \tag{4}$$

where $f(t, y)$ is the nonlinear reaction term, A is a block-diagonal matrix representing discrete diffusion operators, L_k denotes the discrete diffusion operator for each component and κ_k is the scaling parameter. Note that, we neglect cross-diffusion effects and assume a block-diagonal diffusion operator so that each species diffuses independently. More complex full (non-block-diagonal) cross-diffusion models will be considered in future work.

We note that the graph Laplacian L is closely related to a discrete diffusion operator with free (no-flux) boundary conditions and therefore preserves the spatial average. Other types of boundary conditions, such as Dirichlet or Robin conditions, can be incorporated by appropriately modifying the graph operator or by introducing constrained or weighted boundary nodes.

3.1 Properties of Graph Laplacian

The weight matrix W is symmetric with nonnegative entries and zero diagonal. The corresponding graph Laplacian L is therefore symmetric and positive semidefinite. In particular, for any $z \in \mathbb{R}^N$,

$$z^\top L\, z = \frac{1}{2} \sum_{i,j=1}^{N} W_{ij}\, (z_i - z_j)^2 \geq 0.$$

Moreover, the Laplacian satisfies the zero row-sum and column-sum properties,

$$L\mathbf{1} = 0, \qquad \mathbf{1}^{\top}L = 0^{\top},$$

where $\mathbf{1} = (1, \ldots, 1)^{\top}$.

Since the graph is connected and the Laplacian L is symmetric, its eigenvalues satisfy

$$0 = \lambda_1(L) < \lambda_2(L) \leq \cdots \leq \lambda_N(L),$$

where the eigenvector associated with $\lambda_1(L)$ is the constant vector

$$\phi_1 = \frac{1}{\sqrt{N}}\mathbf{1}.$$

As a consequence, the nullspace of L is given by $\ker(L) = \text{span}\{\mathbf{1}\}$.

Let $\mathbf{1}^{\perp} = \{w \in \mathbb{R}^N : \mathbf{1}^{\top}w = 0\}$ denote the orthogonal complement of the constant mode. Any vector $w \in \mathbf{1}^{\perp}$ admits the expansion

$$w = \sum_{k=2}^{N} c_k \phi_k,$$

where $\{\phi_k\}_{k=1}^{N}$ is an orthonormal eigenbasis of L. Using the spectral decomposition of L, we obtain

$$w^{\top}L\,\mathrm{w} = \sum_{k=2}^{N} \lambda_k c_k^2 \geq \lambda_2(L) \sum_{k=2}^{N} c_k^2 = \lambda_2(L)\,\|w\|^2,$$

which proves the inequality

$$w^{\top}Lw \geq \lambda_2(L)\,\|w\|^2 \qquad \text{for all } w \in \mathbf{1}^{\perp}.$$

Remark 1. In the graph-based model (4), the effective strength of diffusion is governed jointly by the scaling parameters κ_k multiplying the graph Laplacian and by the spectral properties of the Laplacian itself. These spectral properties depend on the network size N, the number of nearest neighbors k, and the Gaussian kernel parameter σ. In particular, σ controls the spatial interaction range of the weights, while k determines the sparsity and connectivity of the graph; together, they influence the eigenvalues of L, including the spectral gap $\lambda_2(L)$, which plays a central role in averaged solution dynamics. In the numerical experiments, we fix σ and vary N, k, and κ_k independently in order to isolate and illustrate their qualitative effects on the averaged dynamics, rather than to enforce a strict diffusion scaling across different graph configurations.

3.2   Averaged Dynamics

Next we show that, the spatially averaged dynamics of the graph-based system (4) converge to the corresponding well-mixed ODE dynamics as time increases. Define the network average of a nodal vector $z \in \mathbb{R}^N$ by

$$\bar{z} = \frac{1}{N}\mathbf{1}^{\top}z, \qquad \Pi = I - \frac{1}{N}\mathbf{1}\mathbf{1}^{\top},$$

so that $z = \bar{z}\mathbf{1} + z_\perp$ with $z_\perp = \Pi z$ and $\mathbf{1}^\top z_\perp = 0$. For the full state y, define the block projection $\Pi_K = \mathrm{diag}(\Pi, \dots, \Pi)$ and $y_\perp = \Pi_K y$.

We have $\mathbf{1}^\top L_k = 0^\top$ for every k. Therefore, taking componentwise averages of (4) yields

$$\frac{d}{dt}\bar{y}(t) = \overline{f(t, y(\cdot, t))},$$

i.e., the diffusion term does not directly affect the averaged dynamics.

Assume that the reaction term is Lipschitz with constant $K_F > 0$, i.e., $\|f(t, p) - f(t, q)\| \le K_F\|p - q\|$. Then, for the fluctuation energy $E(t) = \frac{1}{2}\|y_\perp(t)\|^2$, one obtains the estimate

$$\dot{E}(t) \le \left(K_F - \kappa_{\min}\lambda_2(L)\right)\|y_\perp(t)\|^2, \qquad \kappa_{\min} = \min_{1 \le k \le K} \kappa_k,$$

and hence, whenever $\kappa_{\min}\lambda_2(L) > K_F$,

$$\|y_\perp(t)\| \le \|y_\perp(0)\|\, e^{-(\kappa_{\min}\lambda_2(L) - K_F)t}.$$

Thus, sufficiently strong diffusion (or sufficiently large spectral gap) enforces spatial homogenization.

As $y_\perp(t) \to 0$, the nodal states become nearly uniform, so smoothness of f implies

$$\overline{f(t, y(\cdot, t))} = f(t, \bar{y}(t)) + \mathcal{O}(\|y_\perp(t)\|).$$

Therefore, after the homogenization transient, the averaged dynamics satisfy

$$\frac{d}{dt}\bar{y}(t) \approx f(t, \bar{y}(t)),$$

Thus, for connected graphs and sufficiently strong diffusion, the spatially averaged Graph ODE solution approaches the corresponding well-mixed ODE behavior at large times. At the same time, our primary interest lies in the full temporal evolution, as spatial coupling and network structure can significantly influence the averaged dynamics at early and intermediate times.

4   Discrete System

Explicit schemes, such as forward Euler or Runge–Kutta methods, are straightforward to implement and allow direct evaluation of the nonlinear reaction terms. Their stability, however, is limited by a CFL-type condition that depends on the diffusion strength. Implicit methods provide enhanced stability but require solving large coupled nonlinear systems at each time step. A practical compromise is offered by implicit–explicit schemes or operator-splitting approaches, which treat diffusion implicitly while updating the nonlinear terms explicitly.

To numerically solve the time-dependent problems (2) and (4) for $0 \le t \le T$, we discretize the time interval $[0, T]$ using a sequence of time steps Δt_n, which can vary depending on the solution behavior. The discrete time levels are given by

$$t^0 = 0, \quad t^{n+1} = t^n + \Delta t_n, \qquad n = 0, 1, 2, \dots, N_t - 1,$$

where $\{t^n\}_{n=0}^{N_t}$ denotes the sequence of time levels, Δt_n may vary adaptively depending on the solution behavior and N_t is the number of time steps. We denote by y^n the numerical approximation of $y(x,t)$ at time t^n, i.e.,

$$y^n \approx y(x, t^n).$$

A general s-stage Runge–Kutta scheme takes the form

$$Y_i = y^n + \Delta t_n \sum_{j=1}^{s} a_{ij}\, \tilde{f}(t^n + c_j \Delta t_n, Y_j), \qquad i = 1, \ldots, s, \tag{5}$$

$$y^{n+1} = y^n + \Delta t_n \sum_{i=1}^{s} b_i\, \tilde{f}(t^n + c_i \Delta t_n, Y_i), \tag{6}$$

with $\tilde{f}(t,y) = f(t,y)$ (ODEs) and $\tilde{f}(t,y) = f(t,y) - Ay$ (Graph ODEs). Here a_{ij}, b_i, c_i are the coefficients of the chosen Runge–Kutta method. We employ adaptive Runge–Kutta solvers, which adjust the step sizes Δt_n according to local error estimates. This provides higher-order accuracy in time and ensures efficient resolution of the rapid temporal variations characteristic of Lotka–Volterra type dynamics, while allowing larger time steps when the solution evolves smoothly.

In our computations, we employ the classical adaptive Runge–Kutta method of order 4 shown in Eq. (5). This scheme embeds a fourth- and a fifth-order method, allowing an efficient estimate of the local truncation error at each step. Specifically, for every trial step of length Δt_n, the solver computes two approximations, $y_{(4)}^{n+1}$ and $y_{(5)}^{n+1}$, corresponding to the fourth- and fifth-order formulas. The difference between these approximations provides an a posteriori estimate of the local error, which is compared against a prescribed tolerance. If the estimated error exceeds the tolerance, the step is rejected and recomputed with a smaller time step; otherwise, the step is accepted, and the next step size is increased or decreased according to standard controller strategies. This adaptive mechanism ensures that small time steps are taken when the solution exhibits rapid temporal variations, while significantly larger steps are allowed during slowly varying phases of the dynamics. As a result, the Runge–Kutta scheme achieves high accuracy and efficiency without the need for problem-specific tuning.

Next, after numerically solving the graph-based system using a Runge–Kutta time-integration scheme, we compute spatially averaged solutions by averaging the nodal values over all graph nodes at each time step. The spatially averaged state is defined as the componentwise arithmetic mean over the graph nodes

$$\bar{y}_k^n := \frac{1}{N} \sum_{\ell=1}^{N} y_{k,\ell}^n, \qquad \ell = 1, \ldots, K, \tag{7}$$

where $y_{k,\ell}^n$ denotes the k-th component of y_ℓ^n at node ℓ. In particular, for $y_\ell^n = (u_{s,\ell}^n, u_{i,\ell}^n, v_{s,\ell}^n, v_{i,\ell}^n)^\top$, we define

$$\bar{u}_s^n = \frac{1}{N} \sum_{\ell=1}^{N} u_{s,\ell}^n, \quad \bar{u}_i^n = \frac{1}{N} \sum_{\ell=1}^{N} u_{i,\ell}^n, \quad \bar{v}_s^n = \frac{1}{N} \sum_{\ell=1}^{N} v_{s,\ell}^n, \quad \bar{v}_i^n = \frac{1}{N} \sum_{\ell=1}^{N} v_{i,\ell}^n.$$

for all $n = 1, 2, \ldots, N_t$.

5 Numerical Experiments

We investigate how the spatial resolution of the network and the number of nearest–neighbor connections influence the average dynamics of the system. We consider four spatial resolutions, corresponding to graphs with $N = 10$, 100, 1000 nodes randomly generated in the unit square. The interaction strength between two nodes is determined by a Gaussian kernel with $\sigma = 0.2$, which assigns larger weights to closer nodes. To enforce sparsity and control the locality of the interactions, we retain only the k nearest neighbors of each node and remove all other edges. We consider $k = 5$, 10, 50, representing increasingly dense local connectivity patterns. We remark that if $k > N$, then every node is connected to all others, and no pruning occurs. For example, when $N = 100$ and $k = 10$, each node is linked to its ten closest neighbors, whereas for $N = 10$ the choices $k = 10$ and $k = 50$ both result in the same fully connected graph, since all nodes lie within the nearest–neighbor threshold. This setup allows us to systematically examine how both the graph resolution (N) and the interaction locality (k) influence the qualitative behavior of the resulting dynamical system.

Figure 1 illustrates the underlying k-nearest neighbor graph structures used in the simulations for all combinations of $N = 5, 10, 100, 1000$ and $k = 5, 10, 50$. Each row corresponds to a fixed value of k, representing the number of nearest-neighbor connections per node, while each column corresponds to a different network resolution N, representing the total number of nodes. These visualizations highlight how both the network size and the connectivity pattern influence the sparsity and local structure of the graphs, which in turn can affect the dynamics of the system.

We consider a coupled prey–predator system with infection dynamics, described by the variables u_s, u_i, v_s, v_i, representing susceptible and infected prey and predators, respectively. The system is initialized with

$$u_s(0) = u_s^0 = 0.7, \quad u_i(0) = u_i^0 = 0.3, \quad v_s(0) = v_s^0 = 0.8, \quad v_i(0) = v_i^0 = 0.2,$$

which correspond to the initial densities of each population component.

The intrinsic growth, mortality, predation, reproduction, and infection rates are chosen as representative dimensionless values to illustrate the qualitative behavior of the model. These parameters are not calibrated to any specific biological system or empirical dataset; incorporating biologically realistic parameter values based on real data will be considered in future work. Specifically, the prey birth and death rates are $\alpha_s = 0.02$, $\alpha_i = 0.01$, $\gamma_s^u = 0.02$, $\gamma_i^u = 0.02$, while predator death rates are $\gamma_s^v = 0.01$ and $\gamma_i^v = 0.01$. The interaction terms are controlled by the predation rate $\beta = 0.02$ and the reproduction rate $\delta = 0.04$ for Lotka–Volterra, and by the disease transmission rates $\sigma = 0.01$ and recovery rates $\omega = 0.01$ for the SIS infection dynamics. The full model, given in Eq. (1), combines standard Lotka–Volterra predator–prey interactions with SIS-type infection terms. Prey populations grow intrinsically and are reduced by predation and density-dependent mortality. Predators reproduce proportionally to consumed prey and die naturally. The parameters δ control predator reproduction from consuming prey, σ governs infection transmission between populations,

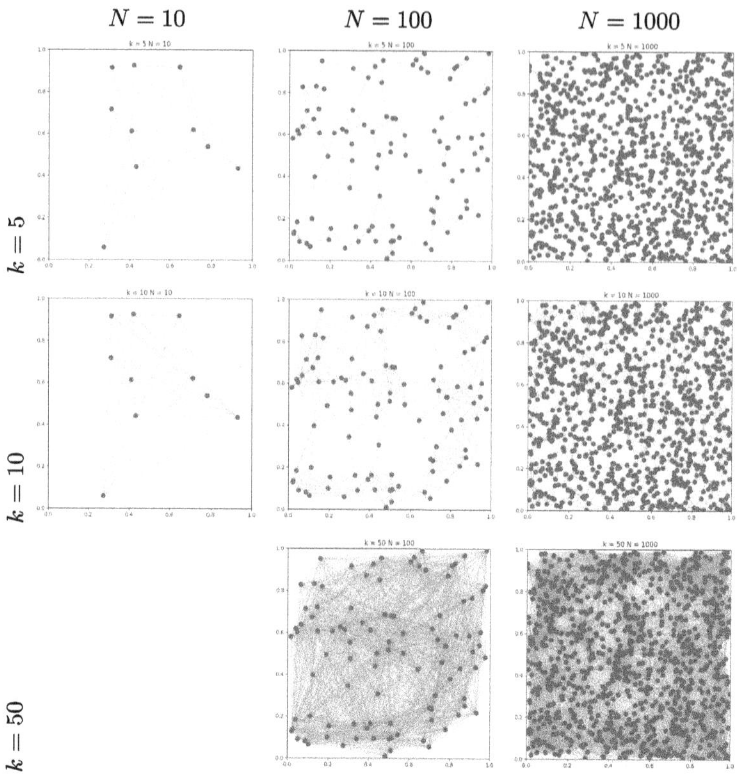

Fig. 1. K-nearest neighbor graph structures for varying $N = 5, 10, 100, 1000$ and $k = 5, 10, 50$, where N denotes the number of nodes and k represents the number of nearest-neighbor connections per node.

and ω describes recovery of infected individuals back to the susceptible state. The system is then evolved in time to explore the interplay between predation, reproduction, and disease transmission in both prey and predator populations.

Figure 2 illustrates the comparison between the temporal (ODE) model and the corresponding spatio-temporal (Graph ODEs) dynamics by showing the spatially averaged solutions for the four population components u_s, u_i, v_s, and v_i. The Graph ODEs averaged solutions are computed on a graph with $N = 100$ nodes and $k = 10$ nearest neighbors, and we consider multiple diffusion scales to assess their impact on the system behavior. The base diffusion coefficients are set to $(10^{-3}, 10^{-4}, 10^{-4}, 10^{-5})$, corresponding to the four population components. To explore the effect of diffusion strength, we also consider cases with diffusion values 100 times smaller and 100 times larger than the base case. Specifically, we consider the following diffusion settings:

1. *ODEs:* $\kappa = (0, 0, 0, 0)$ (no diffusion).
2. *Graph ODEs:Base:* $\kappa = (10^{-3}, 10^{-4}, 10^{-4}, 10^{-5})$ (baseline spatial coupling).
3. *Graph ODEs:Lower:* $\kappa = (10^{-5}, 10^{-6}, 10^{-6}, 10^{-7})$.

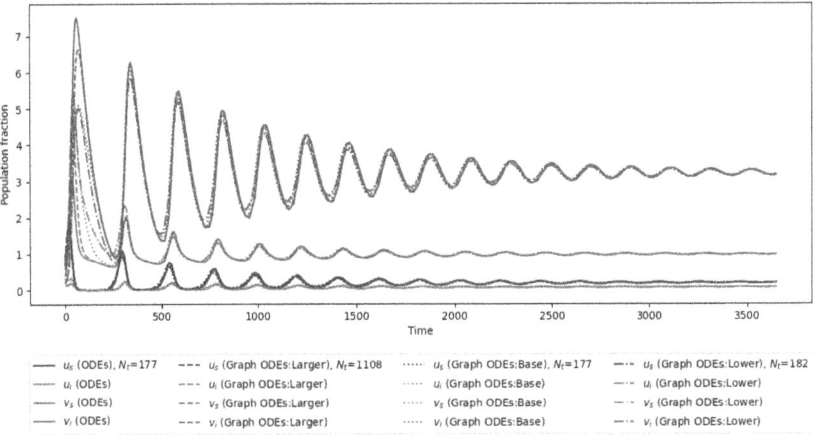

Fig. 2. Comparison of temporal (ODEs) and spatio-temporal (Graph ODEs) solutions under different diffusion regimes: ODEs ($\kappa = 0$); Graph ODEs (Lower) $\kappa = (10^{-5}, 10^{-6}, 10^{-6}, 10^{-7})$; Graph ODEs (Base) $\kappa = (10^{-3}, 10^{-4}, 10^{-4}, 10^{-5})$; and Graph ODEs (Larger) $\kappa = (10^{-1}, 10^{-2}, 10^{-2}, 10^{-3})$. Spatially averaged time evolution of the four population components u_s, u_i, v_s, and v_i is shown. N_t denotes the number of time steps taken by the adaptive time integrator.

4. Graph ODEs:Larger: $\kappa = (10^{-1}, 10^{-2}, 10^{-2}, 10^{-3})$.

These diffusion coefficients represent effective spatial coupling resulting from Gaussian-kernel-based interactions between nodes, capturing the influence of spatial structure on population dynamics. The comparison highlights that, over time, the spatially averaged PDE solutions gradually converge toward the ODE dynamics. However, significant deviations are observed during the early stages of the simulations, particularly for smaller diffusion, where the dynamics are slower and the differences from the ODE solutions are more pronounced. Larger diffusion accelerates the spatial averaging process, leading to faster alignment with the ODE behavior and smaller initial discrepancies. These results emphasize the critical role of spatial coupling and diffusion in shaping the averaged dynamics of eco-epidemic systems. All simulations are performed over a time horizon of $T_{\max} = 3650$ days (10 years). For the simulations with no diffusion or small diffusion (base and lower diffusion cases), the adaptive Runge–Kutta scheme requires approximately $N_t \approx 180$ time steps to advance the solution over the full time horizon. In contrast, when the diffusion strength is increased significantly, the number of adaptive time steps rises substantially, reaching $N_t = 1108$ steps. This increase reflects the greater stiffness of the system induced by stronger diffusion, which forces the adaptive solver to take smaller time steps to maintain stability and accuracy. These observations highlight the pronounced effect of diffusion strength on the temporal resolution required for accurately capturing the spatio-temporal dynamics of the system. We note that the numerical scheme belongs to the class of explicit methods, which imposes a restriction on the time step size;

hence, we focus on cases with relatively small diffusion. Moreover, the dynamics of the considered eco-epidemic system are highly oscillatory, and we employ high-order time-stepping with adaptivity to handle these rapid variations. In general, for larger diffusion coefficients, one could adopt a stable implicit scheme for the diffusion term while retaining an explicit approximation for the reaction term [15, 16, 20, 21]. We plan to investigate such schemes in future work.

From an ecological and epidemiological perspective, the observed convergence of the spatio-temporal (Graph ODE) solutions toward the corresponding ODE dynamics reflects the homogenizing effect of spatial mixing on population and disease dynamics. Smaller diffusion corresponds to limited mobility or weak contact between subpopulations, leading to prolonged transient behavior and delayed synchronization across the spatial network; this manifests as slower convergence and larger early-time deviations from the well-mixed ODE model. In contrast, larger diffusion represents increased movement, contact, or dispersal, which rapidly smooths spatial heterogeneities and drives the system toward a mean-field regime consistent with classical ODE dynamics. The early-time discrepancies between diffusion regimes therefore encode biologically meaningful information about the timescales over which spatial structure influences disease spread and predator–prey interactions. While our primary emphasis is computational, these results suggest that neglecting spatial coupling may lead to misestimation of transient outbreak intensity, timing, and population oscillations in eco-epidemic systems, particularly in settings characterized by weak dispersal or fragmented habitats.

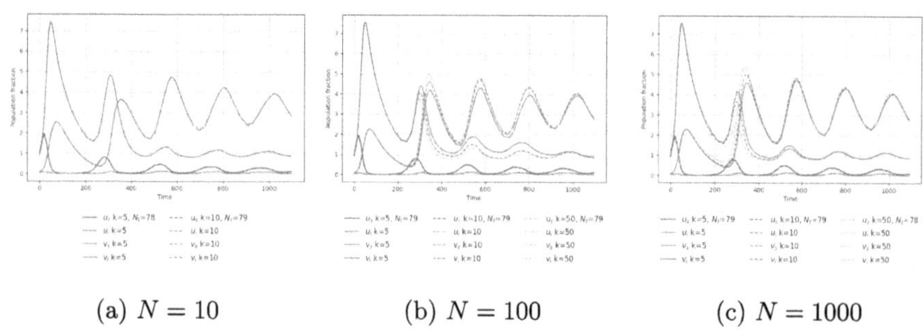

(a) $N = 10$ (b) $N = 100$ (c) $N = 1000$

Fig. 3. Graph ODEs solutions for different graph sizes $N = 10, 100,$ and 1000, illustrating the effect of the number of nearest-neighbor connections ($k = 5, 10,$ and 50). Spatially averaged time evolution of the four population components u_s, u_i, v_s, and v_i is shown. N_t denotes the number of time steps taken by the adaptive time integrator.

Figure 3 illustrates the behavior of the spatio-temporal model for different grid sizes, where we consider $N = 10,$ 100, and 1000 to examine the effect of connection sparsity on the spatially averaged solution. Simulations are performed over a time horizon of $T_{\max} = 3$ years, and the Runge–Kutta scheme executes around $N_t \approx 76 - 70$ adaptive time steps. For $N = 10$, the effect of the number of nearest-neighbor connections ($k = 5$ and 10) is negligible. However, for $N = 100$ and 1000, the influence of the connectivity becomes more pronounced. Larger values of k lead to stronger interactions, which manifest as higher oscillations in the averaged populations. This effect is particularly noticeable at the early stages of the simulation, after which the solutions converge toward behavior similar to the model without spatial variation. These results highlight that the spatial connectivity plays a significant role in shaping the transient dynamics, especially for finer grids and denser networks.

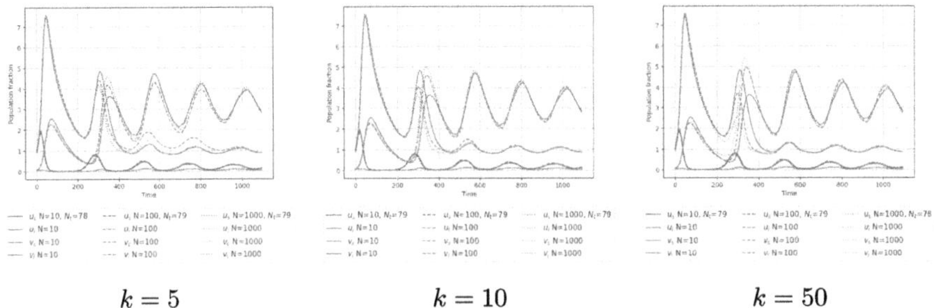

$$k = 5 \qquad\qquad k = 10 \qquad\qquad k = 50$$

Fig. 4. Graph ODEs solutions for different nearest-neighbor connectivity values $k = 5, 10$, and 50, shown for multiple graph sizes $N = 10, 100$, and 1000. Spatially averaged time evolution of the four population components u_s, u_i, v_s, and v_i is shown. N_t denotes the number of time steps taken by the adaptive time integrator.

Figure 4 illustrates how the spatially averaged dynamics of the spatio-temporal model depend on the number of nearest neighbors k, which controls the sparsity of the underlying network. Across all tests, we observe that increasing the number of spatial nodes N leads to more pronounced differences in the averaged solution, resulting in larger amplitudes of oscillation. These differences are particularly significant for larger N when combined with higher values of k, indicating that both the grid resolution and the network connectivity strongly influence the transient dynamics. Overall, the results highlight that network density and resolution play a critical role in shaping the early-time behavior of the system, while the solutions eventually converge toward dynamics similar to the well-mixed model.

From a biological and epidemiological perspective, variation in the connectivity parameter k reflects changes in effective contact rates or mobility between local subpopulations. Larger values of k correspond to denser interaction networks, facilitating faster disease transmission and stronger coupling between predator–prey compartments, which manifests as amplified early-time oscillations, particularly for finer graphs ($N = 100$ and 1000). In contrast, smaller values of k represent limited local interactions and reduced dispersal, leading to more localized dynamics and weaker transient amplification. For coarse networks ($N = 10$), rapid spatial mixing results in minimal sensitivity to connectivity, consistent with a well-mixed population assumption. Across all cases, the convergence of spatially averaged solutions toward similar long-term behavior indicates that spatial structure primarily modulates transient outbreak intensity and timing, while asymptotic population levels are governed by intrinsic ecological and epidemiological processes.

6   Conclusion

In conclusion, our study demonstrates that the spatial structure and connectivity play a pivotal role in shaping the averaged dynamics of eco-epidemic systems. By investigating a coupled Lotka–Volterra–SIS model on various graph structures, we find that both the number of nodes and the density of nearest-neighbor connections strongly influence the transient behavior of the populations. Larger networks with higher connectivity exhibit greater oscillation amplitudes, particularly at early times. Under the free flux boundary conditions considered in this work, the spatially averaged dynamics tend to converge toward the corresponding ODEs solution, indicating that the long-term behavior is largely governed by the intrinsic ecological and epidemiological interactions. These results highlight the need to account for spatial coupling and network structure in shaping early-time dynamics. Moreover, the graph-based formulation provides a flexible foundation for more general settings, including non-homogeneous and alternative boundary conditions (such as Dirichlet or Robin), where spatial effects can persist and qualitatively alter the system behavior, as demonstrated in our earlier work with Dirichlet boundary conditions [22]. Building on this framework, we have developed machine learning approaches to infer hidden eco-epidemiological interactions under known spatial coupling [23]. In future work, we plan to investigate operator-learning methods capable of accurately predicting both spatial and local eco-epidemiological dynamics using real datasets.

References

1. Anderson, R.M.: Populations and infectious diseases: ecology or epidemiology? J. Anim. Ecol. **60**(1), 1–50 (1991)
2. Bairagi, N., Roy, P.K., Chattopadhyay, J.: Role of infection on the stability of a predator-prey system with several response functions–a comparative study. J. Theor. Biol. **248**(1), 10–25 (2007)
3. Chattopadhyay, J., Arino, O.: A predator-prey model with disease in the prey. Nonlinear Anal. **36**, 747–766 (1999)
4. Han, L., Ma, Z., Hethcote, H.: Four predator prey models with infectious diseases. Math. Comput. Model. **34**(7–8), 849–858 (2001)
5. Hethcote, H.W., Wang, W., Han, L., Ma, Z.: A predator-prey model with infected prey. Theor. Popul. Biol. **66**(3), 259–268 (2004)
6. Ji, H., Vasilyeva, M., Mbroh, N.A., Sadovski, A.: Epidemic dynamics in the spatio-temporal predator-prey model. Front. Appl. Math. Stat. **11**, 1642,676 (2025)
7. Kriege, N.M., Johansson, F.D., Morris, C.: A survey on graph kernels. Appl. Netw. Sci. **5**(1), 6 (2020)

8. Malchow, H., Petrovskii, S.V., Venturino, E.: Spatiotemporal Patterns in Ecology and Epidemiology: Theory, Models, and Simulation. Chapman and Hall/CRC (2007)
9. Prenter, J., MacNeil, C., Dick, J.T., Dunn, A.M.: Roles of parasites in animal invasions. Trends Ecol. Evol. **19**(7), 385–390 (2004)
10. Satō, K., Matsuda, H., Sasaki, A.: Pathogen invasion and host extinction in lattice structured populations. J. Math. Biol. **32**(3), 251–268 (1994)
11. Smith, R.L., Smith, R.L., Hickman, G.C., Hickman, S.M.: Elements of ecology (1998)
12. Su, M.: Modeling at the interface of ecology and epidemiology. Comput. Ecol. Softw. **5**(4), 367 (2015)
13. Su, M., Hui, C., Zhang, Y., Li, Z.: Spatiotemporal dynamics of the epidemic transmission in a predator-prey system. Bull. Math. Biol. **70**(8), 2195–2210 (2008)
14. Tilman, D., Kareiva, P.: Spatial Ecology: The Role of Space in Population Dynamics and Interspecific Interactions, vol. 30. Princeton University Press (1997)
15. Vabishchevich, P.N.: Additive Operator-Difference Schemes: Splitting Schemes. Walter de Gruyter (2013)
16. Vasilyeva, M.: Implicit-explicit schemes for decoupling multicontinuum problems in porous media. J. Comput. Phys. **519**, 113,425 (2024)
17. Vasilyeva, M.: Generalized multiscale finite element method for discrete network (graph) models. J. Comput. Appl. Math. **457**, 116,275 (2025)
18. Vasilyeva, M., Brannick, J., Southworth, B.S.: Multiscale graph reduction for heterogeneous and anisotropic discrete diffusion processes. arXiv preprint arXiv:2510.10894 (2025)
19. Vasilyeva, M., Sadovski, A., Palaniappan, D.: Multiscale solver for multicomponent reaction–diffusion systems in heterogeneous media. J. Comput. Appl. Math. **427**, 115,150 (2023)
20. Vasilyeva, M., Southworth, B.S., He, Y., Wang, M.: Implicit-explicit scheme with multiscale Vanka two-grid solver for heterogeneous unsaturated poroelasticity. arXiv preprint arXiv:2508.12197 (2025)
21. Vasilyeva, M., Stepanov, S., Sadovski, A., Henry, S.: Uncoupling techniques for multispecies diffusion-reaction model. Computation **11**(8), 153 (2023)
22. Vasilyeva, M., Wang, Y., Stepanov, S., Sadovski, A.: Numerical investigation and factor analysis of the spatial-temporal multi-species competition problem. arXiv preprint arXiv:2209.02867 (2022)
23. Vasilyeva, M., Wei, Z., Gajamannage, K., Ji, H., Krasnikov, A., Sadovski, A.: Learning ecological and epidemic processes using neural odes, kolmogorov-arnold network odes and sindy. arXiv preprint arXiv:2601.09811 (2026)
24. Wang, Y., Vasilyeva, M., Sadovski, A.: Prediction of the survival status for multispecies competition system. In: AIP Conference Proceedings, vol. 2872, p. 120083. AIP Publishing LLC (2023)

Predicting Equilibrium in a Predator–Prey System with Disease

Hyangim Ji[(✉)], Maria Vasilyeva, and Alexey Sadovski

Department of Mathematics, Texas A&M University – Corpus Christi,
Corpus Christi, TX, USA
`hyangim.ji@tamucc.edu`
`https://vmasha.github.io/mvasilyeva/`

Abstract. We investigate the use of supervised machine learning models to predict equilibrium outcomes in a predator–prey system with disease. Epidemics in this system are described by a coupled model that integrates Lotka–Volterra (LV) predator–prey dynamics with a Susceptible–Infected–Susceptible (SIS) epidemic framework. A synthetic dataset is generated by randomly varying model parameters within prescribed ranges and recording the final susceptible and infected populations as equilibrium states. This dataset is then used to train a wide suite of supervised models, including a linear model, tree ensembles, boosting methods, and neural networks. The results demonstrate that machine learning provides fast and reliable equilibrium prediction, enabling large-scale scenario exploration without repeatedly integrating the underlying system of ordinary differential equations.

Keywords: Lotka–Volterra–SIS model · Equilibrium analysis · Supervised machine learning

1 Introduction

Predator–prey systems with disease naturally arise at the intersection of population ecology and infectious disease dynamics. Classical predator–prey models, beginning with the Lotka–Volterra equations [18,26], have long been used to describe trophic interactions, coexistence, extinction, and nonlinear population oscillations [6,11,16,19,22]. When infectious disease is introduced into prey or predator populations, the resulting eco–epidemiological system can fundamentally alter equilibrium outcomes, convergence behavior, and long–term persistence [2,3,7,10,12,24].

SIS–type diseases, in which individuals return to the susceptible class after infection, provide a natural modeling framework for wildlife and environmental pathogens [9,15,20]. Coupling SIS epidemic dynamics with predator–prey interactions yields a system of nonlinear ordinary differential equations (ODEs) that captures the interplay between trophic processes, population fluctuations, and

© The Author(s), under exclusive license to Springer Nature Switzerland AG 2026
K. L. Kabir et al. (Eds.): BICOB 2026, CCIS 2977, pp. 76–92, 2026.
https://doi.org/10.1007/978-3-032-26028-4_6

disease transmission [14]. Such models allow the study of disease–driven extinction, oscillatory equilibria, threshold conditions for pathogen persistence, and other emergent ecological phenomena.

Despite their theoretical importance, eco–epidemiological predator–prey models pose significant computational challenges. Accurately resolving long–term dynamics requires time integration schemes capable of capturing transient oscillations, stiffness arising from multiscale interactions, and slow convergence toward equilibrium [2,8,25]. As models incorporate additional demographic structure, disease processes, or spatial interactions, the dimensionality of the parameter space grows rapidly, rendering systematic exploration of equilibrium states via repeated numerical integration increasingly time-consuming and computationally prohibitive [1,13,17]. This limitation becomes particularly acute when large-scale sensitivity analysis, uncertainty quantification, or scenario screening is required.

To address these challenges, we propose a machine learning (ML) framework for fast and accurate prediction of equilibrium states in eco–epidemiological predator–prey systems. Surrogate ML models enable rapid exploration of high-dimensional parameter spaces and support efficient sensitivity and uncertainty analyses that would otherwise be prohibitive with direct numerical integration [23,27,28]. Moreover, ML models can capture complex nonlinear dependencies between model parameters and equilibrium outcomes, offering strong predictive accuracy across ecological and epidemiological settings [4,5,21].

In this work, we use a coupled ODE model, referred to as LVSIS [14], to generate a synthetic dataset by randomly varying model parameters within prescribed ranges and recording the final susceptible and infected populations (u_s, u_i, v_s, v_i) as equilibrium states. This dataset is then used to train a diverse suite of supervised machine learning models, including a linear regressor, tree ensembles, boosting methods, and neural networks. The primary objective of this study is to assess the feasibility and accuracy of supervised machine learning models for predicting equilibrium outcomes of this highly nonlinear eco–epidemiological system, rather than to develop fully interpretable surrogate models. We demonstrate that machine learning enables rapid and reliable prediction of equilibrium states without repeatedly integrating the underlying system of ODEs. By capturing the nonlinear mapping from parameters to equilibria, ML models serve as powerful surrogates for ecological and epidemiological forecasting, offering both computational efficiency and mechanistic insight into population–disease dynamics.

The main contributions of this work are threefold: (i) the construction of a large synthetic equilibrium dataset for a coupled LVSIS predator–prey system using systematically verified numerical simulations; (ii) a comparative evaluation of a broad class of supervised machine learning models for direct prediction of equilibrium states from model parameters; and (iii) the demonstration that neural network–based surrogate models achieve high predictive accuracy while dramatically reducing computational cost relative to repeated numerical integration.

The paper is organized as follows. Section 2 presents the coupled ODE formulation governing the temporal evolution of the predator–prey system with disease, together with its numerical discretization using the forward Euler method. We further illustrate the model behavior by computing solutions for four representative test cases. Section 3 presents the dataset generation process, in which model parameters are randomly sampled over prescribed intervals to capture a given range of equilibrium behaviors. In Sect. 4, we examine machine learning techniques for predicting the equilibrium states, focusing on supervised learning models that approximate the nonlinear parameters-to-equilibrium mapping inherent in the LVSIS system. Finally, conclusions are presented in Sect. 5.

2   Predator–Prey System with Disease

In the present study, we consider a mathematical model for a predator–prey system with disease. The model couples classical Lotka–Volterra (LV) predator–prey dynamics with a Susceptible–Infected–Susceptible (SIS) epidemic framework. This coupling captures the interplay between population fluctuations and disease transmission. We refer to this model as the LVSIS model. For further details on the model formulation, parameterization, and underlying assumptions we refer the reader to our previous work [14].

We denote the susceptible and infected prey populations by u_s and u_i, and the susceptible and infected predator populations by v_s and v_i. The temporal evolution of the system is described by the following system of ordinary differential equations:

$$
\begin{aligned}
\frac{du_s}{dt} &= \alpha_s u_s + \alpha_i u_i - \beta_s^s u_s v_s - \beta_i^s u_s v_i - \gamma_s^u u_s^2 - \sigma_i^s u_s u_i - \sigma_{vi}^s u_s v_i + \omega_i u_i, \\
\frac{du_i}{dt} &= -\beta_s^i u_i v_s - \beta_i^i u_i v_i - \gamma_i^u u_i^2 + \sigma_i^s u_s u_i + \sigma_{vi}^s u_s v_i - \omega_i u_i, \\
\frac{dv_s}{dt} &= -\gamma_s^v v_s + \delta_s^s v_s u_s + \delta_i^s v_s u_i - \sigma_i^{vs} v_s u_i - \sigma_{vi}^{vs} v_s v_i + \omega_{vi} v_i, \\
\frac{dv_i}{dt} &= -\gamma_i^v v_i + \delta_s^i v_i u_s + \delta_i^i v_i u_i + \sigma_i^{vs} v_s u_i + \sigma_{vi}^{vs} v_s v_i - \omega_{vi} v_i,
\end{aligned}
\tag{1}
$$

with the following initial conditions

$$
u_s(0) = u_s^0, \quad u_i(0) = u_i^0, \quad v_s(0) = v_s^0, \quad v_i(0) = v_i^0, \quad t = 0.
$$

The terms in system (1) represent processes including intrinsic growth, mortality, predation, disease transmission, and recovery. Disease transmission occurs both within species (prey–prey and predator–predator) and across trophic levels, allowing infection to spread through ecological interactions. Recovery terms return infected individuals to the susceptible class, consistent with the SIS framework. A schematic illustration is included to visualize the disease transmission and recovery pathways of system (1) is shown in Fig. 1.

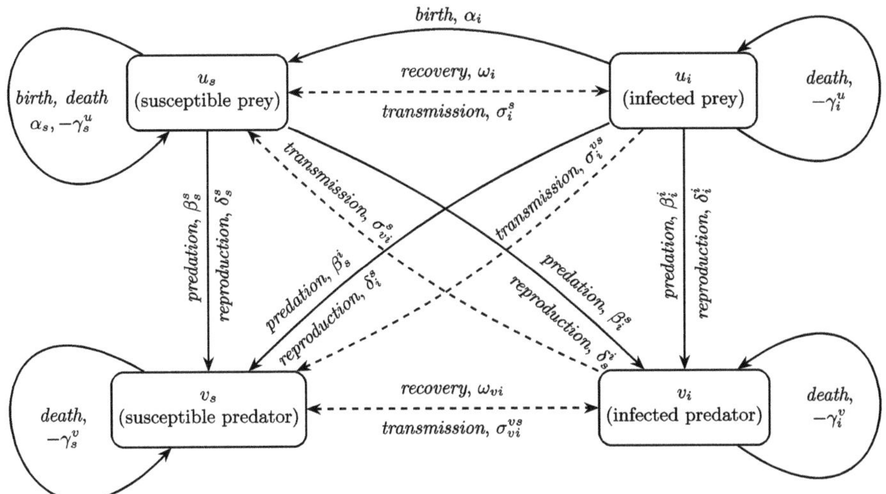

Fig. 1. Compartmental flow diagram for the LVSIS predator–prey disease model (1). Boxes denote susceptible and infected prey (u_s, u_i) and predators (v_s, v_i). Dashed arrows indicate disease transmission and recovery, including within- and cross-species pathways, while solid arrows represent ecological interactions and demographic processes.

All model parameters, including demographic rates, predation coefficients, transmission intensities, and recovery factors, are summarized in Table 1. The table provides their biological roles as well as baseline values together with biologically plausible parameter ranges used for dataset generation in Sect. 3. These ranges are selected to reflect moderate ecological variability while allowing for a wide spectrum of epidemic intensities.

To numerically solve system (1), we discretize time using a fixed step size $\Delta t > 0$. Let $t_n = n\,\Delta t$ for $n = 0, 1, \ldots, N$, where N denotes the total number of time steps. We denote the discrete approximations of the populations at time t_n by $u_s^n \approx u_s(t_n)$, $u_i^n \approx u_i(t_n)$, $v_s^n \approx v_s(t_n)$, and $v_i^n \approx v_i(t_n)$. Applying the forward Euler method yields the update scheme:

$$
\begin{bmatrix} u_s^{n+1} \\ u_i^{n+1} \\ v_s^{n+1} \\ v_i^{n+1} \end{bmatrix} = \begin{bmatrix} u_s^n \\ u_i^n \\ v_s^n \\ v_i^n \end{bmatrix} + \Delta t \begin{bmatrix} \alpha_s u_s^n + \alpha_i u_i^n - \beta_s^s u_s^n v_s^n - \beta_i^s u_s^n v_i^n - \gamma_s^u (u_s^n)^2 - \sigma_i^s u_s^n u_i^n - \sigma_{vi}^s u_s^n v_i^n + \omega_i u_i^n \\ -\beta_s^i u_i^n v_s^n - \beta_i^i u_i^n v_i^n - \gamma_i^u (u_i^n)^2 + \sigma_i^s u_s^n u_i^n + \sigma_{vi}^s u_s^n v_i^n - \omega_i u_i^n \\ -\gamma_s^v v_s^n + \delta_s^s v_s^n u_s^n + \delta_i^s v_s^n u_i^n - \sigma_i^{vs} v_s^n u_i^n - \sigma_{vi}^{vs} v_s^n v_i^n + \omega_{vi} v_i^n \\ -\gamma_i^v v_i^n + \delta_s^i v_i^n u_s^n + \delta_i^i v_i^n u_i^n + \sigma_i^{vs} v_s^n u_i^n + \sigma_{vi}^{vs} v_s^n v_i^n - \omega_{vi} v_i^n \end{bmatrix}.
$$

The forward Euler scheme is adopted for its simplicity and computational efficiency, making it well suited for large-scale dataset generation. Numerical stability and accuracy are ensured by using a sufficiently small time step ($\Delta t = 0.1$ days). Over the parameter ranges considered, the LVSIS system consistently converges to stable equilibrium states, and the explicit Euler scheme is adequate for capturing the long–term dynamics of interest.

Table 1. Model parameters and their biological interpretations.

Parameter	Description	Baseline	Range
α_s	Intrinsic growth rate of susceptible prey.	0.02	0.016–0.024
α_i	Intrinsic growth rate of infected prey.	0.01	0.008–0.012
β_s^s	Predation rate of susceptible predators on susceptible prey.	0.02	0.016–0.024
β_i^s	Predation rate of infected predators on susceptible prey.	0.02	0.016–0.024
β_s^i	Predation rate of susceptible predators on infected prey.	0.02	0.016–0.024
β_i^i	Predation rate of infected predators on infected prey.	0.02	0.016–0.024
γ_s^u	Density-dependent mortality rate of susceptible prey.	0.02	0.016–0.024
γ_i^u	Density-dependent mortality rate of infected prey.	0.02	0.016–0.024
γ_s^v	Intrinsic mortality rate of susceptible predators.	0.01	0.008–0.012
γ_i^v	Intrinsic mortality rate of infected predators.	0.01	0.008–0.012
δ_s^s	Reproduction rate of susceptible predators from consuming susceptible prey.	0.04	0.032–0.048
δ_i^s	Reproduction rate of susceptible predators from consuming infected prey.	0.04	0.032–0.048
δ_s^i	Reproduction rate of infected predators from consuming susceptible prey.	0.04	0.032–0.048
δ_i^i	Reproduction rate of infected predators from consuming infected prey.	0.04	0.032–0.048
σ_i^s	Rate of disease transmission from infected prey to susceptible prey.	0.01	0.001–0.1
σ_{vi}^s	Rate of disease transmission from infected predators to susceptible prey.	0.01	0.001–0.1
σ_i^{vs}	Rate of disease transmission from infected prey to susceptible predators.	0.01	0.001–0.1
σ_{vi}^{vs}	Rate of disease transmission from infected predators to susceptible predators.	0.01	0.001–0.1
ω_i	Recovery rate at which infected prey return to the susceptible class.	0.01	0.001–0.1
ω_{vi}	Recovery rate at which infected predators return to the susceptible class.	0.01	0.001–0.1

Numerical Experiments with Varying Parameters

We present numerical tests designed to evaluate the behavior of the predator–prey system with infection dynamics under varying parameter settings. These numerical experiments explore how changes in model parameters influence the equilibrium populations of susceptible and infected prey (u_s, u_i) and predators (v_s, v_i). Our goal is to illustrate key dynamical patterns that emerge in the system and the resulting equilibrium states. We consider four representative test cases in which one or two key parameters are varied while the remaining coefficients are held fixed. The initial condition is prescribed as $(u_s(0),\, u_i(0),\, v_s(0),\, v_i(0)) = (0.7,\, 0.3,\, 0.8,\, 0.2)$, and is held constant across all experiments.

The system is numerically integrated over a time horizon of $T = 10$ years to ensure that long-term transient dynamics decay and the trajectories approach their corresponding equilibrium states. Each test case isolates the effect of a specific ecological or epidemiological mechanism. We consider the following test cases:

– *Test Case 1 (Baseline).* The baseline parameter set is

$$\alpha_s = 0.02, \qquad \alpha_i = 0.01, \qquad \beta_s^s = \beta_i^s = \beta_s^i = \beta_i^i = 0.02,$$

$$\gamma_s^u = \gamma_i^u = 0.02, \qquad \gamma_s^v = \gamma_i^v = 0.01, \qquad \delta_s^s = \delta_i^s = \delta_s^i = \delta_i^i = 0.04,$$

$$\sigma_i^s = \sigma_{vi}^s = \sigma_i^{vs} = \sigma_{vi}^{vs} = 0.01, \qquad \omega_i = \omega_{vi} = 0.01.$$

This case serves as a reference scenario, representing balanced demographic processes, predation, and disease transmission.

- *Test Case 2 (Increased birth and transmission).* The prey birth rates and disease transmission rates are increased to

$$\alpha_s = 0.024, \quad \alpha_i = 0.012, \quad \sigma_i^s = \sigma_{vi}^s = \sigma_i^{vs} = \sigma_{vi}^{vs} = 0.015,$$

while all other parameters remain at their baseline values. This case illustrates how enhanced population inflow and stronger transmission affect transient dynamics and long–term equilibria.

- *Test Case 3 (Increased mortality and recovery).* The mortality and recovery rates are increased to

$$\gamma_s^u = \gamma_i^u = 0.024 \qquad \gamma_s^v = \gamma_i^v = 0.012, \qquad \omega_i = \omega_{vi} = 0.015,$$

with all other coefficients fixed at their baseline values. This case examines how increased removal and faster recovery influence population persistence.

- *Test Case 4 (Increased ecological coupling).* The predator–prey interaction strength is increased to

$$\beta_s^s = \beta_i^s = \beta_s^i = \beta_i^i = 0.024, \qquad \delta_s^s = \delta_i^s = \delta_s^i = \delta_i^i = 0.048,$$

while all other parameters remain unchanged. This scenario highlights the effect of stronger trophic coupling on convergence rates and equilibrium structure.

Figure 2 shows the solution dynamics and phase portraits for all four test cases. For each case, the left panel displays the temporal trajectories of (u_s, u_i, v_s, v_i) together with their equilibrium values, while the right panel presents phase portraits of (u_s, v_s) and (u_i, v_i), illustrating stability, convergence behavior, and oscillatory dynamics.

Across all cases, the system initially exhibits characteristic predator–prey oscillations driven by trophic interactions and disease transmission. These oscillations gradually decay as demographic turnover, predation, and recovery processes balance, and the trajectories approach their equilibrium states. In most cases, except for Test Case 2, both infected prey and predator populations decline toward extinction, indicating that infection cannot be sustained under the corresponding parameter regimes. The susceptible prey and predator populations stabilize at positive equilibrium levels, reflecting coexistence in the absence of endemic disease.

For the baseline parameters in Test Case 1, all populations exhibit damped oscillations and converge to a disease–free coexistence equilibrium $(u_s, u_i, v_s, v_i) \approx (0.25, 0, 0.75, 0)$. The phase portraits show spiral trajectories toward an interior fixed point for (u_s, v_s) and decay toward the origin for (u_i, v_i).

In Test Case 2, increasing prey birth rates and transmission rates lead to a qualitatively different outcome. The system converges to an endemic equilibrium $(u_s, u_i, v_s, v_i) \approx (0.211, 0.04, 0.594, 0.326)$, in which all four populations persist.

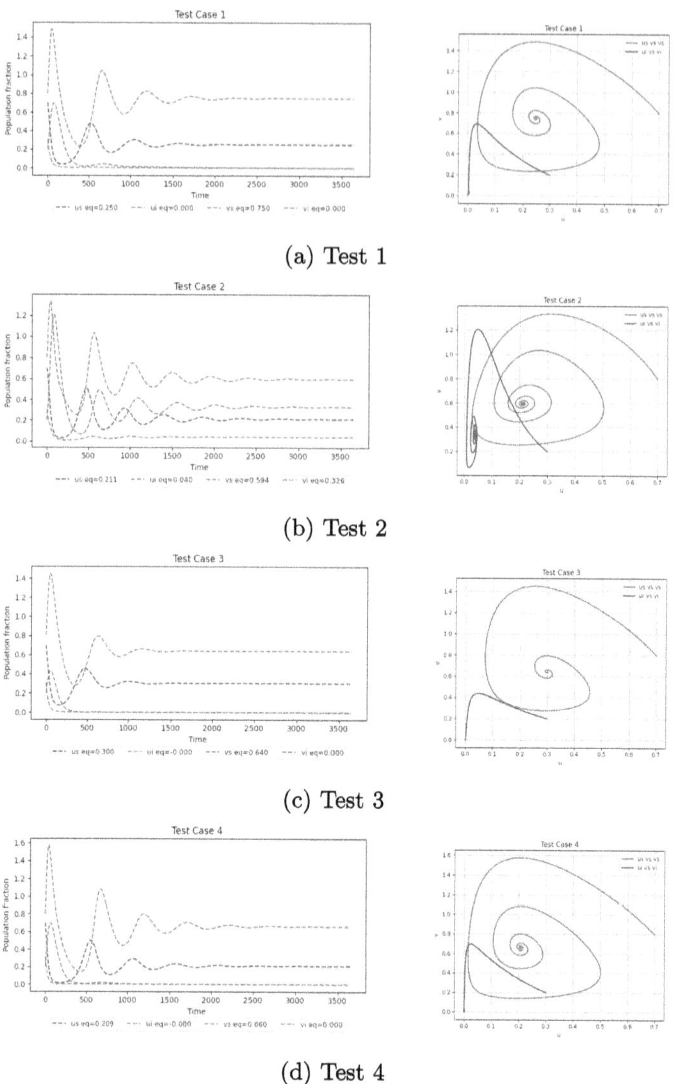

(a) Test 1

(b) Test 2

(c) Test 3

(d) Test 4

Fig. 2. Solution dynamics (left) and phase portraits (right) for four test cases.

The phase portraits exhibit interior spirals in both the susceptible and infected subspaces, indicating sustained infection in both species.

In Test Case 3, increasing mortality and recovery rates drive the system back to a disease–free equilibrium after a short transient, $(u_s, u_i, v_s, v_i) \approx (0.3, 0, 0.64, 0)$. Compared with Test Case 1, prey populations settle at a slightly higher level and predator populations at a slightly lower level, reflecting reduced predator survival and faster clearance of infection.

Finally, increasing ecological coupling in Test Case 4 leads to another disease-free equilibrium, $(u_s, u_i, v_s, v_i) \approx (0.209, 0, 0.66, 0)$, characterized by lower prey density and higher predator density relative to the baseline case. The phase portraits display stronger initial oscillations but the same stable-focus convergence.

Overall, Tests 1, 3, and 4 converge to disease-free coexistence equilibria that differ in prey–predator balance, while Test 2 produces a stable endemic equilibrium. These results demonstrate how modest changes in demographic or epidemiological parameters can shift the LVSIS system between disease–free and endemic regimes while substantially altering long–term population distributions.

3   Dataset Generation

To generate the datasets, we systematically sample parameters from both Lotka–Volterra (LV) predator–prey dynamics and the Susceptible–Infected–Susceptible (SIS) disease processes of the LVSIS model. Simulations are initialized with fixed population proportions, $(u_s(0), u_i(0), v_s(0), v_i(0)) = (0.7, 0.3, 0.8, 0.2)$, representing the susceptible and infected prey and predator populations. Each simulation is carried out over a total time horizon of $T_{\max} = 100$ years using a small uniform time step of $\tau = 0.1$ days with the forward Euler scheme. This extended integration window is chosen to accommodate the wide range of demographic, ecological, and epidemiological parameter values considered.

Parameter variability is introduced to explore a prescribed range of ecological and epidemiological scenarios, as summarized in Table 1. The LV parameters, including birth, death, and predation rates, are sampled from uniform distributions centered around their baseline values (0.02 for most parameters and 0.04 for reproduction rates) and perturbed multiplicatively within $[0.8, 1.2]$ to maintain biological variability. SIS disease parameters, such as transmission and recovery rates, are drawn from uniform distributions centered at 0.01 and scaled over a wider factor range of $[0.1, 10]$ to capture diverse epidemic intensities.

While these parameter ranges are not extremely broad, they are sufficient to assess predictive performance within the sampled parameter regime. Consequently, the trained models are expected to generalize primarily within similar parameter ranges. Extending the parameter space, particularly for ecological parameters, will be considered in future work to further evaluate model robustness and extrapolation capability. We generate N_{sim} simulation samples

$$\mathcal{D}_{\text{LVSIS}} = \left\{ \left(x^{(\ell)}, y^{(\ell)}\right) \right\}_{\ell=1}^{N_{\text{sim}}},$$

where each sample is represented as a pair $(x^{(\ell)}, y^{(\ell)})$.

The input vector $x^{(\ell)}$ contains the sampled LVSIS parameter values

$$x^{(\ell)} = \{\alpha_s, \alpha_i, \beta_s^s, \beta_i^s, \beta_s^i, \beta_i^i, \gamma_s^u, \gamma_i^u, \gamma_s^v, \gamma_i^v, \delta_s^s, \delta_i^s, \delta_s^i, \delta_i^i, \sigma_i^s, \sigma_{vi}^s, \sigma_i^{vs}, \sigma_{vi}^{vs}, \omega_i, \omega_{vi}\},$$

and the output vector $y^{(\ell)}$ records the final population sizes and the number of time steps, N_t, required to reach equilibrium

$$y^{(\ell)} = \{u_s, u_i, v_s, v_i, N_t\}.$$

For each sampled parameter set, the ODE system is numerically integrated until an equilibrium state is detected. Equilibrium is determined by the convergence criterion

$$\|u_s^{n+1} - u_s^n\| < \varepsilon, \quad \|u_i^{n+1} - u_i^n\| < \varepsilon, \quad \|v_s^{n+1} - v_s^n\| < \varepsilon, \quad \|v_i^{n+1} - v_i^n\| < \varepsilon,$$

with tolerance $\varepsilon = 10^{-7}$. This criterion ensures that the recorded states correspond to true equilibria rather than slowly decaying oscillations or long-lived transients. If the equilibrium condition is satisfied before $T_{\max}$, the simulation is terminated early.

This formulation establishes a clear mapping from model parameters to observable outcomes, capturing variability in equilibrium population sizes and the number of time steps. The resulting dataset provides a foundation for training and evaluating machine learning models to predict system equilibria from parameter inputs.

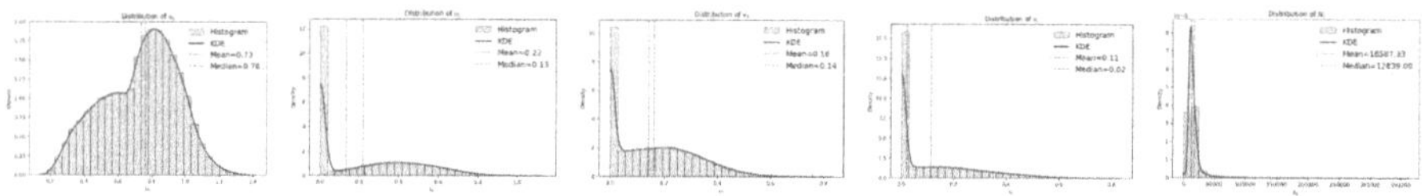

Fig. 3. Empirical distributions of equilibrium populations and convergence measures across all simulation scenarios. From left to right: susceptible prey (u_s), infected prey (u_i), susceptible predators (v_s), infected predators (v_i), and the number of time steps required to reach equilibrium (N_t). Each panel shows a histogram (shaded blue) with a kernel density estimate (solid blue). The red dashed line indicates the sample mean, and the green dashed line indicates the sample median.

The distributions are visualized in Fig. 3, which summarizes the distributions of the four equilibrium populations and the number of time steps generated from the LVSIS model. The susceptible prey populations u_s exhibit a broad, approximately unimodal distribution centered around moderate values, indicating that prey typically persist at relatively high equilibrium densities across the sampled parameter space. In contrast, the infected populations u_i are heavily right-skewed, with most equilibria yielding very small infection levels and only a minority producing moderate infection prevalence. Predator populations, v_s and v_i, show even stronger right–skewness: susceptible predator densities remain low in most scenarios, and infected predator densities are near zero for the overwhelming majority of parameter combinations. These patterns reflect the ecological asymmetry inherent in predator–prey dynamics. Prey maintain larger populations and therefore support frequent infection, whereas predators persist at much lower densities, making sustained infection in the predator population rare. The distribution of N_t indicates that the majority of trajectories converge well before the maximum integration horizon, while a small subset exhibits slower convergence associated with weakly damped or near–threshold dynamics. A more detailed statistical and mechanistic analyses of these equilibrium distributions are deferred to a separate manuscript. Here, these results are presented primarily to characterize the dataset and motivate the use of flexible machine learning models for equilibrium prediction.

4   Machine Learning Prediction Algorithms

We formulate the prediction task as a supervised learning problem, mapping model parameters to their corresponding equilibrium outcomes. The input features consist of the raw LVSIS parameters, while the target variables are the four equilibrium population components (u_s, u_i, v_s, v_i). To assess the generalization performance of each model, the dataset is split into training and test subsets. Before training, all input features are standardized to have zero mean and unit variance, ensuring that differences in parameter scales do not bias the learning process. We implement a diverse collection of supervised learning algorithms to predict equilibrium outcomes from the LVSIS parameters, enabling a consistent and comparative evaluation across model families.

The evaluated methods are organized into four categories:

- *Linear model.* Linear Regression is used as a baseline model, capturing only global linear relationships between the input parameters and population outcomes. Scikit–learn's LinearRegression implementation is employed, which computes the ordinary least–squares solution analytically using linear algebra rather than iterative optimization. As a result, no learning rate or stopping criterion is required.
- *Neural networks (NN).* We employ several multilayer perceptrons (MLPs) with ReLU activation functions and progressively increasing architectural depth and width. The models NN1, NN2, NN3, and NN4 correspond to hidden-layer configurations (100), $(100, 50)$, $(100, 100, 50)$, and $(100, 100, 100, 50)$, respectively. All neural network models are implemented using the MLPRegressor from scikit–learn. The networks are trained with the Adam optimizer for a maximum of 3000 training iterations, together with ℓ_2 regularization $(\alpha = 0.01)$ and the default initial learning rate of 0.001 to mitigate overfitting. The loss function is the default squared error (mean squared error), which measures the average squared difference between predicted and target values. Training is terminated either when the maximum number of iterations is reached or when convergence is achieved, as determined by the optimizer's stopping criterion based on the relative improvement of the loss between successive iterations falling below a prescribed tolerance of 10^{-4}. A fixed random seed is applied to ensure reproducibility across runs.
- *Tree-based models.* The Decision Tree Regressor serves as a nonparametric baseline capable of capturing hierarchical feature interactions and nonlinear dependencies. Trees are grown using greedy variance reduction and are limited to a maximum depth of 8 to control model complexity and reduce overfitting. Tree construction is non-iterative and does not involve learning

rates or convergence–based stopping criteria. A fixed random seed is used to ensure reproducibility.

- *Ensemble methods.* We employ several ensemble algorithms that improve predictive robustness through aggregation. Extra Trees and Random Forest models are implemented as ensembles of 300 decision trees. In Random Forests, the maximum depth of each tree is limited to 10 to control model complexity and reduce overfitting. Bagging aggregates 200 base estimators trained on bootstrap samples of the data. Gradient Boosting employs 300 boosting stages with a learning rate of 0.05 and regression trees of maximum depth 5, while Histogram-Based Gradient Boosting uses the same learning rate with trees of maximum depth 10. All ensemble models are implemented using the corresponding algorithms from the scikit–learn library.

Finally, we present prediction results obtained using various machine learning algorithms on the dataset described above.

Prediction Results

The ML models were evaluated using a unified prediction and scoring pipeline to ensure consistent comparison across all eleven algorithms. Numerical experiments were performed on an Apple M4 CPU with 16 GB of RAM running macOS Tahoe 26.2, using a single CPU core, and machine learning models were implemented with scikit-learn version 1.7.2.

The evaluation computed several error metrics, including the root mean squared error (RMSE) and mean absolute error (MAE), expressed as relative errors in percentage

$$RMSE^{\%} = \sqrt{\frac{\sum_i (Y_i - \tilde{Y}_i)^2}{\sum_i Y_i^2}} \times 100\%, \quad MAE^{\%} = \frac{\sum_i |Y_i - \tilde{Y}_i|}{\sum_i Y_i} \times 100\%.$$

In addition, we report the coefficient of determination, R^2, which measures the proportion of variance in the reference data explained by the predictions:

$$R^2 = 1 - \frac{\sum_i (Y_i - \tilde{Y}_i)^2}{\sum_i (Y_i - \bar{Y})^2}.$$

Here, Y_i denotes the reference equilibrium value, $\tilde{Y}_i$ the corresponding predicted value, and $\bar{Y}$ the mean of the reference values for each population component.

Tables 2, 3 and 4 and 5 summarize the predictive performance of all machine learning models considered in this study. For each model, we report RMSE and MAE expressed as percentages, together with the coefficient of determination (R^2), computed over both the training and test sets. The tables also report computational cost in terms of training and prediction time. To assess the impact of data availability, each model was trained and evaluated under four different training–test splits: 80–20, 60–40, 40–60, and 20–80.

Across all splits, clear performance trends emerge. Linear Regression consistently provides the fastest training and prediction times, but exhibits limited

Table 2. Performance comparison of machine learning models for predicting the four LVSIS equilibrium populations under an 80–20 training-test split. Reported metrics include percentage RMSE, percentage MAE, coefficient of determination (R^2), and computational cost measured by training and prediction time.

Model	Train			Test			Time(s)	
	RMSE$^\%$	MAE$^\%$	R^2	RMSE$^\%$	MAE$^\%$	R^2	Train	Pred
Linear	49.85	37.16	0.698	49.65	37.02	0.700	0.04	0.009
NN1	16.22	9.47	0.969	15.96	9.44	0.970	5.53	0.061
NN2	12.50	6.18	0.982	12.33	6.20	0.983	7.81	0.104
NN3	11.48	5.77	0.984	11.40	5.81	0.985	12.45	0.186
NN4	12.00	6.18	0.984	12.00	6.21	0.984	17.82	0.240
DecisionTree	43.10	28.34	0.783	45.78	29.90	0.756	3.47	0.019
ExtraTrees	0.00	0.00	1.000	25.85	16.54	0.924	251.94	32.574
GradientBoosting	24.97	16.86	0.927	27.51	18.47	0.911	760.02	1.940
HistGradientBoosting	30.06	20.65	0.895	31.33	21.49	0.886	3.36	0.280
Bagging	10.61	6.59	0.987	28.25	17.70	0.909	805.76	15.340
RandomForest	28.22	18.70	0.905	33.52	21.88	0.869	772.10	6.558

Table 3. Performance comparison of machine learning models for predicting the four LVSIS equilibrium populations under a 60–40 training–test split. Reported metrics include percentage RMSE, percentage MAE, coefficient of determination (R^2), and computational cost measured by training and prediction time.

Model	Train			Test			Time(s)	
	RMSE$^\%$	MAE$^\%$	R^2	RMSE$^\%$	MAE$^\%$	R^2	Train	Pred
Linear	49.86	37.16	0.698	49.65	37.01	0.700	0.06	0.007
NN1	16.67	9.71	0.968	16.43	9.71	0.969	4.62	0.069
NN2	12.50	6.33	0.982	12.35	6.35	0.983	6.41	0.109
NN3	11.75	5.82	0.984	11.57	5.82	0.985	10.23	0.170
NN4	11.67	6.00	0.984	11.64	6.04	0.984	14.52	0.246
DecisionTree	43.69	28.65	0.777	45.65	29.83	0.757	2.56	0.020
ExtraTrees	13.46	4.53	0.980	26.61	17.16	0.920	206.76	22.531
GradientBoosting	25.64	17.27	0.923	27.73	18.64	0.910	557.74	1.928
HistGradientBoosting	30.25	20.73	0.894	31.35	21.47	0.886	2.62	0.287
Bagging	17.70	9.98	0.964	29.08	18.37	0.903	578.96	12.894
RandomForest	29.23	19.23	0.899	33.73	22.04	0.868	561.36	6.340

predictive accuracy, with test errors exceeding 35% and R^2 values around 0.7. Decision Trees offer only modest improvements over the linear baseline and display poor generalization, particularly as the training set size decreases, reflecting

Table 4. Performance comparison of machine learning models for predicting the four LVSIS equilibrium populations under a 40–60 training–test split. Reported metrics include percentage RMSE, percentage MAE, coefficient of determination (R^2), and computational cost measured by training and prediction time.

Model	Train			Test			Time(s)	
	RMSE%	MAE%	R^2	RMSE%	MAE%	R^2	Train	Pred
Linear	49.87	37.17	0.698	49.67	37.03	0.700	0.06	0.005
NN1	18.27	10.93	0.962	18.01	10.92	0.963	3.47	0.059
NN2	13.20	6.74	0.980	12.86	6.69	0.981	4.88	0.101
NN3	12.10	5.96	0.983	11.95	5.99	0.984	7.53	0.171
NN4	12.36	6.45	0.982	12.28	6.51	0.983	11.24	0.238
DecisionTree	44.43	28.97	0.770	46.64	30.13	0.748	1.63	0.019
ExtraTrees	19.89	9.30	0.956	27.78	18.12	0.913	117.52	17.651
GradientBoosting	26.24	17.66	0.919	28.05	18.87	0.908	359.44	1.884
HistGradientBoosting	30.45	20.85	0.893	31.40	21.48	0.886	2.65	0.276
Bagging	23.28	13.54	0.939	30.41	19.39	0.894	365.35	11.608
RandomForest	30.35	19.88	0.892	34.03	22.31	0.866	359.75	6.265

Table 5. Performance comparison of machine learning models for predicting the four LVSIS equilibrium populations under a 20–80 training–test split. Reported metrics include percentage RMSE, percentage MAE, coefficient of determination (R^2), and computational cost measured by training and prediction time.

Model	Train			Test			Time(s)	
	RMSE%	MAE%	R^2	RMSE%	MAE%	R^2	Train	Pred
Linear	49.80	37.12	0.698	49.61	36.98	0.701	0.05	0.006
NN1	22.43	14.45	0.942	22.27	14.52	0.943	2.20	0.069
NN2	14.92	8.21	0.974	14.77	8.25	0.975	2.82	0.102
NN3	12.57	6.21	0.982	12.30	6.19	0.983	5.92	0.183
NN4	13.00	6.54	0.980	12.73	6.53	0.981	7.45	0.243
DecisionTree	46.36	29.83	0.751	48.00	30.88	0.733	0.76	0.019
ExtraTrees	26.10	15.14	0.923	29.73	19.78	0.900	56.07	14.332
GradientBoosting	27.80	18.65	0.910	29.03	19.57	0.902	170.79	1.808
HistGradientBoosting	31.04	21.24	0.889	31.73	21.72	0.883	2.34	0.248
Bagging	28.90	17.96	0.905	32.24	20.97	0.881	164.34	9.771
RandomForest	32.27	21.18	0.880	34.53	22.83	0.862	166.94	6.052

their susceptibility to overfitting and limited capacity to represent the strongly nonlinear LVSIS parameter–equilibrium mapping.

Neural network models achieve the highest predictive accuracy across all training–test splits. In particular, NN2 and NN3 consistently yield the lowest

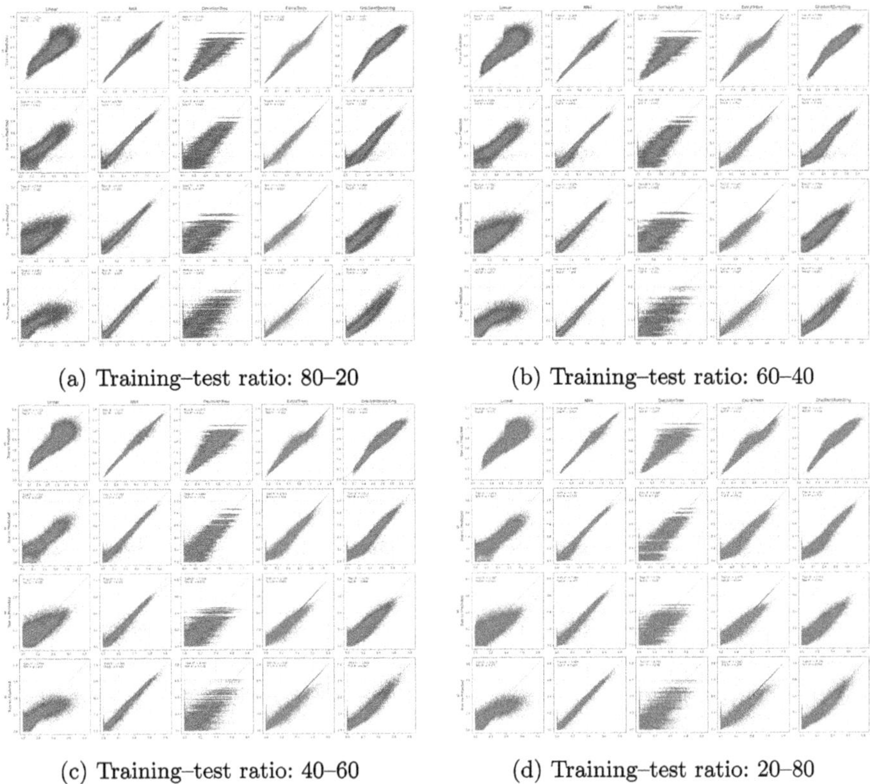

(a) Training–test ratio: 80–20

(b) Training–test ratio: 60–40

(c) Training–test ratio: 40–60

(d) Training–test ratio: 20–80

Fig. 4. Cross-plots (true vs predicted) for selected machine learning algorithms (Linear Regression, NN4, DecisionTree, ExtraTrees, GradientBoosting). Blue points represent the training data, and red points represent the test data. (Color figure online)

test RMSE% and MAE% values, together with R^2 values exceeding 0.98 in most cases. Increasing network depth beyond NN3 does not result in further performance gains; NN4 exhibits comparable or slightly worse generalization accuracy while incurring higher training costs. This behavior indicates diminishing returns from additional network depth and highlights the importance of balancing model capacity and generalization.

Ensemble methods show more heterogeneous behavior. The Extra Trees model achieves perfect or near-perfect training accuracy, but its substantially degraded test performance indicates pronounced overfitting. This behavior is accompanied by high computational cost, as predictions are obtained by averaging over a large ensemble of randomized trees. Other ensemble approaches, including Random Forests, Bagging, and Gradient Boosting improve generalization relative to single-tree models, but at the expense of significantly increased training time. Among these approaches, Gradient Boosting provides the most balanced trade-off between accuracy and generalization, albeit with the highest overall training cost.

Focusing on the representative 80–20 training–test split (Table 2), the neural networks consistently outperform all other models. While deeper architectures require longer training times (up to approximately 18 s for NN4), prediction times remain low across all neural network configurations (approximately 0.1–0.3 s). In contrast, ensemble methods require training times on the order of several hundred seconds. Overall, NN3 offers the best combination of predictive accuracy, robustness, and computational efficiency for this problem.

We note that percentage–based error metrics should be interpreted with caution due to the strong skewness of equilibrium distributions, particularly for infected populations, where near-zero equilibrium values are common. In such cases, relative errors may obscure model performance on rare endemic regimes, making complementary measures such as R^2 and qualitative diagnostic plots (Fig. 4) essential for assessing predictive performance.

Figure 4 presents cross-plots comparing true versus predicted equilibrium populations for a selection of machine learning models: Linear Regression, NN4, Decision Tree, Extra Trees, and Gradient Boosting. Each subplot corresponds to one of the four LVSIS population components (u_s, u_i, v_s, v_i), with the models arranged horizontally. Blue points indicate training-set predictions, while red points show test-set predictions, allowing visual assessment of both model fit and generalization. The 1:1 dashed line in each plot represents perfect prediction. We observe that a deeper neural network (NN4) demonstrates the closest alignment with the 1:1 line, reflecting high accuracy on both training and test data. Ensemble methods such as Extra Trees and Gradient Boosting also capture overall trends effectively but exhibit slightly higher variance, whereas simpler models such as Linear Regression and Decision Tree show larger deviations, particularly for test data, highlighting their limited capacity to model the underlying nonlinear dynamics.

5   Conclusion

In this work, we investigated the long–term dynamics and equilibrium structure of a coupled LVSIS system and assessed the ability of supervised machine learning models to predict its equilibrium outcomes. Numerical experiments across a range of representative parameter regimes demonstrated how demographic rates, transmission mechanisms, and trophic interactions govern transitions between disease–free and endemic states. Using the numerically generated dataset, we trained a variety of machine learning models to predict the four equilibrium population components directly from model parameters. Across all training–test splits, neural network models consistently achieved the highest predictive accuracy, with moderately deep architectures (NN3) providing the best balance between accuracy and generalization. Ensemble methods yielded reasonable predictions but often incurred substantially higher computational cost and, in some cases, exhibited overfitting or reduced generalization performance. In contrast, linear regression and single decision tree models performed poorly across all settings, highlighting the strongly nonlinear mapping between LVSIS parameters and equilibrium states.

References

1. Allen, L.J.: An introduction to stochastic epidemic models. In: Mathematical Epidemiology. LNM, pp. 81–130. Springer (2007). https://doi.org/10.1007/978-3-540-78911-6_3
2. Anderson, R.M., May, R.M.: Infectious diseases of humans: dynamics and control. Oxford university press (1991)
3. Banerjee, M., Petrovskii, S.: Infected predator–prey dynamics: a review. J. Theoret. Biol. **515**, 110,594 (2021)
4. Carrella, E., Miller, J., Mumford, J.: Using machine learning to unravel ecological and epidemiological interactions. Methods Ecol. Evol. **11**(11), 1378–1390 (2020)
5. Dietze, M.C.: Ecological Forecasting. Princeton University Press (2017)
6. Freedman, H.I.: Deterministic Mathematical Models in Population Ecology. Marcel Dekker (1980)
7. Greenman, J., Benton, T.: The impact of population dynamics on disease transmission in animal predator-prey communities. Theor. Popul. Biol. **68**, 75–90 (2005)
8. Haerter, J.O., Mitarai, N., Sneppen, K.: Predator–prey oscillations in unstable environments. Phys. Rev. E **94**(2), 022,406 (2016)
9. Hethcote, H.W.: Three basic epidemiological models. In: S.A. Levin, T.G. Hallam, L.J. Gross (eds.) Applied Mathematical Ecology, pp. 119–144. Springer, Berlin, Heidelberg (1989). https://doi.org/10.1007/978-3-642-61317-3_5. Introduces the SIS framework for diseases without permanent immunity, where individuals transition repeatedly between susceptible and infected states
10. Hethcote, H.W.: The mathematics of infectious diseases. SIAM Rev. **42**(4), 599–653 (2000)
11. Holt, R.D.: Predation, apparent competition, and the structure of prey communities. Theor. Popul. Biol. **12**, 197–229 (1977)
12. Hsieh, Y.H.: Predator–prey model with disease infection in both populations. Math. Med. Biol. **25**(3), 247–266 (2008). https://doi.org/10.1093/imammb/dqn017. Advance Access publication on August 12, 2008
13. Iacus, S.M.: Simulation and Inference for Stochastic Differential Equations. Springer (2008)
14. Ji, H., Vasilyeva, M., Mbroh, N.A., Sadovski, A.: Epidemic dynamics in the spatio-temporal predator-prey model. Front. Appli. Mathem. Statist. **11**, 1642,676 (2025)
15. Keeling, M., Rohani, P.: Modeling Infectious Diseases in Humans and Animals. Princeton University Press (2008)
16. Kot, M.: Elements of Mathematical Ecology. Cambridge University Press (2001)
17. Liu, Z., Zhang, Z., Sun, G.: Dynamics of a predator-prey model with disease in both populations. Appl. Math. Model. **65**, 689–706 (2019)
18. Lotka, A.J.: Elements of physical biology. Williams & Wilkins (1925)
19. Murray, J.D.: Mathematical Biology I: An Introduction. Springer (2002)
20. Pastor-Satorras, R., Castellano, C., Van Mieghem, P., Vespignani, A.: Epidemic processes in complex networks. Rev. Mod. Phys. **87**(3), 925 (2015)
21. Pathak, J., Hunt, B.R., Girvan, M., Lu, Z., Ott, E.: Using machine learning to replicate chaotic attractors and forecast dynamical systems. Chaos **27**(12), 121,102 (2017)

22. Rosenzweig, M.L.: Graphical analysis of the logistic growth curve and the paradox of enrichment. Science **157**, 754–755 (1963)
23. Vasilyeva, M., Wang, Y., Stepanov, S., Sadovski, A.: Numerical investigation and factor analysis of the spatial-temporal multi-species competition problem. arXiv preprint arXiv:2209.02867 (2022)
24. Venturino, E.: The influence of diseases on lotka-volterra systems. Rocky Mountain J. Math. **24**(1), 381–402 (1994)
25. Venturino, E.: Epidemics in predator-prey models: disease in the predators. IMA J. Math. Appl. Med. Biol. **12**(4), 287–302 (1995). https://doi.org/10.1093/imammb/12.4.287
26. Volterra, V.: Fluctuations in the abundance of a species considered mathematically. Nature **118**, 558–560 (1926)
27. Wang, Y., Vasilyeva, M., Sadovski, A.: Prediction of the survival status for multi-species competition system. In: AIP Conference Proceedings, vol. 2872. AIP Publishing (2023)
28. Wang, Y., Vasilyeva, M., Stepanov, S., Sadovski, A.: Numerical investigation and factor analysis of two-species spatial-temporal competition system after catastrophic events. WSEAS Trans. Syst. **22**(1), 423–436 (2023)

A Pilot Feasibility Study of an Interpretable Clinical Decision Support Framework for ICU Readmission Risk After Postoperative Breast Cancer Surgery

Yashasvi Nijampurkar$^{(\boxtimes)}$ ⓘ and Sanja Avramovic ⓘ

George Mason University, Fairfax, VA 22030, USA
{ynijampu,savramov}@gmu.edu

Abstract. **Background:** Unplanned Intensive Care Unit (ICU) readmission after breast cancer surgery is associated with increased morbidity, mortality, and healthcare utilization [1, 2]. Existing ICU readmission models are typically developed for heterogeneous populations and may not generalize to narrowly defined postoperative oncology cohorts [3–5].

Objective: This study presents a pilot feasibility analysis of an interpretable, cohort-specific clinical decision support framework for identifying postoperative breast cancer patients at elevated risk of ICU readmission using routinely collected electronic health record (EHR) data.

Methods: A retrospective cohort was curated from the MIMIC-IV database [6] using reproducible phenotyping and transfer-based ICU readmission logic. Thirteen physiologic, laboratory, and medication-related variables recorded within 24 h after surgery or index ICU discharge were extracted in raw clinical units. Exploratory modeling using logistic regression, random forest, and k-nearest neighbors models was performed solely to assess feature stability and relevance rather than predictive performance. In parallel, a transparent rule-based alert module categorized abnormalities using established clinical thresholds [7–9] and generated color-coded (Green/ Yellow/ Red) severity alerts.

Results: The final analytic cohort consisted of 51 postoperative breast cancer patients of whom 40 (78.4%) experienced ICU readmission during the index hospitalization. Recurrent abnormalities were observed in electrolytes (sodium, potassium), blood pressure indices, creatinine, and temperature, consistent across exploratory modelling approaches. Cross-validation analyses demonstrated substantial variability, consistent with small-cohort feasibility conditions. Within the rule-based alert framework, higher alert severity qualitatively aligned with physiologic instability patterns.

Conclusion: This study demonstrates the feasibility of constructing a cohort-specific, interpretable ICU readmission decision support framework for postoperative breast cancer patients. The framework is intended as a proof of concept to inform future large-scale validation and prospective evaluation.

Keywords: ICU Readmission · Breast Cancer Surgery · Clinical Decision Support · MIMIC-IV · Pilot study

K. L. Kabir et al. (Eds.): BICOB 2026, CCIS 2977, pp. 93–100, 2026.
https://doi.org/10.1007/978-3-032-26028-4_7

1   Introduction

Unplanned ICU readmission is associated with increased mortality, prolonged hospitalization, and increased healthcare utilization [1, 2]. Surgical oncology patients, including those undergoing breast cancer surgery, may be particularly vulnerable due to the combined physiologic stress of malignancy and postoperative recovery.

Numerous approaches have been proposed to predict ICU readmissions, including severity scores and machine learning models trained on heterogeneous ICU populations [3–5]. While informative at a population level, these methods often underperform when applied to narrowly defined postoperative oncology cohorts with distinct recovery trajectories.

The MIMIC-IV database enables reproducible investigation of cohort-specific clinical decision support frameworks using real-world electronic health record (EHR) data [6]. However, restricting analyses to clinically homogeneous surgical oncology populations typically yields small sample sizes. In such settings, feasibility, interpretability, and methodological transparency are more appropriate objectives than establishing predictive accuracy.

Accordingly, this study presents a pilot feasibility analysis of an interpretable ICU readmission decision support framework tailored to postoperative breast cancer patients, emphasizing cohort construction, feature behavior, and transparent rule-based alerting under realistic data constraints.

2   Materials and Methods

2.1   Data Source and Cohort Construction

This retrospective study utilized the Medical Information Mart for Intensive Care IV (MIMIC-IV, v3.1), a publicly available, de-identified critical care database containing detailed electronic health record data from ICU admissions at Beth Israel Deaconess Medical Centre between 2008 and 2019 [6]. Access to MIMIC-IV requires completion of human subjects training and agreement to data use policies. The database includes demographic information, diagnoses, procedures, medication orders, laboratory results, charted vital signs, and detailed transfer records.

Data extraction was performed using structured query language (SQL) queries applied to relevant hospital and ICU tables, including diagnoses_icd, procedure_icd, admissions, patients, transfers, icustays, chartevents, labevents, and prescriptions.

Postoperative breast cancer patients were identified through a multi-stage cohort construction pipeline designed to ensure clinical specificity and reproducibility. Hospital admissions associated with breast cancer were identified using ICD-10 (C50*) and corresponding legacy ICD-9 diagnosis codes recorded during the index hospital admission. Admissions were restricted to adult patients ($\geq$18 years) with a documented surgical procedure during the same hospital encounter, with surgical timing defined as the earliest recorded procedure date per admission.

To focus on postoperative critical care trajectories, only ICU admissions occurring after the recorded surgical time were considered. ICU stays were identified using transfer level records to capture true transitions into and out of ICU care rather than relying solely on ICU stay counts. Transfers were restricted to ICU care units, and sequential ICU episodes were ordered chronologically within each hospital admission.

Unplanned ICU readmission was operationalized using a transfer-based definition to avoid misclassification of intra-ICU movements or administrative transfers. ICU readmission was defined as a subsequent transfer into an ICU care unit occurring more than six hours after discharge from a prior ICU stay within the same hospital admission. Transfers separated by six hours or less were considered part of a continuous ICU episode and were excluded from readmission classification. Readmission status was determined at the patient-admission level.

Patients lacking complete physiologic, laboratory, or medication data within the defined observation window were excluded from the final cohort to maintain feature consistency. Application of these criteria resulted in a final analytic cohort of 51 unique postoperative breast cancer patients. Figure 1 Summarizes the cohort selection process, and Table 1 reports cohort characteristics.

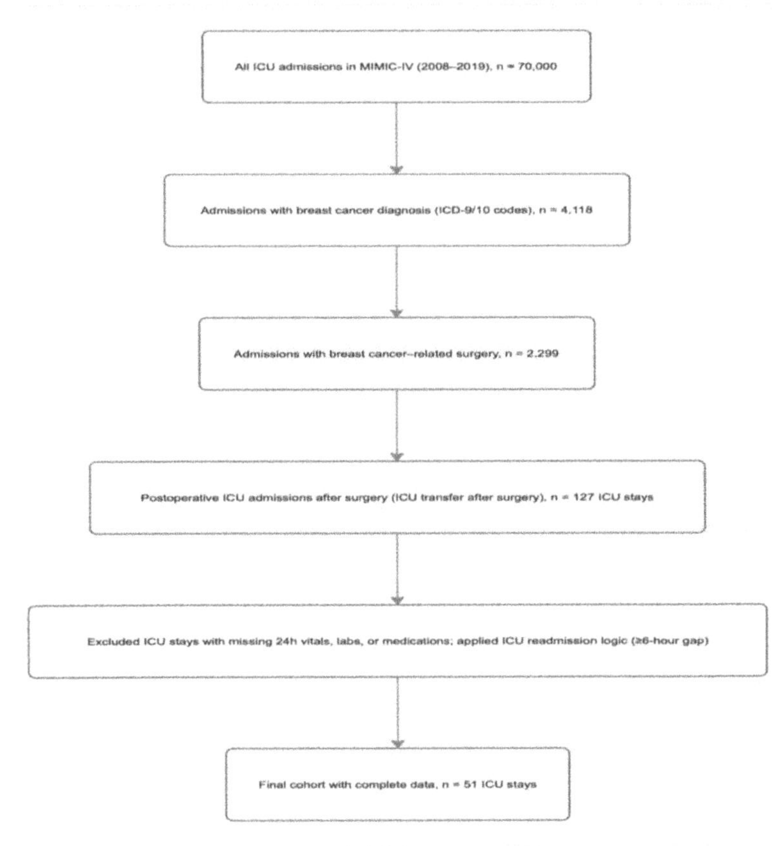

Fig. 1. Cohort selection flowchart

Table 1. Cohort characteristics

Characteristics	Value
Total patients	51
ICU readmission	40 (78.4%)
ICU stays analyzed	127
Complete feature availability	Yes
Data source	MIMIC-IV (2008–2019)

2.2 Feature Extraction and Outcome

Thirteen routinely collected physiologic, laboratory, and medication-related variables were extracted within a 24-h observation window following surgery or index ICU discharge. The 24-h window was selected to capture early postoperative physiologic instability while maintaining a consistent temporal reference across patients (Table 2).

Physiologic variables were obtained from ICU charted vital sign records (chartevents table) and included heart rate, systolic blood pressure, diastolic blood pressure, mean arterial pressure, respiratory rate, temperature, and oxygen saturation. Laboratory values were extracted from the hospital laboratory event table and included hemoglobin, creatinine, glucose, sodium, and potassium.

When multiple measurements were available within the 24-h window, summary values were computed at the ICU stay level. To capture potential physiologic instability, the maximum recorded value within the observation window was retained for each variable. All measurements were preserved in their original clinical units to maintain interpretability and alignment with established bedside reference ranges. No feature scaling or normalization was applied.

Medication exposure was derived from prescription records within the same observation window. A binary indicator was generated to reflect whether any ICU-relevant medication was administered during this period.

The primary outcome was binary ICU readmission status, defined using the transfer-based logic described in Sect. 2.1.

2.3 Exploratory Modeling and Analysis

Exploratory machine learning analyses were conducted to assess the stability, consistency, and clinical relevance of extracted features within this small, specialized postoperative cohort. The objective of this analysis was not to optimize predictive performance or develop a deployable clinical prediction model, but rather to evaluate whether routinely collected physiologic and laboratory variables exhibited coherent signal behavior consistent with known clinical risk factors for ICU deterioration and readmission.

Three commonly used classification approaches were examined: logistic regression, random forest, and k-nearest neighbors (k-NN). Logistic regression was selected to provide a transparent baseline model with interpretable coefficients. Random forest was included to explore potential non-linear relationships and interaction effects among physiologic variables. The k-NN classifier was evaluated as a distance-based method sensitive to local neighborhood structure in feature space. Model hyperparameters were intentionally constrained to reduce overfitting risk given the limited cohort size.

Five-fold stratified cross-validation was used to summarize variability in model behavior while preserving the proportion of ICU readmission cases across folds. No independent hold-out test set evaluation was performed, and no claims of predictive accuracy, generalizability, or clinical performance are made.

Feature relevance was assessed using multiple complementary approaches, including standardized regression coefficients for logistic regression models, impurity-based feature importance measures for random forest models, SHAP (Shapley Additive exPlanations) value decomposition [10], and univariate statistical comparisons between readmitted and non-readmitted patients. Across modeling approaches, electrolyte abnormalities (sodium and potassium), blood pressure indices, renal function (creatinine), and temperature consistently emerged as influential variables.

This exploratory analysis emphasizes methodological transparency and interpretability in a data-sparse setting, highlighting clinically plausible signals without overinterpreting predictive metrics.

3 Rule-Based Alert Framework

A parallel rule-based alert system was developed to translate physiologic and laboratory abnormalities into clinically interpretable signals. Each variable was categorized as low, normal, or high using established clinical thresholds derived from standard reference ranges and guidelines [7–9]. Patient-level alert severity was determined by the cumulative burden of abnormalities and mapped to color-coded categories (Green, Yellow, Red). Figure 2 illustrates the alert framework. Clinical thresholds were selected from established ICU reference ranges and were not optimized for predictive performance.

Table 2. Clinical Variables, Units, Thresholds, and Reference Sources

Variable	Unit	Normal range/ threshold used	Alert condition	Source
Heart rate	beats/min	≤ 120	High > 120	[7, 9]
Systolic blood pressure	mmHg	≤ 140	High > 140	[9]
Diastolic blood pressure	mmHg	≤ 90	High > 90	[9]
Respiratory rate	breaths/min	12–20	Low < 12 or High > 20	[9]
Temperature	°C	≤ 38.0	High > 38.0	[9]
SpO$_2$	%	≥ 95	Low < 90	[9]
Hemoglobin	g/dl	12.0–15.0	Low < 12.0 or High > 15.0	[9]

(continued)

Table 2. (*continued*)

Variable	Unit	Normal range/ threshold used	Alert condition	Source
Creatinine	mg/dl	0.7–1.2	Low < 0.7 or High > 1.2	[9]
Glucose	mg/dl	70–140	Low < 70 or High > 140	[9]
Potassium	mmol/L	3.6–5.2	Low < 3.6 or High > 5.2	[9]
Sodium	mmol/L	135–145	Low < 135 or High > 145	[9]
Medications	binary	Any administered	Presence = abnormal context	[7]

The alert framework was intentionally designed as an additive burden model rather than a weighted scoring system in order to maximize transparency and bedside interpretability. Each abnormal variable contributed equally to overall alert severity. This design prioritizes explainability and ease of clinical communication over algorithmic complexity. While weighted scoring approaches may improve discrimination in larger datasets, equal-weight aggregation was considered appropriate for this feasibility study.

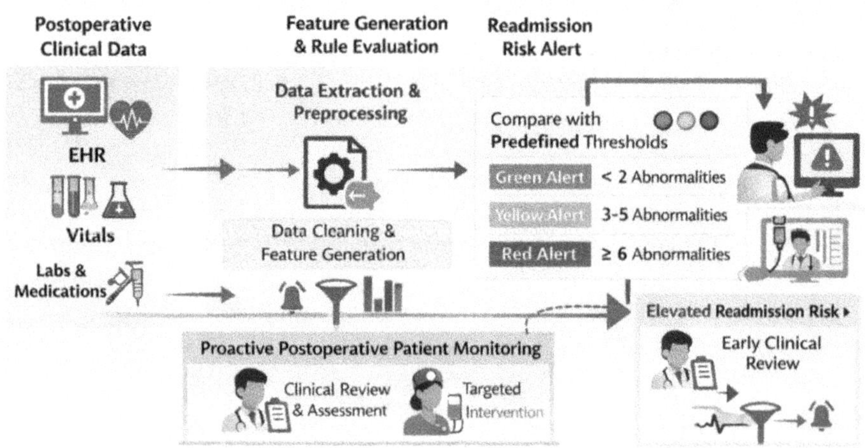

Fig. 2. Schematic overview of the rule-based alert framework. Physiologic, laboratory, and medication-related variables are evaluated against established clinical thresholds and aggregated to generate patient-level alert severity categories (Green, Yellow, Red)(Colour Figure Online).

4   Results (Pilot Feasibility Observations)

The final analytic cohort consisted of 51 unique postoperative breast cancer patients encompassing 127 ICU stays, of whom 40 patients (78.4%) experienced at least one ICU readmission during the index hospitalization. This elevated readmission prevalence reflects the cohort's restriction to postoperative ICU trajectories and the application of a transfer-based readmission definition designed to capture true ICU returns.

Frequent abnormalities were observed in electrolyte measures (sodium and potassium), blood pressure indices, renal function (creatinine), and temperature. These variables were consistently highlighted across exploratory modeling approaches and within the rule-based alert framework.

Exploratory cross-validation analyses exhibited substantial variability in model outputs, consistent with heterogeneous postoperative ICU courses and small cohort size. With the alert framework, patients were categorized as Green (≤ 2 abnormalities), Yellow (3–5 abnormalities), or Red (≥ 6 abnormalities). Higher alert severity qualitatively aligned with abnormal electrolyte values, blood pressure instability, and renal dysfunction.

5 Discussion

This pilot feasibility study demonstrates the practicality of constructing a cohort-specific, interpretable ICU readmission decision support framework for postoperative breast cancer patients.

Although predictive modeling in such a small cohort is inherently unstable, consistent feature patterns across multiple analytical approaches suggest clinically meaningful signals worthy of further investigation.

The primary contribution of this study lies in methodological transparency in a small, specialized cohort; interpretable feature analysis; and a rule-based alert system providing actionable bedside guidance.

6 Limitations

Several limitations must be acknowledged. First, the final analytic cohort consisted of 51 patients, which limits statistical power and restricts the stability of performance estimates. Reported cross-validation variability reflects the inherent uncertainty associated with small sample modeling. Accordingly, the exploratory modeling results should be interpreted as hypothesis-generating rather than confirmatory.

Second, this study was derived from a single-center dataset, which may limit generalizability to other institutions or surgical populations. External validation in independent cohorts is necessary before any clinical deployment.

Third, the modeling framework relied on static features extracted within a 24-h window and did not incorporate temporal dynamics or longitudinal physiologic trajectories. Future work may explore time-aware modeling approaches.

Finally, the rule-based alert framework utilized fixed threshold values derived from established references ranges and did not incorporate variable weighting or individualized risk calibration. While this simplicity enhances interpretability, future refinement may evaluate adaptive or weighted scoring mechanisms.

7   Conclusion

This study presents a pilot feasibility analysis of an interpretable ICU readmission decision support framework tailored to postoperative breast cancer patients. Rather than claiming clinical accuracy, the study emphasizes methodological transparency, feature interpretability, and alert design under realistic data constraints. The framework serves as a proof of concept to guide future multicenter validation.

AI Disclosure. Generative AI tools were used for language editing and structural refinement only. All study design, data extraction, analysis, and interpretation were performed by the authors.

References

1. Pattalung, T.N., Chaichulee, S.: Comparison of machine learning algorithms for mortality prediction in intensive care patients on multi-centre critical care databases. IOP Conf. Ser. Mater. Sci. Eng. **1163**(1), 012027 (2021). https://doi.org/10.1088/1757-899x/1163/1/012027
2. Ruppert, M. M., et al.: Predictive modeling for readmission to intensive care: a systematic review. *Crit. Care Explor.* **5**(1), e084 (2023). https://pubmed.ncbi.nlm.nih.gov/36699252/
3. Hempel, L., Sadeghi, S., Kirsten, T.: Prediction of intensive care unit length of stay in the MIMIC-IV dataset. *Appl. Sci.* **13**(12), 6930 (2023). https://www.mdpi.com/2076-3417/13/12/6930
4. Alghatani, K., Ammar, N., Rezgui, A., Shaban-Nejad, A.: Predicting intensive care unit length of stay and mortality using patient vital signs: machine learning model development and validation. JMIR Med. Inform. **9**(5), e21347 (2021). https://doi.org/10.2196/21347
5. Steiner, C. A., Weiss, A. J., Barrett, M. L., Fingar, K. R., Davis, P. H.: Trends in bilateral and unilateral mastectomies in hospital inpatient and ambulatory settings, 2005–2013. *HCUP Statistical Briefs.* Agency Healthcare Res. Qual. (2016). https://www.ncbi.nlm.nih.gov/books/NBK367629/
6. ICD-10-CM Codes C50: Malignant neoplasm of breast. ICD10Data.com. Available at: https://www.icd10data.com/ICD10CM/Codes/C00-D49/C50-C50/C50-
7. Cleeland, C. S., et al.: Automated symptom alerts reduce postoperative symptom severity after cancer surgery: a randomized controlled clinical trial. J. Clin. Oncol. **29**(8), 994–1000 (2011). https://doi.org/10.1200/JCO.2010.29.8315
8. Jacobs, F., et al.: Digital innovations in breast cancer care: exploring the potential and challenges of digital therapeutics and clinical decision support systems. Digital Health (2024). https://doi.org/10.1177/20552076241288821
9. Lee, H., et al.: Efficacy of the APACHE II score at ICU discharge in predicting post-ICU mortality and ICU readmission in critically ill surgical patients. Anaesth. Intensive Care **43**(2), 175–186 (2015). https://doi.org/10.1177/0310057x1504300206
10. Rojas, J.C., Carey, K.A., Edelson, D.P., Venable, L.R., Howell, M.D., Churpek, M.M.: Predicting intensive care unit readmission with machine learning using electronic health record data. Ann. Am. Thorac. Soc. **15**(7), 846–853 (2018). https://doi.org/10.1513/annalsats.201710-787oc

Decoding the Latent Space: A Comparative Analysis of Autoencoder Representations for Protein Fold Recognition

Shraddha Patre[2], Fardina Fathmiul Alam[1]($\boxtimes$), and Aarnav Tare[1]

[1] University of Maryland, College Park, MD 20742, USA
`fardina@umd.edu`, `atare@terpmail.umd.edu`
[2] Cornell Tech – Cornell University, New York, NY 10044, USA
`sp2666@cornell.edu`

Abstract. Representation learning is central to structural bioinformatics, enabling shared embeddings for protein fold recognition, structural similarity search, and function prediction. However, the organization and interpretability of latent representations learned from protein distance or contact maps remain poorly understood. We present a representation-focused comparative analysis of two of our previously developed geometric autoencoders, SuperFoldAE and its contractive variant ConSOLAE, trained on C_α distance maps. To characterize latent representations, we examine latent organization using visualizations, clustering alignment, perturbation sensitivity, resolution studies, and transfer tasks. While both models capture coarse fold structure, their latent spaces differ markedly in geometry and robustness. Representations learned with contractive regularization exhibit smoother, more compact, and more stable latent geometry, with stronger unsupervised alignment to fold structure (ARI = 0.87, NMI = 0.91). These geometric advantages are reflected in downstream behavior, including consistent Top-1 and Top-5 accuracy and support for unsupervised protein length prediction (R^2 = 0.64). Overall, this study provides a concise comparative characterization of latent spaces learned by distance-map autoencoders, offering a geometric perspective that links latent geometry and stability to robust protein fold representations.

Keywords: structural bioinformatics · latent space · representation learning · distance maps · autoencoder · regularization · interpretability

1 Introduction

Protein fold recognition remains a central problem in structural bioinformatics, and recent advances in deep representation learning have enabled models to learn low-dimensional embeddings that capture meaningful structural patterns. Autoencoders trained on C_α distance maps are particularly attractive because

K. L. Kabir et al. (Eds.): BICOB 2026, CCIS 2977, pp. 101–118, 2026.
https://doi.org/10.1007/978-3-032-26028-4_8

distance maps compactly encode global and local geometric relationships [14,22]. These approaches address limitations of sequence-based methods such as PSI-BLAST, HHblits, and DeepSF [10], which often struggle to capture structural similarity at low sequence identity. Despite this progress, latent representations learned by distance-map autoencoders remain largely unexplored. Most prior work [1,2,20] emphasizes predictive accuracy rather than latent-space geometry or structural signals encoded. Consequently, robust generalization in protein structure analysis depends not only on model capacity but also on whether learned embeddings meaningfully organize fold- and family-level structure.

In our earlier work, we introduced SuperFoldAE [15], a supervised convolutional autoencoder that applied latent representation learning to protein fold recognition and achieved strong in-distribution performance on SCOP 1.75. However, its accuracy declined sharply on the out-of-distribution SCOP 2.06 dataset, indicating limited robustness to unseen structural variation. To address this, we proposed ConSOLAE [16], which incorporates contractive regularization to encourage smoother latent manifolds and demonstrated improved generalization to unseen proteins. Both models were previously benchmarked against established methods such as DeepSF [10], a deep learning-based fold classifier, and PSI-BLAST, a classical sequence alignment tool, providing context for their performance and highlighting the complementary strengths of learned structural embeddings. These results suggest that classification accuracy alone does not guarantee robust generalization; rather, latent geometry plays a critical role in a model's ability to generalize across diverse protein folds. This observation motivates a fundamental open question: what structural information do distance-map autoencoders encode, and how do architectural choices shape latent geometry?

Addressing this question is important for three reasons that motivate our research questions. First, understanding how regularization influences latent geometry, including smoothness, stability, and organization, helps explain differences in embedding generalization (RQ1). Second, because protein fold recognition depends on global structural topology, it is essential to assess whether latent spaces organize proteins by underlying geometry rather than superficial patterns; examining their discriminative structure reveals alignment with biologically meaningful fold categories, even when accuracy appears high (RQ2). Third, for embeddings to be useful beyond classification, they must encode transferable biological signals that support downstream structural transfer tasks, which accuracy metrics alone cannot capture (RQ3). Guided by these considerations, we formulate three research questions.

- **RQ1: Latent Geometry Under Regularization.** How does contractive regularization influence the geometry of latent representations learned from protein distance maps, particularly in terms of smoothness, stability, and manifold organization?
- **RQ2: Influence of Input Resolution and Model Design.** How do distance-map resolution and architectural differences between SuperFoldAE and ConSOLAE affect classification performance and the discriminative organization of fold classes?

- **RQ3: Biological Signal and Representational Transferability.** To what extent do learned latent embeddings capture continuous protein-level properties beyond fold identity, and how well do they support downstream structural tasks?

To address these questions, we conduct a qualitative and quantitative comparative analysis of the latent spaces learned by SuperFoldAE and ConSOLAE. We examine fold separability, geometric smoothness, perturbation stability, and the encoding of global structural properties using UMAP projections, clustering metrics, principal component perturbations, and transfer tasks. Our analysis additionally includes family-level classification on SCOP 2.06, enabling evaluation of how latent geometry supports finer-grained structural relationships beyond fold identity. Our results show that representations learned with contractive regularization form smoother and more compact latent manifolds with reduced variance, while both models capture global structural cues such as protein length. Together, these findings provide a representation-level characterization of latent space organization in distance-map autoencoders and clarify how geometric structure and stability are associated with structurally informative protein representations.

2   Related Works

Classical protein fold recognition often relies on sequence- and template-based methods using profile alignment or handcrafted sequence descriptors. Approaches such as PSI-BLAST-style profile construction and block-based representations [9], HMM–HMM alignment for fold recognition [13], and ensemble classifiers combining template assignment with support vector machines [21] perform well when sequence identity is moderate to high. Deep sequence models, notably DeepSF [10], map protein sequences directly to fold labels using 1D convolutional networks, demonstrating that sequence-based deep learning can scale to large fold dictionaries. However, these methods often struggle when remote homologs share structural similarity but little sequence identity [8].

To overcome these limitations, recent work has shifted toward structure-based representation learning. Autoencoders applied to protein conformations, contact maps, and distance maps produce low-dimensional embeddings that preserve essential geometric information [6,7,12]. These embeddings have been used for decoy selection [2], tertiary structure generation [3,17], and secondary structure prediction [14]. Prior studies show that structure-derived latent embeddings can outperform traditional hand-engineered features and, in many cases, surpass sequence-only models on structurally driven tasks. Supervised autoencoders combine reconstruction with a predictive objective to improve generalization [11], while contractive autoencoders further promote robustness by penalizing the encoder Jacobian and encouraging smooth latent manifolds [18]. Our prior work, SuperFoldAE [15], applied a supervised autoencoder to $C\alpha$ distance maps for fold recognition, outperforming sequence-based baselines on

SCOP 1.75; ConSOLAE [16] extended this approach with contractive regularization, yielding smoother latent manifolds and improved generalization to SCOP 2.06. Parallel efforts have also combined structural embeddings with protein language models to integrate geometric and sequence information [5].

Despite these advances, most studies treat the latent space as a black box, using embeddings for classification, retrieval, or generation without examining internal organization. In particular, it remains unclear how different autoencoder objectives shape latent geometry, how fold- and family-level information is organized, and to what extent embeddings capture structural properties beyond fold identity. This work addresses this gap through a systematic, interpretability-focused analysis of the latent spaces learned by SuperFoldAE and ConSOLAE, emphasizing geometry, robustness, and transferability to downstream structural tasks.

3   Methods

3.1   Datasets

We conduct all experiments using the SCOP 1.75 benchmark, which provides experimentally determined protein structures annotated at the fold, superfamily, and family levels. Following prior work [15,16], each protein is represented as a C_α distance map, where the (i, j) entry corresponds to the Euclidean distance between residues i and j. Distance maps capture both local backbone geometry and global topological relationships, providing an image-like representation suitable for convolutional architectures. To obtain fixed-size inputs for the neural network, variable-length distance maps are resized to standardized resolutions of 64×64, 128×128, and 256×256 using the PIL library with nearest-neighbor interpolation, following the preprocessing of [4]. This process normalizes protein length and ensures that input dimensionality does not explicitly encode protein size, enabling scale-invariant learning and controlled analysis of the effects of input resolution on model performance and latent geometry. We use the same train-validation-test split as in [15,16], with 88% of proteins for training and 12% for testing, and a standard 80%–20% split of the training set for validation. The SCOP 1.75 dataset is filtered to 95% sequence identity and focuses on the 27 most populous fold classes [15,16] among the 1,195 folds in the dataset, avoiding folds with insufficient training data.

3.2   Models Evaluated

We compare two previously developed geometric autoencoder architectures (visualized in Fig. 1), each trained for 30 epochs with a batch size of 32 for consistency:

SuperFoldAE. SuperFoldAE [15] is a convolutional autoencoder that applies a supervised loss at the latent layer to jointly learn structural representations and fold classifications from Cα distance maps. The model places a classification head on the latent layer and is trained using a balanced combination of cross-entropy loss for fold prediction and mean squared error for distance-map reconstruction. This joint objective encourages the latent space to contain information useful both for reconstruction and for discriminative classification. The model is trained using the SGD optimizer with a learning rate of $7.5e-3$, momentum of 0.8, and weight decay of $5e-4$, as used in [15]. The encoder consists of two convolutional layers with 32 and 64 filters followed by a fully connected layer which produces a 256-dimensional latent vector, and the decoder mirrors this structure using a fully connected layer and two transposed convolution layers. Leaky ReLU activations with $\alpha = 0.08$ are used throughout. As reported in [15], the reconstruction term acts as a regularizer linked to uniform stability, helping the model generalize within SCOP 1.75. However, prior results indicated that the latent representations learned by SuperFoldAE exhibited limited generalization to the out-of-distribution SCOP 2.06 dataset. This limitation motivates our comparative analysis of latent geometry, specifically the effect of the contractive regularization used in ConSOLAE [16].

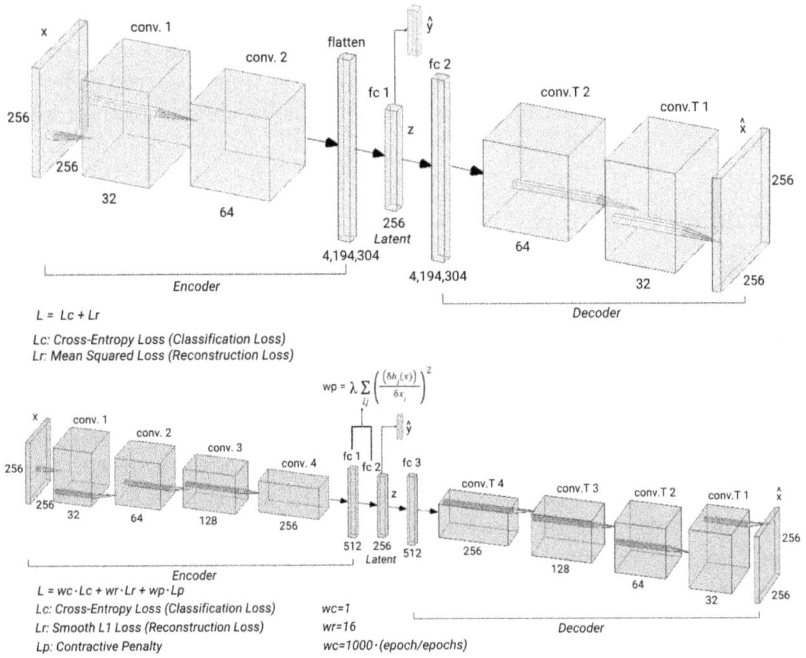

Fig. 1. Architectural comparison of SuperFoldAE [15] (blue) and ConSOLAE [16] (green). ConSOLAE extends the baseline autoencoder with a deeper network and a contractive penalty which encourages smooth, stable latent representations. (Color figure online)

ConSOLAE. ConSOLAE [16] extends the SuperFoldAE framework by calculating a contractive penalty term between the two fully connected layers of the encoder, in addition to dynamic weighting of loss objectives, a smooth L1 loss objective for reconstructions, and a deeper network. The model is trained using the Adam optimizer with a learning rate of $1e-4$ and a weight decay of $1e-5$, as used in [16]. The deeper encoder is made up of four convolutional layers with 32, 64, 128, and 256 channels, followed by two fully connected layers, resulting in a more compact latent space compared to SuperFoldAE. The decoder mirrors this structure with fully connected layers and four transposed convolutions. The contractive penalty term, parametrized by $\lambda = 1e-6$, minimizes the Frobenius norm of the encoder Jacobian, promoting a locally smooth latent manifold that is less sensitive to small perturbations in the input. Prior work showed that this regularization improves generalization on out-of-distribution datasets by producing more stable latent representations. This study focuses on comparing the latent geometry of ConSOLAE with that of SuperFoldAE rather than solely evaluating classification performance.

ConSOLAE-0 (Baseline Variant without Contractive Penalty). To isolate the effect of contractive regularization, we evaluate a variant of ConSOLAE with the contractive weight λ set to zero. This baseline shares the same architecture and latent dimensionality as ConSOLAE, ensuring that observed differences in latent-space behavior arise solely from the presence or absence of the contractive penalty.

3.3   Training Procedure

The goal of this comparative characterization is not to reproduce or modify the models in [15,16], but to evaluate the latent representations they learn under a consistent experimental setting. All models are trained on the SCOP 1.75 training set and evaluated on the corresponding validation and test sets. C_α distance maps are normalized and resized to fixed resolutions (64×64, 128×128, 256×256), ensuring identical inputs across models. Both SuperFoldAE and ConSOLAE optimize joint reconstruction and fold-classification objectives, using mean squared error (SuperFoldAE) or smooth L1 loss (ConSOLAE) for reconstruction and cross-entropy loss for fold prediction. ConSOLAE further incorporates a contractive regularization term, $\lambda \| J_\phi(x) \|_F^2$, while the ConSOLAE-0 baseline sets $\lambda = 0$ to isolate its effect. Evaluation on SCOP 2.06 is performed without retraining, enabling assessment of out-of-distribution generalization.

3.4   Statistical Analysis

We perform paired t-tests between our selected models across ten training runs to assess the effects of architectural choices and the contractive penalty, separating meaningful representational differences from variations due to random initialization or stochastic training. Effect sizes are reported using Cohen's d to complement p-values. Cohen's d is a standardized measure of the difference between two group means, quantifying the magnitude of an effect (e.g., the impact of adding a contractive penalty), rather than its existence (indicated by p-value $\prec 0.05$). Conventionally, $d \approx 0.2$ represents a small effect, $d \approx 0.5$ a medium effect, and $d \geq 0.8$ a large effect.

3.5 ConSOLAE Latent-Space Characterization

To address our research questions, we employ several complementary methods to obtain a more detailed analysis of the geometry, separability, and stability of ConSOLAE's learned latent space.

Latent Norms, Variances, and Smoothness. To assess latent geometry and encoder stability, we compute metrics characterizing the learned manifold. We measure the ℓ_2 norm and per-dimension variance of latent vectors to quantify embedding scale and dispersion, and compute mean activation magnitudes across encoder layers to assess stability. Latent smoothness and local robustness are evaluated by examining how small perturbations to input distance maps propagate through the encoder. Perturbed inputs are generated as $x' = x + \varepsilon \eta$, where $\eta \sim \mathcal{N}(0, I)$ and ε controls perturbation magnitude. We then compute the latent displacement ratio

$$L(x) = \frac{\|f(x') - f(x)\|_2}{\|x' - x\|_2}. \tag{1}$$

This quantity provides an empirical estimate of the local Lipschitz constant of the encoder f, measuring sensitivity to input perturbations. Lower values indicate smoother and more stable encodings, while higher values reflect greater sensitivity and less constrained latent geometry. We average $L(x)$ over the test set to estimate overall latent smoothness. For protein distance maps, smoothness is desirable, as minor structural variations due to noise or conformational flexibility should not induce large latent shifts. Together, these measures enable direct comparison of latent robustness between SuperFoldAE and ConSOLAE.

UMAP (Uniform Manifold Approximation and Projection) Visualizations. We use UMAP to project high-dimensional latent vectors into 2D, revealing cluster compactness, fold separability, and global manifold structure.

Clustering Metrics. We compute the Adjusted Rand Index (ARI) and Normalized Mutual Information (NMI) between latent-space k-means clusters and true fold labels. These metrics complement UMAP visualizations by quantitatively assessing how well the latent geometry reflects fold-level organization.

Principal Component Perturbation. We perform controlled perturbations along the first two principal components of the latent space. By monitoring reconstruction error and fold prediction changes with varying perturbation magnitudes, we identify the components with the most influence on these tasks.

3.6   Transfer Task: Protein Length Prediction

To assess whether latent spaces capture global structural properties beyond fold identity, we evaluate family-level prediction on the SCOP 2.06 dataset and train a linear regression model to predict protein sequence length from latent vectors. Protein length reflects overall structural size and complexity, and successful recovery from latent space indicates preservation of meaningful global geometric information without direct supervision. We report mean squared error, mean absolute error, and R^2 on the test set, using this task to probe whether latent representations encode biologically meaningful global-scale structure.

4   Results and Discussion

We now address three research questions by analyzing latent geometry, discriminative structure, and biological signal in SuperFoldAE and ConSOLAE.

RQ1: Latent Geometry Under Regularization

A meaningful latent space should organize proteins into stable, well-structured regions that reflect underlying biological relationships. We examine how contractive regularization influences this geometric organization and whether its inclusion in ConSOLAE yields a more coherent manifold than SuperFoldAE.

Baseline Comparison Models. We compare the performance of our selected models against two external baseline methods evaluated using the same evaluation metrics and test splits. As dimensionality-reduction baseline, we apply PCA to the flattened distance maps of the SCOP 1.75 and SCOP 2.06 test datasets. We additionally compare our models against ESM-2 [19], a state-of-the-art pretrained protein language model. For this comparison, we evaluate ESM-2 embeddings generated from the PDB sequences of proteins used in the SCOP 1.75 and SCOP 2.06 test datasets.

Latent Smoothness and Local Sensitivity. To quantify local robustness, we introduce controlled Gaussian perturbations to the input and measure latent smoothness using the latent displacement ratio (Eq. 1). We compare models using a paired t-test to assess whether the smoothness differs reliably. For the SCOP 1.75 test dataset, at small noise levels ($\varepsilon = 0.01$), ConSOLAE achieves a mean $L = 0.8129$ versus 0.6901 for SuperFoldAE ($t = 4.16$, $p = 0.0024$, Cohen's $d = 1.316$), indicating significantly lower local stability. At larger perturbations ($\varepsilon = 0.05$), the difference is not significant (ConSOLAE $L = 0.8763$, SuperFoldAE $L = 0.8949$, $t = 0.3305$, $p = 0.748$, Cohen's $d = 0.1045$). Performing the same t-test with the out-of-distribution SCOP 2.06 dataset, we see that at small noise levels ($\varepsilon = 0.01$), ConSOLAE achieves a mean $L = 0.6503$ versus 0.3982 for SuperFoldAE ($t = 12.07$, $p = 7.33e - 7$, Cohen's $d = 3.816$), indicating significantly lower local stability. At larger perturbations ($\varepsilon = 0.05$), the difference is significant, but with a very low effect size (ConSOLAE $L = 0.7490$, SuperFoldAE $L = 0.6372$, $t = -2.2727$, $p = 0.0491$, Cohen's $d = 0.719$). This pattern shows greater local sensitivity in ConSOLAE at very small perturbation magnitudes, which weakens at larger scales. This reflects preserved local responsiveness rather than instability, indicating that contractive regularization primarily improves large-scale latent organization, which we analyze next.

Table 1. Summary of latent-space and encoder stability metrics. Arrows ($\downarrow$) indicate that lower values are preferable (more compact and stable representations). Bold values represent the best performance in each category. Reconstruction MSE is not defined for ESM-2 since the original pretrained transformer model lacks a decoder component.

Metrics	PCA	ESM-2	SuperFoldAE	ConSOLAE-0	ConSOLAE
SCOP 1.75 Dataset					
Latent L_2 norm $\downarrow$	211.85	156.54	29.80	76.80	**10.30**
Latent var (mean across dims) $\downarrow$	244.77	0.94	2.85	18.84	**0.12**
Latent var (max dim) $\downarrow$	33222.48	98.78	19.26	61.80	**0.16**
Reconstruction MSE $\downarrow$	–	–	0.00225	0.00049	**0.00048**
SCOP 2.06 Dataset					
Latent L_2 norm $\downarrow$	217.14	154.24	24.23	59.37	**9.30**
Latent var (mean across dims) $\downarrow$	242.50	0.72	1.72	11.80	**0.07**
Latent var (max dim) $\downarrow$	32298.48	74.06	15.02	38.44	**0.15**
Reconstruction MSE $\downarrow$	–	–	0.00585	0.00177	**0.00178**

Global Stability. To further confirm the global stability and smoothness of the latent space in ConSOLAE over SuperFoldAE, we compute the ℓ_2 norm and per-dimension variance of latent vectors in these models, as detailed in Table 1. ConSOLAE has the most compact and stable latent space, with the smallest vector norms and lowest per-dimension variance. Its maximum-to-mean variance ratio ($1.3\times$) is much lower than the other evaluated methods (PCA: $135.6\times$, ESM-2: $105.1\times$, SuperFoldAE: $6.8\times$, ConSOLAE-0: $3.3\times$), indicating evenly distributed variance across latent dimensions. This demonstrates that contractive regularization produces well-regularized, balanced representations, yielding a smooth and stable latent manifold. In contrast, SuperFoldAE exhibits moderately scaled latent vectors, while ConSOLAE-0, ESM-2, and PCA have extremely large latent norms which, in combination with increased per-dimension variance, indicate higher instability. We see very similar trends in the results for the SCOP 2.06 dataset. This pattern reflects a more dispersed latent space with weaker geometric constraints. Notably, both ConSOLAE and ConSOLAE-0 achieve lower reconstruction error than SuperFoldAE, showing that contractive regularization improves latent stability without sacrificing reconstruction fidelity. These latent-space differences due to the contractive penalty, will be further analyzed next.

Ablation Study: Isolating Effect of the Contractive Penalty. To isolate the effect of the contractive penalty, we directly compare ConSOLAE (with penalty, $\lambda > 0$)

to its non-contractive baseline ConSOLAE-0 ($\lambda = 0$). We perform paired t-tests using the SCOP 1.75 dataset and find that, across both perturbation levels ($\varepsilon = 0.01$ and $\varepsilon = 0.05$), the baseline model exhibits extremely large latent displacement ratios ($L \approx 10$–12), whereas the contractive model remains below 1 ($L \approx 0.8$–0.9). Effect sizes are very large (Cohen's $d = 2.8$–6.6) with $p < 10^{-5}$ in both cases, demonstrating that the contractive penalty dramatically improves latent stability. Performing the above t-test using the SCOP 2.06 dataset we find that, across both perturbation levels ($\varepsilon = 0.01$ and $\varepsilon = 0.05$), the baseline model again exhibits extremely large latent displacement ratios ($L \approx 7.39$–9.18), whereas the contractive model remains below 1 ($L \approx 0.65$–0.75). Effect sizes are very large (Cohen's $d = 2.7$–6.0) with $p < 10^{-5}$ in both cases. Importantly, this stabilization does not significantly change classification accuracy, a point we quantify further in RQ2, indicating that the penalty primarily reshapes the latent geometry rather than the predictive decision boundary.

Table 2. Comparison of activations at each encoder layer for ConSOLAE (with contractive penalty) and ConSOLAE-0 (baseline without contractive penalty)

Metric	ConSOLAE [1.75]	ConSOLAE-0 [1.75]	ConSOLAE [2.06]	ConSOLAE-0 [2.06]
Conv. 1 activation range	{−0.2888, 0.2322}	{−0.3450, 0.3105}	{−0.9953, 0.8547}	{−1.1668, 1.0573}
Conv. 2 activation range	{−8.1126, 6.0235}	{−7.5168, 7.5391}	{−39.1529, 28.5069}	{−28.8564, 29.8864}
Conv. 3 activation range	{−9.5662, 8.4299}	{−9.9955, 5.3873}	{−52.3569, 34.9549}	{−39.4008, 19.1549}
Conv. 4 activation range	{−13.2198, 7.7032}	{−8.4439, 9.2431}	{−77.6688, 30.0478}	{−38.3461, 24.6179}
fc1 activation range	{−61.04, 34.68}	{−127.96, 113.07}	{−72.1576, 23.1527}	{−149.4453, 118.1135}
fc2 activation range	{−0.7113, 1.6672}	{−23.5579, 23.6408}	{−0.7990, 2.2313}	{−22.9753, 22.6539}

To further verify this effect, we compare activation distributions layer by layer between ConSOLAE and ConSOLAE-0 (Table 2). Activation ranges in early convolutional layers (Conv1 and Conv2) are similar across both models. In deeper convolutional layers (Conv3 and Conv4), ConSOLAE exhibits comparable or slightly broader activation ranges, indicating that contractive regularization does not uniformly suppress intermediate feature responses. In contrast, the fully connected layers show substantial differences: ConSOLAE-0 displays more than a $2\times$ larger activation range in fc1 and over a $20\times$ amplification in fc2. Because these layers directly map convolutional features into the latent space, the uncontrolled activation growth in ConSOLAE-0 explains the large latent norms and variances observed in Table 1. Together, these results show that the contractive penalty restructures activation distributions, strongly constraining latent layers while preserving stable feature representations upstream.

Cluster Structure and Separation. We first quantify fold-level organization in the learned representations using unsupervised clustering metrics.

As shown in Table 3, ConSOLAE achieves substantially higher agreement with ground-truth SCOP fold labels (ARI $= 0.87$, NMI $= 0.91$) than both PCA

Table 3. Comparison of ARI and NMI scores for PCA with flattened distance maps, ESM-2, and ConSOLAE methods. Scores are calculated between true labels and KMeans natural clusterings with $k = 27$ (number of fold classes).

Method	PCA	ESM-2	ConSOLAE
ARI (Adjusted Rand Index)	0.04	0.34	**0.87**
NMI (Normalized Mutual Information)	0.21	0.61	**0.91**

on flattened distance maps (ARI = 0.04, NMI = 0.21) and sequence-based ESM-2 embeddings (ARI = 0.34, NMI = 0.61). This indicates that ConSOLAE's latent geometry aligns more strongly with fold structure than either a linear dimensionality-reduction baseline or a sequence-only pretrained language model.

To qualitatively illustrate this result, we visualize ConSOLAE's latent space using a two-dimensional UMAP projection (Fig. 2). The projection reveals well-separated, compact clusters corresponding to individual SCOP folds with minimal intermixing. The strong alignment between the true-label coloring and unsupervised KMeans assignments shows that the latent geometry alone is sufficiently discriminative to recover fold-level structure, supporting the quantitative clustering analysis above. Overall, these results demonstrate that contractive regularization enables ConSOLAE to learn a class-separable, interpretable, and structurally meaningful latent manifold.

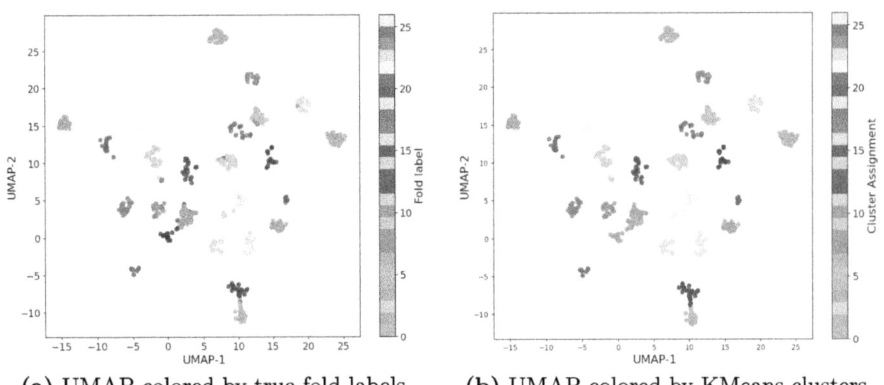

(a) UMAP colored by true fold labels.   (b) UMAP colored by KMeans clusters.

Fig. 2. UMAP projections of ConSOLAE's latent space on the SCOP 1.75 validation set. Compact clusters emerge for each fold, with strong alignment between KMeans clusters and true labels (ARI = 0.87, NMI = 0.91).

Summary (RQ1). Although contractive regularization can exhibit higher local responsiveness at very small perturbation scales, it reshapes the global latent space to be more compact, balanced, and geometrically constrained.

RQ2: Influence of Input Resolution and Model Design

We next examine how distance-map resolution and architectural differences between SuperFoldAE and ConSOLAE affect fold-recognition performance. We begin with a statistical analysis to observe the difference in overall classification performance between the models with the original 256×256 resolution. We then move on to test various resolutions (64×64, 128×128, 256×256), where we observe systematic differences in performance trends between SuperFoldAE and ConSOLAE across resolutions.

Statistical Analysis of Classification Performance Across Models. We use paired t-tests to compare fold classification performance across SuperFoldAE, ConSOLAE, and ConSOLAE-0. Across 10 trials on SCOP 1.75, SuperFoldAE achieves a mean accuracy of 0.71, significantly lower than ConSOLAE's 0.76 ($t = -17.55$, $p = 2.86 \times 10^{-8}$, Cohen's $d = 5.55$). In contrast, no significant difference is observed between ConSOLAE (0.7597) and its non-contractive baseline ConSOLAE-0 (0.7588) ($t = -0.152$, $p = 0.882$, Cohen's $d = 0.048$). Similar trends hold under out-of-distribution evaluation on SCOP 2.06: SuperFoldAE attains a mean accuracy of 0.2406, significantly lower than ConSOLAE's 0.4382 ($t = -8.37$, $p = 1.54 \times 10^{-5}$, Cohen's $d = 2.65$), while ConSOLAE and ConSOLAE-0 again show no significant difference ($t = -0.598$, $p = 0.564$, Cohen's $d = 0.189$). These results indicate that ConSOLAE's performance gains over SuperFoldAE arise from architectural and training differences, whereas the contractive penalty primarily shapes the compact and smooth latent geometry observed in RQ1.

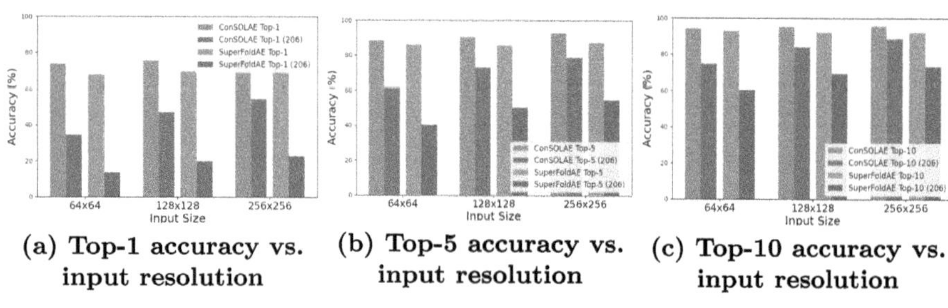

(a) **Top-1 accuracy vs. input resolution**   (b) **Top-5 accuracy vs. input resolution**   (c) **Top-10 accuracy vs. input resolution**

Fig. 3. Classification accuracy across input resolutions on SCOP 1.75 and out-of-distribution (OOD) SCOP 2.06 test sets. Subfigures show (a) Top-1, (b) Top-5, and (c) Top-10 accuracy for distance-map sizes 64×64, 128×128, and 256×256. ConSOLAE (green bars) consistently achieves higher Top-1 and Top-5 accuracy than SuperFoldAE (blue bars), while the methods show comparable Top-10 performance on SCOP 1.75. (Color figure online)

Classification Trends Across Varying Input Resolutions. Figure 3 shows that Top-1 accuracy improves consistently with increasing resolution, with ConSOLAE achieving a stable 5–7%-point gain over SuperFoldAE on SCOP 1.75 and a 2.4–2.6× improvement on the out-of-distribution SCOP 2.06 test set, highlighting its stronger fine-grained ranking ability. This trend aligns with the well-separated latent structure learned by ConSOLAE (RQ1; UMAP, ARI = 0.87, NMI = 0.91). A similar but smaller advantage is observed for Top-5 accuracy. By Top-10, the performance gap narrows on SCOP 1.75, indicating that both models capture coarse fold-level similarity; however, ConSOLAE still achieves 13–15% point gains on SCOP 2.06, reflecting superior generalization.

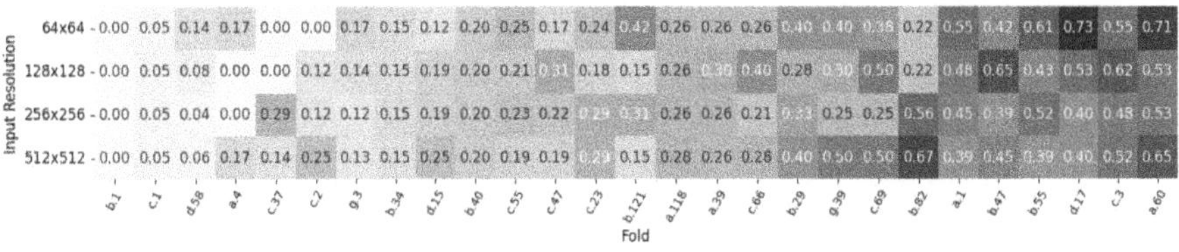

Fig. 4. Classification error by fold and input resolution for ConSOLAE on the SCOP 1.75 test set. Darker values indicate higher error. Four resolutions (64 × 64 to 512 × 512) are shown. Resolution effects are heterogeneous: some folds improve with higher detail, others remain difficult, and some degrade due to increased local noise.

Fold-Wise Discriminability. Figure 4 shows classification error by fold and resolution for ConSOLAE across four input sizes (64 × 64 to 512 × 512). Several folds (d.58, g.3, g.39, a.1, d.17) improve as resolution increases up to 256 × 256, indicating that moderate resolution captures discriminative structural patterns. For some folds (e.g., a.1, d.17), performance remains stable or improves at 512 × 512, while others (e.g., g.39) peak at intermediate resolution and degrade at the highest resolution. In contrast, folds such as c.37, c.2, d.15, and b.82 show increased error at higher resolutions, and others exhibit non-monotonic trends (e.g., b.47, c.3), reflecting fold-specific trade-offs between global topology and fine-grained detail. Overall, resolution interacts with fold geometry in a non-uniform manner: higher resolution often helps, but gains are fold-dependent and may reverse at the highest resolution.

Summary (RQ2). Model architecture and input resolution affect performance in complementary ways: architectural depth yields consistent accuracy gains, while resolution modulates fold-specific discriminability. ConSOLAE shows higher accuracy than SuperFoldAE across resolutions, and as shown in RQ1, the contractive term primarily stabilizes latent geometry rather than improving classification accuracy.

RQ3: Biological Signal and Representational Transferability

To assess whether the learned representations encode biologically meaningful structure beyond fold identity, we evaluate ConSOLAE's latent space on downstream transfer tasks that probe both fine-grained structural discrimination and continuous global properties. In particular, we consider family-level classification on an out-of-distribution dataset and a comparison of root mean squared error (RMSE) results as primary tests of biological relevance, and protein length prediction as a complementary continuous signal.

Table 4. Family-level classification accuracy on the SCOP 2.06 test set, evaluating transferability of latent representations to fine-grained structural discrimination.

Method	Top1	Top5	Top10
SuperFoldAE	15.67%	27.46%	31.90%
ConSOLAE	**34.38%**	**55.89%**	**60.63%**

Family-Level Classification on SCOP 2.06. To evaluate transferability under a more challenging and biologically meaningful label space, we benchmark family-level classification on the SCOP 2.06 dataset. Family prediction requires distinguishing proteins with subtle structural and evolutionary differences and therefore provides a stronger test of representational quality than fold-level classification. As shown in Table 4, ConSOLAE achieves a Top-1 accuracy of 34.38%, substantially outperforming SuperFoldAE (15.67%). Similar gains are observed for Top-5 and Top-10 accuracy, indicating improved retrieval performance under stricter family-level discrimination. These results demonstrate that the latent representations learned by ConSOLAE generalize beyond the training objective and capture biologically meaningful structure that transfers to finer-grained classification tasks.

Distance Map RMSE Comparison. We use distance-map reconstruction error as a diagnostic to assess whether proximity in the learned latent space corresponds to geometric similarity in structure space. On the SCOP 1.75 test set, ConSOLAE achieves a lower reconstruction RMSE than SuperFoldAE (0.0228 vs. 0.0423), while the reconstruction-focused baseline ConSOLAE-0 attains the lowest RMSE (0.0182), as expected. This confirms that the contractive penalty is not optimized to minimize reconstruction error alone, but instead reshapes the latent geometry. To evaluate latent-space consistency, we compare distance-map RMSE between proteins that are nearest neighbors in latent space and random or cross-fold protein pairs. For ConSOLAE, latent neighbors exhibit markedly lower RMSE (0.038) than random or cross-fold pairs (both $\approx$0.053), indicating that proteins that are close in latent space are also geometrically similar in distance-map space. This result directly validates the claim that ConSOLAE's latent neighborhoods reflect meaningful geometric structural similarity beyond fold identity, even when reconstruction error is not the primary optimization objective.

Protein Length Prediction as a Continuous Transfer Task. Beyond discrete classification, we evaluate whether ConSOLAE's latent space encodes continuous global biological properties by predicting protein sequence length (range 17–1146) on SCOP 1.75. Although trained only for fold classification and distance-map reconstruction, a simple two-layer regressor on ConSOLAE's latent representations achieves $R^2 = 0.8141$ on the training set and $R^2 = 0.6380$ on the validation set, with a mean absolute error of 39.87 residues. In contrast, a naive mean predictor yields $R^2 \approx 0$ and a substantially larger error of approximately 121 residues, confirming that the task is non-trivial. As shown in Fig. 5, predictions closely follow the identity line, indicating that the latent space preserves coarse-grained information correlated with protein size. While protein length is not a functional label, this result demonstrates that ConSOLAE captures meaningful continuous global structural variation without explicit supervision.

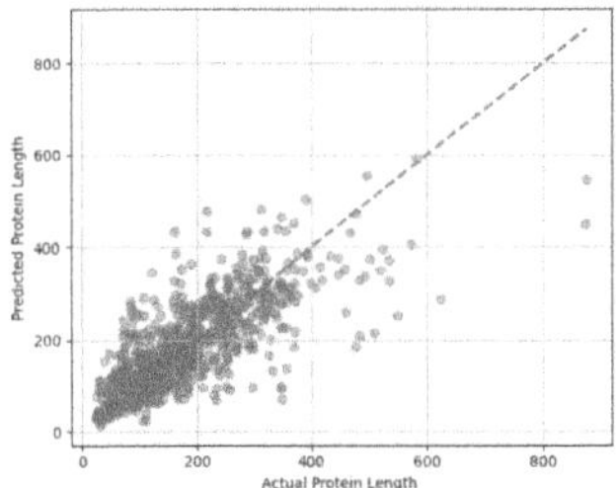

Fig. 5. Predicted vs. actual protein length for the SCOP 1.75 validation set using a regression model trained on ConSOLAE latent representations. Each point corresponds to a protein, and the dashed red line ($y = x$) indicates perfect prediction. (Color figure online)

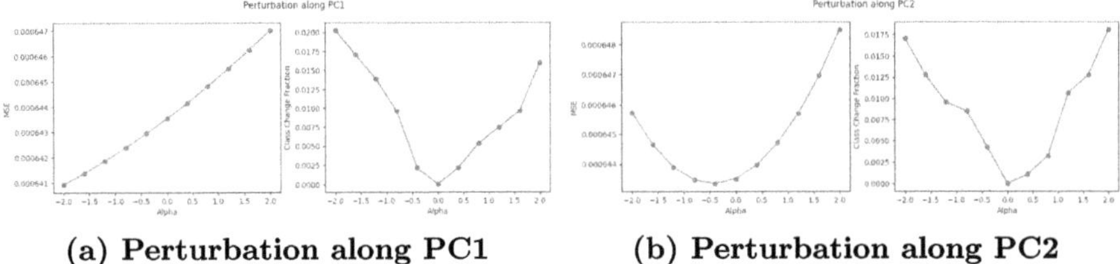

(a) **Perturbation along PC1**   (b) **Perturbation along PC2**

Fig. 6. Latent perturbation analysis of ConSOLAE latent representations generated from SCOP 1.75 test set. Each subplot shows reconstruction MSE (blue) and class-change fraction (red) as latent vectors are shifted by a scalar α along major principal directions. MSE varies smoothly with α across all curves, and class changes are minimal near $\alpha = 0$ and rise symmetrically outward. (Color figure online)

Global Perturbation Analysis Along Principal Latent Directions. To further understand the latent structure underlying this transferability, we move beyond the local smoothness demonstrated by random ε-perturbations and examine the global geometric organization of ConSOLAE's latent space. We perturb latent vectors from the SCOP 1.75 test set along the top principal components (PC1 and PC2) and measure changes in reconstruction error (MSE) and classification consistency. This analysis tests whether large, directed movements in latent space remain predictable and well organized rather than producing unstable behavior. As shown in Fig. 6, reconstruction MSE varies smoothly with perturbation magnitude, and class-change fractions remain near zero around $\alpha = 0$, indicating stable behavior near the original embedding. As perturbations increase, fold predictions change gradually and symmetrically, suggesting that fold information is encoded in a distributed manner across the latent space. Together, these results demonstrate that ConSOLAE's latent space is globally organized and supports continuous, biologically meaningful variation.

Summary (RQ3). ConSOLAE's latent embeddings encode biologically meaningful structure beyond fold identity. Family-level classification, nearest-neighbor distance-map RMSE, and protein length prediction together show that the latent space captures both fine-grained structural similarity and global geometric properties.

Overall Discussion

Across RQ1–RQ3, a coherent representational narrative emerges. Through contractive regularization, ConSOLAE learns a smoother and more stable latent geometry than SuperFoldAE, yielding compact manifolds with low variance and strong robustness to perturbations. Architectural design drives consistent gains in classification performance, while input resolution modulates fold-specific discriminability. This trend extends to the more challenging SCOP 2.06 family-level task, where ConSOLAE improves Top-1/ Top-5/ Top-10 accuracy from 15.67/27.46/31.90 (SuperFoldAE) to 34.38/ 55.89/ 60.63 on average (Table 4). Beyond classification, the latent space captures continuous biological properties, enabling transferable tasks such as protein length prediction. Supporting geometry-consistency analysis further confirms that latent proximity corresponds to structural similarity: nearest neighbors in ConSOLAE's latent space exhibit approximately 29% lower distance-map error than random or cross-fold pairs (mean DM-RMSE ≈ 0.038 vs. ≈ 0.053). Together, these results show that representation-level analysis reveals structure not evident from accuracy alone, establishing ConSOLAE's latent manifold as structured, biologically meaningful, and robust, providing a foundation for next-generation protein structural embeddings.

5   Conclusion and Future Work

In this paper, we characterized the structural information encoded in the latent spaces of distance-map autoencoders and examined how latent geometry relates to generalization in protein fold recognition. Using SuperFoldAE and its contractive variant ConSOLAE as comparative baselines, we showed that architectural and regularization choices fundamentally shape latent representations. ConSOLAE learns a smoother, more compact, and more class-separable latent manifold, exhibiting strong unsupervised alignment with SCOP fold structure, improved robustness to perturbations, and preservation of continuous biological signals such as protein length beyond discrete fold labels. Although both models achieve comparable Top-10 accuracy, suggesting similar coverage of the global fold landscape, their latent geometries differ markedly in quality and stability, underscoring the limitations of accuracy-based evaluation alone. Together, these findings highlight the importance of representation-level analysis for geometric protein embeddings. Future work will explore how specific structural features (e.g., local versus long-range contacts) are encoded across latent dimensions, extend transfer analyses to additional structural and functional tasks, incorporate features such as secondary structure composition, and apply this framework to other structural encodings and model families to support robust, interpretable, and biologically meaningful protein representations.

Acknowledgements. The authors acknowledge the University of Maryland supercomputing resources (http://hpcc.umd.edu) made available for conducting the research reported in this paper.

Disclosure of Interests. The authors declare that they have no competing interests relevant to the content of this article.

References

1. Alam, F.F., Rahman, T., Shehu, A.: Learning reduced latent representations of protein structure data. In: Proceedings of the ACM International Conference on Bioinformatics, Computational Biology and Health Informatics, pp. 592–597 (2019)
2. Alam, F.F., Rahman, T., Shehu, A.: Evaluating autoencoder-based featurization and supervised learning for protein decoy selection. Molecules **25**(5), 1146 (2020)
3. Alam, F.F., Shehu, A.: Generating physically-realistic tertiary protein structures with deep latent variable models. In: Proceedings of the IEEE International Conference on Bioinformatics and Biomedicine, pp. 2463–2470 (2021)
4. Alam, F.F., Shehu, A.: Data size and quality matter: generating physically-realistic distance maps of protein tertiary structures. Biomolecules **12**(7), 908 (2022)
5. Ali, S., Chourasia, P., Patterson, M.: When protein structure embedding meets large language models. Genes **15**(1), 25 (2024)
6. Brown, W.M., Martin, S., Pollock, S.N., Coutsias, E.A., Watson, J.P.: Algorithmic dimensionality reduction for molecular structure analysis. J. Chem. Phys. **129**(6) (2008)

7. Chen, W., Tan, A.R., Ferguson, A.L.: Collective variable discovery and enhanced sampling using autoencoders. J. Chem. Phys. **149**(7) (2018)
8. Cheng, J., Tegge, A.N., Baldi, P.: Machine learning methods for protein structure prediction. IEEE Rev. Biomed. Eng. **1**, 41–49 (2008)
9. Henikoff, S., Henikoff, J.G., Alford, W.J., Pietrokovski, S.: Automated construction and graphical presentation of protein blocks from unaligned sequences. Gene **163**(2), GC17–GC26 (1995)
10. Hou, J., Adhikari, B., Cheng, J.: DeepSF: deep convolutional neural network for mapping protein sequences to folds. Bioinformatics **34**(8), 1295–1303 (2017)
11. Le, L., Patterson, A., White, M.: Supervised autoencoders: improving generalization performance with unsupervised regularizers. In: Bengio, S., Wallach, H., Larochelle, H., Grauman, K., Cesa-Bianchi, N., Garnett, R. (eds.) Advances in Neural Information Processing Systems, vol. 31. Curran Associates, Inc. (2018)
12. Lemke, T., Peter, C.: EncoderMap: dimensionality reduction and generation of molecule conformations. J. Chem. Theory Comput. **15**(2), 1209–1215 (2019)
13. Lyons, J., Paliwal, K.K., Dehzangi, A., Heffernan, R., Tsunoda, T., Sharma, A.: Protein fold recognition using hmm-hmm alignment and dynamic programming. J. Theor. Biol. **393**, 67–74 (2016)
14. Manzoor, U., Halim, Z., et al.: Protein encoder: an autoencoder-based ensemble feature selection scheme to predict protein secondary structure. Expert Syst. Appl. **213**, 119081 (2023)
15. Patre, S., Kanani, R., Alam, F.: SuperFoldAE: Enhancing Protein Fold Classification with Autoencoders, pp. 1–15 (2025)
16. Patre, S., Kanani, R., Tare, A., Vallabhaneni, P., Alam, F.F.: ConSOLAE: learning smooth and generalizable representations for protein fold recognition. In: Companion Proceedings of the 16th ACM International Conference on Bioinformatics, Computational Biology and Health Informatics (2025)
17. Rahman, T., Alam, F.F., Shehu, A.: Equivariant encoding based GVAE (EqEn-GVAE) for protein tertiary structure generation. In: Proceedings of the IEEE International Conference on Bioinformatics and Biomedicine, pp. 3470–3477 (2022)
18. Rifai, S., Vincent, P., Muller, X., Glorot, X., Bengio, Y.: Contractive auto-encoders: explicit invariance during feature extraction. In: Proceedings of the International Conference on Machine Learning, pp. 833–840 (2011)
19. Rives, A., et al.: Biological structure and function emerge from scaling unsupervised learning to 250 million protein sequences. Proc. Natl. Acad. Sci. (2019)
20. Rostami, Z., Mukund, K., Masnadi-Shirazi, M., Subramaniam, S.: An autoencoder learning method for predicting breast cancer subtypes. PLoS ONE **20**(7), e0327773 (2025)
21. Xia, J., Peng, Z., Qi, D., Mu, H., Yang, J.: An ensemble approach to protein fold classification. Bioinformatics **33**(6), 863–870 (2017)
22. Yuan, M., et al.: ProteinMAE: masked autoencoder for protein surface self-supervised learning. Bioinformatics **39**(12), btad724 (2023)

Integration of Supervised Machine Learning Algorithms and MALDI-TOF MS for Identifying Vancomycin-Resistant *Enterococcus Faecium* Isolates

Tetyana King and Hisham Al-Mubaid[✉]

University of Houston – Clear Lake, Houston, TX 77058, USA
hisham@uhcl.edu

Abstract. Matrix-assisted laser desorption ionization–time of flight (MALDI-TOF) mass spectrometry (MS) is a rapid and widely utilized method for microbial identification and detecting their antibiotic-resistance profiles. This method extends to the detection of clinically significant organisms, such as vancomycin-resistant *Enterococcus faecium* (VRE), an opportunistic pathogen of the gut microbiome. Machine learning is well-suited and has been increasingly applied for analyzing MALDI-TOF MS data. However, the MALDI-TOF studies that focus on VRE are limited. In binary *resistant* vs. *susceptible E. faecium* classification using a feature matrix with 36,000 features created from preprocessed spectra without signal-to-noise ratio subtraction (SNR), *random forest* (RF) achieves the highest accuracy (78.7%, F1 79.4%) among other algorithms. In the vanA vs. vanB gene classification of VRE, RF achieves 82.6% accuracy (F1 86.7%). At the matrix with 18,000 features created from the preprocessed spectra with SNR subtraction, *support vector machines* (SVM) algorithm is more balanced in sensitivity and specificity and can accurately predict more VRE/VSE isolates than RF. Using lower-resolution data, with 18,000 features, *convolutional neural network* (CNN) outperforms RF highlighting its potential for accurate VRE/VSE prediction from MALDI-TOF MS spectra.

Keywords: machine learning · MALDI-TOF MS · *Enterococcus faecium* · vancomycin resistance

1 Introduction

As a mass spectrometry (MS) technique, the matrix-assisted laser desorption/ionization time-of-flight (MALDI-TOF) is one of the most important resources for proteomics of bacteria and disease research. MALDI-TOF MS is a simple, low-cost, and rapid method that is widely used for bacterial identification (Du et al. 2002; Josten et al. 2014; Idelevich et al. 2018; Florio et al. 2020). MALDI-TOF MS produces mass spectra that reflect the proteomic profiles of bacteria and, when used with additional analytical tools, allow the identification of specific biomarkers. The most used range is from 2000 to 20,000 m/z, which contains the most robust, abundant and reproducible proteomic signatures of bacteria (Yoon and Jeong 2021).

K. L. Kabir et al. (Eds.): BICOB 2026, CCIS 2977, pp. 119–128, 2026.
https://doi.org/10.1007/978-3-032-26028-4_9

There are two ways to determine the resistance of bacteria to antibiotics using MALDI-TOF spectra. The first is to compare the area under the curve of mass spectra of bacteria grown in standard medium and medium containing antibiotics, or to determine the peak shift in bacteria grown with 13C isotope media (Yoon and Jeong 2024; Florio et al. 2020). General computational biology techniques and tools as FlexAnalysis and MALDI Biotyper software (Bruker Daltonics Inc.) are used for this purpose (Idelevich et al. 2018). The analysis typically begins with baseline subtraction and peak intensity normalization. Peak detection and alignment correct minor mass shifts between biological and technical replicates (Candela et al. 2022). After alignment, spectra are often binned into fixed m/z intervals (e.g., 0.5 or 1 Da) to create homogeneous feature matrices (Candela et al. 2022; Wang et al. 2022). The data can be simplified using principal component analysis (PCA) and processed by the MALDIquant R package (Astudillo et al. 2024; Gibb and Strimmer 2012). Integrating genomic databases (e.g., CARD, ResFinder) allows linking detected spectral features to known resistance genes.

The second main approach to antimicrobial resistance analysis using MALDI-TOF MS is spectral identification of resistance mechanisms (Yoon and Jeong 2024), which can generally be accomplished using direct or machine learning methods. Direct analysis examines specific spectral features to identify changes associated with resistance (Calderaro et al. 2021; Du et al. 2002). For example, MALDI-TOF MS can show changes in peaks (e.g., β-lactamases) in an antimicrobial drug after it is tested on bacteria. Bruker MBT STAR-Carba assay and custom-created scripts can identify changes in peak position, intensity, or disappearance, indicating enzyme inactivation and confirming bacterial resistance (Florio et al. 2020; Idelevich et al. 2018). Machine learning algorithms are applied to find more complex patterns between spectra. One application is classifying spectra associated with resistance phenotypes (Astudillo et al. 2024; Chung et al. 2023; Weis et al. 2022). Unsupervised methods (e.g., *hierarchical clustering*) and supervised algorithms (e.g., *random forest*) can make predictions based on peak matrices (Calderaro et al. 2021). Another machine learning strategy includes the identification of peaks specific to the expression of resistance genes (e.g., vanA or mecA) (Candela et al. 2022; Du et al. 2002). In this case, the spectra of resistant and susceptible isolates are compared using statistical tests (e.g., t-tests, ANOVA) or feature importance metrics derived from trained models (Wang et al. 2022; Candela et al. 2022).

Machine learning being a well-suited technique for analyzing and interpreting the data generated by MALDI-TOF MS (López-Cortés *et al.* 2024; Wang et al. 2022) is actively used to identify antibiotic-resistant bacteria (Ma et al. 2024; Yu et al. 2023), one of which is vancomycin-resistant *Enterococcus faecium* (VRE). *E. faecium* a Gram-positive facultative anaerobe bacterium that is one of the most common representatives of the human gut microbiota. VRE can sometimes cause serious infections such as endocarditis and septicemia (Gao et al. 2024; Arredondo-Alonso et al. 2020; Lebreton et al. 2017).

There are more than 15 different algorithms used to classify antibiotic-resistant bacteria, and most research include several algorithms at a time (López-Cortés *et al.* 2024; Ren et al. 2024; Candela et al. 2022). Among the most used algorithms are *support vector machines* (SVM), *random forest* (RF), and *logistic regression* (LR) based algorithms (López-Cortés *et al.* 2024). SVM and LR belong to the traditional type of supervised machine learning algorithms, in which a model is trained on labeled data (*i.e.,* the input data is labelled based on the class of bacteria as *resistant* or *susceptible*). The algorithms trains and learns from this data to predict the most accurate class label (López-Cortés *et al.* 2024). RF belongs to ensemble methods of supervised machine learning and combines the predictive power of multiple *decision trees* to improve accuracy, robustness, and generalization (Ren et al. 2024). RF is less sensitive to noise and feature scaling than SVM or LR.

The *partial least squares discriminant analysis* (PLS-DA) and *neural networks* (NN) are also actively used for antimicrobial studies (Godmer et al. 2024; Calderaro et al. 2021). NN structures resemble the connections of neurons in the human brain and function similarly. They can determine the most complex patterns in mass spectral data (Lopez-Cortes *et al.* 2024). The *convolutional neural networks* (CNN) is a type of NN with several layers of neuron nodes which surpass traditional supervised machine learning algorithms in forecasting accuracy and have been actively used (Godmer et al. 2024; López-Cortés *et al.* 2024; Wang et al. 2022; Mortier et al. 2021).

Despite the active usage and high potential of machine learning in the prediction of antibiotic-resistant bacteria, only a few studies have focused on applying machine learning to identify VRE (Lopez-Cortes *et al.* 2024; Chung et al. 2023; Wang et al. 2022; Candela et al. 2022). This study aims to compare the accuracy of four algorithms, RF, SVM, PLS-DA, and CNN, in the classification of VRE.

2 Methods

2.1 Mass Spectra Preprocessing

With 178 isolates, the raw spectra (n = 1547 for 178 isolates) were preprocessed using two procedures. The *first procedure* was described by Candela et al. (2022). Briefly, the quality of raw spectra and internal calibration of the genus-specific peak at 4428 m/z were performed using *FlexAnalysis* v3.4 software (Bruker Daltonics Inc.), then spectra were preprocessed using a custom R script including *MALDIquant* v1.22.3 and *MALDIquant-Foreign* v0.14.1 packages (Gibb and Strimmer 2012). Preprocessing parameters were as follows: range 2,000 – 20,000 m/z; smoothing – Savitzky-Golay filter (half-window size 5; polynomial order 3). Since the *MALDIquant* package does not support an argument "factor = 0.02" for the TopHat filter, this value was mimicked using "halfWindowSize = round (0.02 * (20000–2000))". Clean spectra were aligned per spot. Average spectra per spot was aligned to create an average spectrum per isolate.

The *second procedure* of spectra preprocessing was described by Wang et al. (2022). Briefly, all the raw spectra (n = 1547 for 178 isolates) were preprocessed using *FlexAnalysis* v.3.4 with default parameters, which included smoothing with Savitsky-Golay method (width = 2 m/z, cycles = 10), baseline subtraction using TopHat filter, and peak detection with a relative intensity threshold of 0%, signal-to-noise ratio of 2, minimal intensity threshold of 300, peak width of 2 m/z, height of 90%, and baseline subtraction using of TopHat filter. Preprocessed spectra were converted to a peak position (mass-to-charge ratio, m/z) and intensity (the signal strength at each peak) matrix using a custom R script.

2.2 Feature Matrices Preparation

The preprocessed spectra by the first method were used to prepare feature matrices as Candella *et al.* (2022) described. The first matrix was created based on the *full spectrum method*. The spectra were binned by 0.5 Da, and peaks were included regardless of intensity. Then, the intensity was normalized by the total ion current (TIC). The other two matrices were constructed using the *threshold method* where only peaks with an intensity higher than 1.0% of the maximum peak intensity were selected. The difference between the last two matrices was in the normalization procedure. For one matrix created using the "threshold method" (TICp), intensity normalization (TIC) was applied before peak selection, for the other (pTIC), normalization was applied after peak selection.

The feature matrix derived from spectra preprocessed using the second method was used to construct the "full-spectrum" matrix described above and was vectorized using preprocess.py script from published CNN pipeline by Wang et al. (2022) (https://github.com/p568912/CNN_MALDI-TOF_VREfm/tree/main).

2.3 Machine Learning Algorithms: Selection and Training

We used RF, SVM, PLS-DA, and CNN learning techniques to classify VRE and VSE isolates. Three of these algorithms also were applied to differentiate isolates within the VRE class. The vancomycin resistance of most clinical isolates of E. faecium is mediated by *vanA* or *vanB* gene clusters, and RF, SVM and PLS-DA have been used to classify VRE isolates based on *vanA*- and *vanB*-associated resistance (Candela et al. 2022). The RF model was based on 100 decision trees. SVM implemented with the *radial basis function* (RBF) kernel, the *StandardScaler* function for feature normalization.

$$z - score \text{ standardization}: \quad z = \frac{x - \mu}{\sigma}, \tag{1}$$

and regularization parameter $C = 1$. All these algorithms were applied with *10-fold cross validation, 10-FCV*. PLS-DA was used with a number of latent variables (components) equal to five. The training step was done using 178 isolates with 10FCV for VRE and VSE classification, and 86 isolates for VanA VRE and VanB VRE classification.

We used *CNN* (https://github.com/p568912/CNN_MALDI-TOF_VREfm/tree/main) for the classification and prediction step with 177 isolates: 92 VSE and 85 VRE because isolate *VanA 584* was excluded by the algorithm due to insufficient peak intensity during matrix-to-files conversion. CNN was trained using 50 epochs and 100 epochs.

3 Results

3.1 VRE vs. VSE and VanA vs. VanB Classifications

Firstly, the VRE vs. VSE classification was done based on the matrix with 36,000 features based on 0.5 Da binning prepared by the procedure of Candela et al. (2022), the RF algorithm showed the highest accuracy (78.65%), followed by SVM (70.22%) and PLS-DA (65.17%) (see Table 1). The RF algorithm could accurately classify 73 VSE (class = 1) and 67 VRE (class = 0), as well as 22 VanA VRE (class = 0) and 49 VanB VRE (class = 1) isolates as shown in Table 2.

Table 1. The results of VRE vs. VSE classification, and VanA vs. VanB classification withing VRE class using the "full-spectrum method" described by Candela et al. (2022) with 0.5 Da binning (36,000 features). K-fold cross validation (k = 10) was applied. All values are shown as percentages.

Metrics/ Algorithm	VRE vs VSE			Van A VRE vs VanB VRE		
	RF	SVM	PLS-DA	RF	SVM	PLS-DA
Accuracy	78.65	70.22	65.17	82.56	65.12	76.74
F1	79.35	71.35	66.67	86.73	77.27	81.82
Sensitivity	79.35	71.74	67.39	89.09	92.73	81.82
Specificity	77.91	68.60	62.79	70.97	16.13	67.74
PPV	79.35	70.97	65.96	84.48	66.23	81.82
NPV	77.91	69.41	64.29	78.57	55.56	61.74

Table 2. Confusion matrix of predicted isolates. The matrix was prepared with "full-spectrum method" described by Candela et al. (2022) with 0.5 Da binning (36,000 features). Three supervised machine learning algorithms were applied.

Algorithm	Actual Classification	Predicted		Actual Classification	Predicted	
		VRE	VSE		VanA	VanB
RF	VRE	67	19	VanA	22	9
	VSE	19	73	VanB	6	49
SVM	VRE	59	27	VanA	5	26
	VSE	26	66	VanB	4	51
PLS-DA	VRE	54	32	VanA	21	10
	VSE	30	62	VanB	10	45

Other key metrics, such as F1 score, sensitivity, specificity, PPV, and NPV, were generally in line with this trend, indicating that RF is the best performing algorithm for this classification task. Similarly, RF performed best in differentiating isolates with *vanA*- and *vanB*-associated resistance in the VRE group with an accuracy of 82.56%, followed by PLS-DA (76.74%) (see Table 1). SVM showed the lowest accuracy, 65.12%, among the algorithms and could correctly classify most VanB VRE isolates as shown in Table 2.

The second evaluation experiment was performed using the matrix prepared according to Wang et al. (2022), and with binning 0.5 Da, 36,000 features in total, full spectra method and the results are in Table 3. The RF showed the highest accuracy in distinguishing VRE and VSE (72.47%). However, SVM with an accuracy of 70.22% was the most balanced in other indicators, Table 3, and correctly identified 68 VSE and 67 VRE against 63 VSE and 66 VRE isolates classified by RF (see Table 4). In the classification of VanA VRE and VanB VRE, RF turned out to be the leader both in accuracy (74.81%) (see Table 3) and in the number of correctly classified isolates: 16 VanA VRE and 48 VanB VRE (see Table 4).

Table 3. The results of VRE vs. VSE classification, and VanA vs. VanB classification withing VRE class using the "full-spectrum method" described by Candela et al. (2022), with spectra preprocessed according to Wang et al. (2022) and 0.5 Da binning (36,000 features). K-fold cross validation (k = 10) was applied. All values are shown as percentages.

Metrics/ Algorithm	VRE vs VSE			Van A VRE vs VanB VRE		
	RF	SVM	PLS-DA	RF	SVM	PLS-DA
Accuracy	72.47	70.22	61.24	74.81	63.95	51.16
F1	72.00	71.96	63.49	81.36	77.04	60.38
Sensitivity	68.48	73.91	65.22	87.27	94.55	58.18
Specificity	76.74	66.28	59.98	51.61	9.68	38.71
PPV	75.90	70.10	61.86	79.19	65.00	62.75
NPV	69.47	70.37	60.49	69.57	50.00	34.29

Table 4. Confusion matrix of predicted isolates. The matrix was prepared with "full-spectrum method" described by Candela et al. (2022), with spectra preprocessed according to Wang et al. (2022) and 0.5 Da binning (36,000 features). Three supervised machine learning algorithms were applied.

Algorithm	Actual Classification	Predicted		Actual Classification	Predicted	
		VRE	VSE		VanA	VanB
RF	VRE	66	20	VanA	16	15
	VSE	29	63	VanB	7	48
SVM	VRE	67	29	VanA	3	28
	VSE	24	68	VanB	3	52
PLS-DA	VRE	49	37	VanA	12	19
	VSE	32	60	VanB	23	32

The VRE vs. VSE classification based on the matrix according to Wang et al. (2022) and with 1 Da binning, 18,000 features, was more accurate when applying RF (74.71%), while SVM and PLS-DA lagged slightly behind as shown in Table 5. Further, RF was able to accurately identify 66 VSE and 67 VRE isolates (see Table 5). Also, RF performed best in VanA vs. VanB prediction producing 76.74% accuracy (see Table 5) and accurately classifying 18 VanA VRE and 48 VanB VRE (see Table 6). When using the pTIC and TICp matrices, the accuracy of the algorithms turned out to be critically low for all three algorithms.

Table 5. The results of VRE vs. VSE classification, and VanA vs. VanB classification withing VRE class using the "full-spectrum method" described by Candela et al. (2022), with spectra preprocessed according to Wang et al. (2022) and 1 Da binning (18,000 features). K-fold cross validation (k = 10) was applied. All values are shown as percentages.

Metrics/ Algorithm	VRE vs VSE			Van A VRE vs VanB VRE		
	RF	SVM	PLS-DA	RF	SVM	PLS-DA
Accuracy	74.71	67.97	65.73	76.74	61.62	73.25
F1	74.57	69.51	67.37	82.75	76.59	79.27
Sensitivity	71.73	70.65	68.47	87.27	96.36	8.0
Specificity	77.90	65.11	62.79	58.06	0	61.29
PPV	77.64	68.42	66.31	78.68	63.09	78.57
NPV	72.04	67.46	65.06	72.00	0	63.33

Table 6. Confusion matrix of predicted isolates. The matrix was prepared with "full-spectrum method" described by Candela et al. (2022), with spectra preprocessed according to Wang et al. (2022) and 0.5 Da binning (18,000 features). Three supervised machine learning algorithms were applied.

Algorithm	Actual Classification	Predicted		Actual Classification	Predicted	
		VRE	VSE		VanA	VanB
RF	VRE	67	19	VanA	18	13
	VSE	26	66	VanB	7	48
SVM	VRE	56	30	VanA	0	31
	VSE	27	65	VanB	2	53
PLS-DA	VRE	54	32	VanA	19	12
	VSE	29	63	VanB	11	44

3.2 VRE vs. VSE Classification Using CNN

The prediction accuracy using CNN increased from 82.49% for 50 epochs to 83.05% for 100 epochs. Specificity increased from 91.30% to 94.57% and AUC increased from 91.32% to 92.31%, respectively. At the same time, the sensitivity of the model decreased from 72.94% to 70.59%, respectively. As a result, the accuracy of the *VSE* isolates detection was 84 for 50 epochs versus 87 for 100 epochs, which increased due to a decrease in the accuracy of VRE isolates detection which was 62 for 50 epochs versus 60 for 100 epochs.

4 Discussion

The MALDI-TOF MS technique is a simple, low-cost, and rapid technique for producing robust bacterial proteomic signatures. Furthermore, it is widely used for the identification of antibiotic-resistant bacteria (Du et al. 2002; Josten et al. 2014; Idelevich et al. 2018; Florio et al. 2020). Integrating MALDI-TOF MS with computational biology tools and machine learning allows discovering hidden complex patterns in mass spectra to classify resistant and susceptible isolates. In this study, several learning algorithms are evaluated for the classification of vancomycin-resistant and vancomycin-susceptible E. faecium, and for distinguishing between resistant isolates containing vanA or vanB gene clusters.

Two preprocessing workflows based on Candela et al. (2022) and Wang et al. (2022) are used to preprocess spectra and construct multiple feature matrices, including full-spectrum and threshold-based methods. The RF classifier consistently outperforms SVM and PLS-DA across all preprocessing strategies, with the highest accuracy of 78.65% achieved on the 36,000-feature matrix obtained using the preprocessed spectra without SNR elimination. RF also demonstrated robust performance in VanA-VanB classification with an accuracy of 82.56%. These results support the advantage of the RF as an ensemble method, as it is less sensitive to noise and feature scaling than SVM or LR models (Ren et al. 2024). Although SVM demonstrates slightly lower accuracy, it provides a more balanced classification profile when using spectra preprocessed with noise reduction (SNR = 2). This suggests that SVM may be more sensitive to small variations in the quality of spectral signals and benefits from cleaner data. PLS-DA consistently demonstrates the lowest accuracy across all scenarios.

The CNN model (training with 50 epochs) achieved a reasonably good accuracy of 83.05% compared to the traditional machine learning models. When the training was extended to 100 epochs, the accuracy further improved to 94.57%. However, this improvement came with a slight drop in sensitivity (70.59%), indicating that CNN may favor susceptible isolates. Furthermore, CNN achieves high accuracy using a lower-resolution matrix (18,000 features) compared to traditional machine learning algorithms tested with 36,000 features.

Interestingly, the accuracy of all models drops significantly when pTIC and TICp matrices are used. The matrices created with the threshold-based method contain such a number of features that it is difficult to systematize them and identify distinguished markers for the classification of VRE/VSE isolates.

Despite the multiple studies on integrating machine learning and MALDI-TOF MS, relatively few studies have focused specifically on VRE (Lopez-Cortes et al. 2024; Candela et al. 2022; Wang et al. 2022). This study can help to narrow the gap by comparing the accuracy of traditional machine learning algorithms and CNN using different spectral preprocessing strategies.

5 Conclusion

This study demonstrates the accuracy of machine learning algorithms in analyzing MALDI-TOF mass spectra for identifying vancomycin-resistant E. faecium. Among traditional machine learning methods, RF consistently demonstrates high accuracy when trained on a matrix of different resolutions. However, it is inferior to SVM in other parameters when using a matrix created on noise-removed spectra. CNN outperforms traditional methods even on lower-resolution feature matrix.

Acknowledgements. We are grateful to the Clinical Microbiology and Infectious Diseases Department of Hospital General Universitario Gregorio Marañón (Madrid, Spain) and especially to Belén Rodríguez-Sánchez, PhD, from Instituto de Investigación Sanitaria Gregorio Marañón (Madrid, Spain), for kindly providing a training data set of *E. faecium* clinical isolates. This project was supported by NSF grant #2320765.

References

Arredondo-Alonso, S., et al.: Plasmids shaped the recent emergence of the major nosocomial pathogen *Enterococcus faecium*. mBio **11**(1), 19–e03284 (2020)

Astudillo, C.A., López-Cortés, X.A., Ocque, E., Manriquez, J.: Multi-label classification to predict antibiotic resistance from raw clinical MALDI-TOF mass spectrometry data. Sci. Rep. **14**(1), 31283 (2024)

Calderaro, A., et al.: Rapid identification of *Escherichia coli* colistin-resistant strains by MALDI-TOF mass spectrometry. Microorganisms **9**(11), 2210 (2021)

Candela, A., et al.: Rapid and reproducible MALDI-TOF-based method for the detection of vancomycin-resistant *Enterococcus faecium* using classifying algorithms. Diagnostics (Basel) **12**(2), 328 (2022)

Chung, C.R., et al.: Towards accurate identification of antibiotic-resistant pathogens through the ensemble of multiple preprocessing methods based on MALDI-TOF spectra. Int. J. Mol. Sci. **24**(2), 998 (2023)

Gao, W., et al.: Machine learning assisted MALDI mass spectrometry for rapid antimicrobial resistance prediction in clinicals. Anal. Chem. **96**(33), 13398–13409 (2024)

Godmer, A., et al.: Contribution of MALDI-TOF mass spectrometry and machine learning including deep learning techniques for the detection of virulence factors of *Clostridioides difficile* strains. Microb. Biotechnol. **17**(6), e14478 (2024)

Du, Z., Yang, R., Guo, Z., Song, Y., Wang, J.: Identification of *Staphylococcus aureus* and determination of its methicillin resistance by matrix-assisted laser desorption/ionization time-of-flight mass spectrometry. Anal. Chem. **74**(21), 5487–5491 (2002)

Florio, W., Baldeschi, L., Rizzato, C., Tavanti, A., Ghelardi, E., Lupetti, A.: Detection of antibiotic-resistance by MALDI-TOF mass spectrometry: an expanding area. Front. Cell. Infect. Microbiol. **10**, 572909 (2020)

Gibb, S., Strimmer, K.: MALDIquant: a versatile R package for the analysis of mass spectrometry data. Bioinf. (Oxford, England) **28**(17), 2270–2271 (2012)

Idelevich, E.A., Sparbier, K., Kostrzewa, M., Becker, K.: Rapid detection of antibiotic resistance by MALDI-TOF mass spectrometry using a novel direct-on-target microdroplet growth assay. Clinical Microbiol. Infect. Offic. Publ. Europ. Soc. Clinic. Microbiol. Infect. Diseas. **24**(7), 738–743 (2018)

Josten, M., et al.: Identification of agr-positive methicillin-resistant *Staphylococcus aureus* harbouring the class a mec complex by MALDI-TOF mass spectrometry. Int. J. Med. Microbiol. IJMM **304**(8), 1018–1023 (2014)

Lebreton, F., Manson, A.L., Saavedra, J.T., Straub, T.J., Earl, A.M., Gilmore, M.S.: Tracing the *Enterococci* from Paleozoic origins to the hospital. Cell **169**(5), 849–861 (2017)

López-Cortés, X.A., Manríquez-Troncoso, J.M., Kandalaft-Letelier, J., Cuadros-Orellana, S.: Machine learning and matrix-assisted laser desorption/ionization time-of-flight mass spectra for antimicrobial resistance prediction: a systematic review of recent advancements and future development. J. Chromatogr. A **1734**, 465262 (2024)

López-Cortés, X.A., Manríquez-Troncoso, J.M., Hernández-García, R., Peralta, D.: MSDeep-AMR: antimicrobial resistance prediction based on deep neural networks and transfer learning. Front. Microbiol. **15**, 1361795 (2024)

Ma, W.H., Chang, C.C., Lin, T.S., Chen, Y.C.: Distinguishing methicillin-resistant *Staphylococcus aureus* from methicillin-sensitive strains by combining Fe_3O_4 magnetic nanoparticle-based affinity mass spectrometry with a machine learning strategy. Mikrochim. Acta **191**(5), 273 (2024)

Mortier, T., Wieme, A.D., Vandamme, P., Waegeman, W.: Bacterial species identification using MALDI-TOF mass spectrometry and machine learning techniques: a large-scale benchmarking study. Comput. Struct. Biotechnol. J. **19**, 6157–6168 (2021)

Ren, M., Chen, Q., Zhang, J.: Repurposing MALDI-TOF MS for effective antibiotic resistance screening in *Staphylococcus epidermidis* using machine learning. Sci. Rep. **14**(1), 24139 (2024)

Yu, J., et al.: Prediction of methicillin-resistant *Staphylococcus aureus* and carbapenem-resistant *Klebsiella pneumoniae* from flagged blood cultures by combining rapid Sepsityper MALDI-TOF mass spectrometry with machine learning. Int. J. Antimicrob. Agents **62**(6), 106994 (2023)

Wang, H.Y., et al.: Clinically applicable system for rapidly predicting *Enterococcus faecium* susceptibility to vancomycin. Microbiology spectrum **9**(3), e0091321 (2021)

Wang, H.Y., et al.: Efficiently predicting vancomycin resistance of *Enterococcus faecium* from MALDI-TOF MS spectra using a deep learning-based approach. Front. Microbiol. **13**, 821233 (2022)

Weis, C., et al.: Direct antimicrobial resistance prediction from clinical MALDI-TOF mass spectra using machine learning. Nat. Med. **28**(1), 164–174 (2022)

Yoon, E.J., Jeong, S.H.: MALDI-TOF mass spectrometry technology as a tool for the rapid diagnosis of antimicrobial resistance in bacteria. Antibiotics (Basel) **10**(8), 982 (2021)

DCJ and Indel Distance Considering Intergenic Region Information

Gabriel Siqueira[1(✉)], Alexsandro Oliveira Alexandrino[1],
Andre Rodrigues Oliveira[2], and Zanoni Dias[1]

[1] Instituto de Computação, Universidade Estadual de Campinas (Unicamp),
Campinas, Brazil
{gabriel.siqueira,alexsandro,zanoni}@ic.unicamp.br
[2] Faculdade de Computação e Informática, Universidade Presbiteriana Mackenzie,
São Paulo, Brazil
andre.rodrigues@mackenzie.br

Abstract. Since its introduction, the double cut and join (DCJ) operation has gained popularity in Computational Biology as a convenient abstraction that generalizes various genome rearrangement operations – mutation events that affect large genome segments. The central problem involving this operation is the DCJ distance problem, which seeks to minimize the number of DCJ operations required to transform one genome into another. In particular, recent studies incorporate intergenic region sizes to obtain more realistic rearrangement distance estimates. When the genomes differ in gene content, indel (insertion or deletion) events must also be considered. We present the first algorithm for the DCJ distance that incorporates intergenic information and supports indels, a 2-approximation applicable to single-copy, co-tailed genomes.

Keywords: Approximation · Double Cut and Join · Indel · Intergenic Regions

1 Introduction

For many years, the study of genome rearrangement distances has been present in the computational biology literature due to its importance in the development of methods for genome comparison. The advances in this topic include improvements on the practical and theoretical aspects of algorithms and the inclusion of more biologically relevant information in the representation of the genomes. The rearrangement distance consists in estimating the minimum number of operations of genome rearrangement, which are mutations that affect large sequences of the genome, necessary to transform one genome into another.

A common first step in the study of a proposed new genome rearrangement distance problem is to develop an algorithm for some restricted case. Many such problems are NP-hard. So the proposed algorithms are in many cases inexact, but with an ensured approximation factor. In some cases, approximation algorithms are proposed even before the NP-hardness proof [2,6].

K. L. Kabir et al. (Eds.): BICOB 2026, CCIS 2977, pp. 129–139, 2026.
https://doi.org/10.1007/978-3-032-26028-4_10

An important rearrangement operation is the double cut and join (DCJ), which was defined to generalize other operations and is widely adopted in the field [14]. For the simpler model where no intergenic region information is used and both genomes have the same set of genes, each gene with a single copy, finding the DCJ distance (rearrangement distance problem considering only the event of DCJ) can be done in polynomial time [14]. If the genomes can have distinct sets of genes and we allow the insertion or deletion of genes, there is also a polynomial time algorithm [4].

One important advancement in the study of the DCJ operation was the inclusion of intergenic region information. Motivated by the fact that the use of these regions tends to improve the distance estimation on real genomic data, Fertin *et al.* [7] proposed a representation of the genomes including the number of nucleotides in the intergenic regions.

With the inclusion of intergenic region information, the problem becomes NP-hard when the model has only the DCJ operation [7]. For that case, there is a 4/3-approximation algorithm [7]. If we allow for the insertion and deletion of intergenic regions, the problem becomes solvable in polynomial time [5]. The DCJ and indel distance problem, where intergenic regions are considered and genes can be inserted or deleted, is still an open problem in terms of its complexity and the existence of algorithms for it.

Simonaitis *et al.* [10] presented a generalization of this representation of intergenic regions, which allows other types of biological information to be represented as well. In their work, they develop a framework to simplify the process of developing exact or approximation algorithms for certain DCJ-based rearrangement problems by ensuring that if the problem can be described in their model, then we only need to prove the exactness or approximation of the algorithm in a particular subset of instances.

To represent a pair of genomes, we can use the breakpoint graph, which encodes the adjacencies between genes of both genomes. We can include weights and labels on the edges of this graph to represent information regarding the intergenic regions [2].

There are two approaches to representing in the breakpoint graph genes that are present in only one of the genomes. One approach is the use of ghost adjacencies [9], which includes some extra edges and vertices. The other approach is the use of labels on the existing edges. We opt for the second approach.

In this work, we propose an approximation algorithm for the DCJ and Indel Distance problem and adopt two common assumptions for the pair of genomes under comparison. The first assumption is that each genome has only one copy of each gene. This assumption simplifies the problem, and it is commonly used as a first step in the exploration of a rearrangement problem. This assumption can be lifted by some common approaches at a computational cost. The simpler approach is to search for the best solution within the correspondences between the copies of the same gene in both genomes. The second assumption is that the genomes are co-tailed, that is, for every pair of extremities in a linear chromosome in the origin genome, there is a corresponding pair of extremities in the target genome. This assumption allows us to consider fewer cases in our proofs.

The remainder of this paper is organized as follows. Section 2 presents the necessary definitions. Section 3 describes the proposed algorithm, Sect. 4 discusses the consequences of our assumptions, and Sect. 5 concludes the paper.

2 Definitions

A genome $\mathcal{G}$ is represented by a set of chromosomes, each one is an interleaved sequence of genes and intergenic regions, all represented by integers. In the case of genes, the sign of the integers represents the orientation of the genes, and the absolute value is used to represent each distinct gene. Once we assume that genes only have one copy in the genome, there is no repetition in the absolute value of the integers representing the genes. In the case of intergenic regions, the integers are non-negative, and their values represent the number of nucleotides in the corresponding region. We call this value the size of the intergenic region. Each chromosome may be linear, which we represent by surrounding the sequence with [], or circular, which we represent by surrounding the sequence with (). For a circular genome $(g_1, r_1, \ldots, g_n, r_n)$ with n genes, the first and last genes of the sequence, g_1 and g_n, are considered adjacent, and the sequence ends with an intergenic region (r_n), which is the region between those genes. For a linear genome $[g_1, r_1, \ldots, g_n]$ with n genes, the sequence starts and ends with genes (g_1 and g_n). Normally, real genomes start and end with intergenic regions, but there is a process called capping that inserts artificial genes in the model to ensure that there are genes on the extremities of the genome.

Example 1. The following set of sequences of integers represents a genome with three chromosomes. The genes are shown in black and the intergenic regions in gray. The first chromosome is linear with the genes $+1$, -2, and $+3$. Between the genes $+1$ and -2, there is an intergenic region with size 10, and between the genes -2 and $+3$ there is an intergenic region with size 28. The second chromosome is circular with the genes $+4$, $+5$, and -6. The genes -6 and $+4$ are adjacent and have an intergenic region with size 13 between them. The last chromosome is linear with genes $+7$ and $+8$.

$$\{[+1, 10, -2, 28, +3], (+4, 18, +5, 21, -6, 13), [+7, 25, +8]\}$$

Given a genome $\mathcal{G}$, the DCJ operation cuts the genome $\mathcal{G}$ between two genes w and x, in an intergenic region with size a, and between two genes y and z, in an intergenic region with size b. Afterward, it joins the genes in a new order, either combining w with y and x with z, or combining w with z and x with y. The newly formed intergenic regions will have sizes a' and b', such that $a+b = a'+b'$. Depending on how the combination is performed, the sign of the elements may be flipped. We use the notation $\mathcal{G} \cdot \delta$ to denote the application of the DCJ δ in the genome $\mathcal{G}$.

Example 2. A sequence of two DCJs applied to a genome. The first DCJ acts on a linear and on a circular chromosome, combining them in a single linear chromosome. Note that a DCJ could also do the opposite and extract a circular chromosome from a linear chromosome. The second DCJ acts on two linear chromosomes.

$$\{[+1, 10, -2, 28, +3], (+4, 18, +5, 21, -6, 13), [+7, 25, +8]\}$$

$$\{[+1, 5|5, -2, 28, +3], (+4, 10|8, +5, 21, -6, 13), [+7, 25, +8]\}$$

$$\{[+1, 5|10, -4, 13, +6, 21, -5, 8|5, -2, 28, +3], [+7, 25, +8]\}$$

$$\{[+1, 15, -4, 13, +6, 21, -5, 13, -2, 28, +3], [+7, 25, +8]\}$$

$$\{[+1, 15, -4, 13, +6, 15|6 - 5, 13 - 2, 28, +3], [+7,18|7, +8]\}$$

$$\{[+1, 15, -4, 13, +6, 15|7, +8], [+7,18|6 - 5, 13 - 2, 28, +3]\}$$

$$\{[+1, 15, -4, 13, +6, 22, +8], [+7, 24, -5, 13, -2, 28, +3]\}$$

Given a genome $\mathcal{G}$, an indel is an operation that inserts or deletes genes and intergenic regions. In the case of the insertion, the operation takes a sequence of genes, and intergenic regions around them, and inserts it between two genes x and y of the same chromosome. In the case of the deletion, the operation cuts the genome in two intergenic regions of the same chromosome and removes all genes and intergenic regions between these cuts. We use the notation $\mathcal{G} \cdot \beta$ to denote the application of the indel β in the genome $\mathcal{G}$.

Example 3. Deletion followed by the insertion of a sequence of genetic material in a genome. The first four lines represent the deletion, and the last three lines represent the insertion. The deleted and inserted material are highlighted with an underline.

$$\{[+1, 10, -2, 28, +3], (+4, 18, +5, 21, -6, 13), [+7, 25, +8]\}$$

$$\{[+1, 5|5, -2, 8|20, +3], (+4, 18, +5, 21, -6, 13), [+7, 25, +8]\}$$

$$\{[+1, 5|20, +3], (+4, 18, +5, 21, -6, 13), [+7, 25, +8]\}$$

$$\{[+1, 25, +3], (+4, 18, +5, 21, -6, 13), [+7, 25, +8]\}$$

$$\{[+1, 25, +3], (+4, 18, +5, 21, -6, 13), [+7, 16|9, +8]\}$$

$$\{[+1, 25, +3], (+4, 18, +5, 21, -6, 13), [+7, 16|5, +9, 14, -10, 11|9, +8]\}$$

$$\{[+1, 25, +3], (+4, 18, +5, 21, -6, 13), [+7, 21, +9, 14, -10, 20, +8]\}$$

The DCJ and Indel Distance problem consists in finding the minimum number of DCJs and indels necessary to transform an origin genome $\mathcal{G}$ into a target genome $\mathcal{H}$. Once we are assuming co-tailed genomes, the extremities of linear chromosomes are the same in both genomes. Another important assumption is that we only insert genes that occur exclusively in $\mathcal{H}$ and only delete genes that occur exclusively in $\mathcal{G}$. Otherwise, it would be possible to turn one genome into the other by deleting all the genes in $\mathcal{G}$ and inserting all the genes in $\mathcal{H}$.

Once we only have one copy of each gene, we can always assume that the genes of $\mathcal{H}$ have positive sign and are in an increasing order in each chromosome. In the case of circular chromosomes, there is one gene where we can start to read the chromosome where we read all genes in an increasing order. For example, if the genome shown in example 1 was the target, we could replace the negative signs with positive signs to ensure this property: $\{[+1, 10, +2, 28, +3],$ $(+4, 18, +5, 21, +6, 13), [+7, 25, +8]\}$. In this case, the circular genome is already shown with the reading order that ensures the increasing sequence of gene numbers.

Our algorithms for the DCJ and Indel Distance problem will require a structure called the breakpoint graph. The benefit of that structure is that it can represent the adjacencies of two genomes in a single structure, and we can use the size of a cycle decomposition of this graph to measure how similar the two genomes are. The version of the breakpoint graph we are going to use was defined by Alexandrino *et al.* [3] and has weights and labels on its edges to represent the size of intergenic regions and the genes exclusive to one of the genomes, respectively.

The breakpoint graph of two genomes $\mathcal{G}$ and $\mathcal{H}$ has a vertex set V, an edge set E, a weight function $w : E \to \mathbb{N}$, and a label function $\lambda : E \to \{\circ, \bullet\}$. To define this graph, we start by choosing an arbitrary order for the chromosomes in $\mathcal{G}$ and a direction to read each one. For circular chromosomes, we also choose a gene to start to read it. These choices will help simplify the definitions but do not impact the results, as any other choice would result in an equivalent graph.

In V, there is one vertex for each gene x in the extremity of the linear chromosomes. If x is the first gene to be read, we denote this vertex by x^h, otherwise, if x is the last gene to be read, we denote this vertex by x^t. For each other gene x in both $\mathcal{G}$ and $\mathcal{H}$ there are two vertices x^t and x^h. We are going to position these vertices according to the reading order chosen before. If a gene x has two vertices and a positive sign, the vertices will be positioned in the order x^t and x^h. If a gene x has two vertices and a negative sign, the vertices will be positioned in the order x^h and x^t.

The set E is formed by two sets E_o (origin edges) and E_t (target edges). In E_o, there is one edge connecting each two consecutive vertices that are in the same chromosome and not from the same gene. There is also in E_o one edge for each circular chromosome of $\mathcal{G}$ connecting the first and last vertex of that chromosome. Those edges represent the adjacencies of $\mathcal{G}$, ignoring genes exclusive to $\mathcal{G}$. In E_t, there is one edge connecting each pair of vertices x^h and y^t, such that y is the smallest value larger than x in a vertex and x is from the same chromosome as y. There is also in E_t one edge for each circular chromosome of $\mathcal{H}$ connecting x^h and y^t, such that x and y are the largest and smallest values of that chromosome, respectively. Those edges represent the adjacencies of $\mathcal{H}$, ignoring genes exclusive to $\mathcal{H}$.

For each edge of E connecting two vertices from genes x and y, the weight function w will attribute the weight corresponding to the sum of the intergenic regions between x and y. The label function λ will attribute the label $\bullet$ if there is a gene between x and y in S, if we are looking at an origin edge, or in ι, if we are looking at a target edge. Otherwise, λ attributes the label $\circ$. Figure 1 shows an example of the breakpoint graph and Fig. 2 shows an example of how rearrangement operations affect the breakpoint graph.

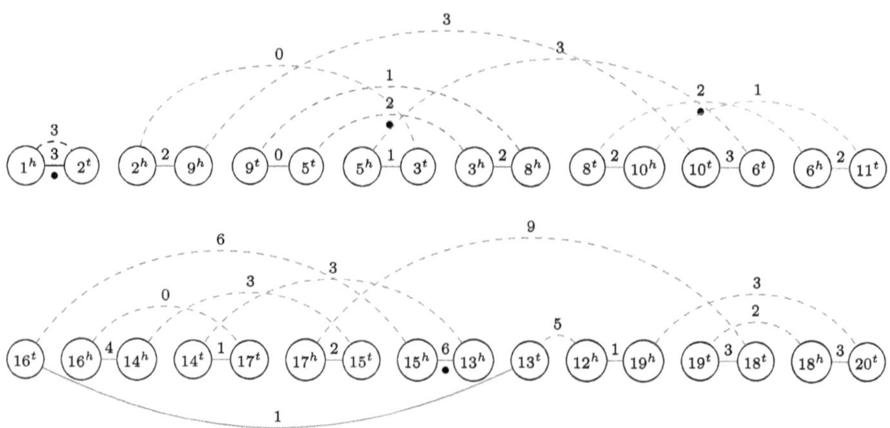

Fig. 1. Breakpoint Graph created from two genomes $\mathcal{G} = \{[+1, 1, +21, 2, +2, 2, -9, 0, +5, 1, +3, 2, -8, 2, -10, 3, +6, 2, +11], (+16, 4, -14, 1, +17, 2, +15, 4, +22, 2, -13, 1), [+12, 1, -19, 3, +18, 3, +20]\}$ and $\mathcal{H} = \{[+1, 3, +2, 0, +3, 1, +4, 1, +5, 3, +6, 0, +7, 2, +8, 1, +9, 3, +10, 1, +11], [+12, 5, +13, 3, +14, 3, +15, 6, +16, 0, +17, 9, +18, 2, +19, 3, +20]\}$. The solid lines are the origin edges, and the dashed lines are the target edges. Each color represents a different cycle in the cycle decomposition of the graph. The weights (representing the intergenic regions) are indicated above each edge, and the labels (indicating if there are genes exclusive to one of the genomes) are indicated below each edge. The labels $\circ$ are omitted from this drawing of the graph.

The breakpoint graph can be decomposed into edge-alternating cycles, which are cycles alternating between origin and target edges. We can categorize the cycles according to the weight and label of their edges. A cycle is balanced if the sum of the weights of its origin edges is equal to the sum of the weights of its target edges. The cycle is unbalanced otherwise. A cycle is labeled if any of its edges have a label $\bullet$. The cycle is clean otherwise. If a cycle is balanced and clean, it is called good, and it is called bad otherwise. A cycle is trivial if it has only one origin and one target edge, and it is non-trivial otherwise.

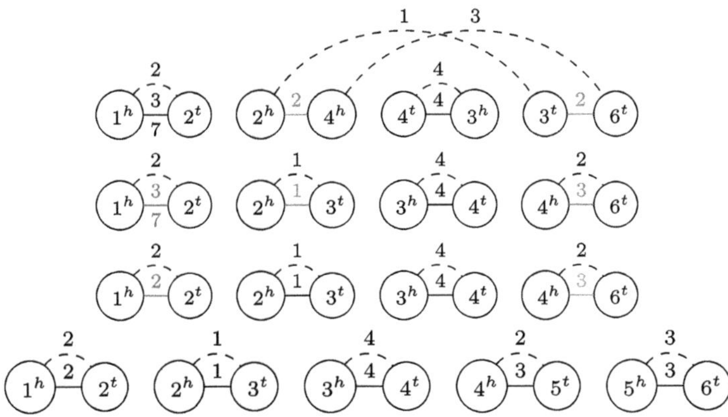

Fig. 2. Given $\mathcal{G} = [+1, 1, +7, 2, +2, 2, -4, 4, -3, 2, +6]$ and $\mathcal{H} = [+1, 2, +2, 1, +3, 4, +4, 3, +5, 3, +6]$. The effect in the breakpoint graph of a sequence of rearrangement operations capable of transforming the genome $\mathcal{G}$ into the genome $\mathcal{H}$. The sequence is composed of: a DCJ affecting the red edges; a deletion in the blue edge (removing the gene $+7$ and one nucleotide in the intergenic regions surrounding it); and an insertion in the cyan edge (inserting the gene $+5$ and three nucleotides in the intergenic regions surrounding it).

We denote by c_g the number of good cycles in the breakpoint graph and by c_{max} the number of genes common to both genomes minus the number of linear genomes in $\mathcal{G}$. The value c_{max} is the maximum number of good cycles we can have after the application of any sequence of indels and DCJs in the genome $\mathcal{G}$ if we do not insert any new gene.

Lemma 1. *Considering the breakpoint graph of two genomes $\mathcal{G}$ and $\mathcal{H}$. The values c_g and c_{max} are equal if and only if $\mathcal{G}$ and $\mathcal{H}$ are equal.*

Proof. If the genomes are equal, we only have trivial cycles, and all cycles are good. So we have $c_g = c_{max}$.

If $c_g = c_{max}$, then all cycles must be good. Consequently, we have the maximum number of genes common to both genomes. In other words, all genes are common. Otherwise, there would be a label on some edge. Additionally, the cycles must be all trivial. Otherwise, there would be fewer than c_{max} cycles. If all genes are common and there are only trivial cycles the genomes must be equal. $\qquad\square$

3   Approximation for DCJ and Indel Distance

The DCJ and Indel Distance problem still has unknown complexity, but there are two important reasons to propose an approximation for it. The first reason is that it would be the first algorithm proposed for that problem, and it could be used in practice with the theoretical guarantee of the approximation. The second reason is that there is a variant of the problem that is NP-hard and the algorithm will also be applicable to it. This variant consists in assuming that an indel always has to affect genes. In that case, the reduction from the problem of calculating the DCJ distance with intergenic regions, which is NP-hard [7], is direct, because any instance of this problem, where the genomes must have the same set of genes, is an instance of the DCJ and Indel Distance, where no indel is used in the solution due to the restriction of the indels only affecting genes exclusive to one of the genomes.

Our algorithm is based on the idea of increasing the number of good cycles in the breakpoint graph until we reach the value of c_{max}. Note that insertion operations are necessary to clean the cycles but can increase the value of c_{max}. Consequently, we must consider the change of the value c_{max} in the algorithm.

The following lemmas are used in the proof for the approximation factor of the algorithm.

Lemma 2. *Considering the breakpoint graph of two genomes $\mathcal{G}$ and $\mathcal{H}$. For a DCJ δ acting on $\mathcal{G}$ we have $(c_{max} - c_g) - (c'_{max} - c'_g) \leq 1$, where c'_{max} and c'_g are the updated values considering the resulting genome $\mathcal{G}' = \mathcal{G} \cdot \delta$.*

Proof. A DCJ will cut two origin edges and reconnect the vertices with new edges. This will either split a cycle into two, combine two cycles into one, or maintain the number of cycles. In the best case, we can create a new good cycle. As this operation cannot change the value of c_{max} the lemma follows.   □

Lemma 3. *[3, Lemma 3] Considering the breakpoint graph of two genomes $\mathcal{G}$ and $\mathcal{H}$. For an indel β acting on $\mathcal{G}$ we have $(c_{max} - c_g) - (c'_{max} - c'_g) \leq 1$, where c'_{max} and c'_g are the values considering the resulting genome $\mathcal{G}' = \mathcal{G} \cdot \beta$.*

Lemma 4. *For two genomes G and H we have $d^{id}_{DCJ}(\mathcal{G}, \mathcal{H}) \geq c_{max} - c_g$.*

Proof. Directly from lemmas 2 and 3, and from the fact that c_{max} is equal to c_g if and only if the genomes are equal (Lemma 1).   □

In the following lemma, we take advantage of a previous result using the same breakpoint graph.

Lemma 5. *[1, Lemma 5] If the breakpoint graph of two genomes $\mathcal{G}$ and $\mathcal{H}$ has a trivial cycle C with an empty origin edge, such that C is unbalanced and the target edge is empty or C is non-negative and the target edge has a non-empty label, then there exists an indel δ that increases by one the value of $c_{max} - c_g$.*

Algorithm 1 presents a pseudocode of our algorithm, and Lemma 6 describes the maximum number of operations needed by the algorithm. Note that at every iteration the number of good cycles increases, so eventually the algorithm finishes. The detection of cycles and application of the operations can be done in quadratic time. As we require a linear number of iterations to complete the algorithm, the total running time of the algorithm is cubic.

Algorithm 1. The input is a pair of genomes $\mathcal{G}$ and $\mathcal{H}$. The output is a sequence S of DCJs and indels capable of transforming $\mathcal{G}$ into $\mathcal{H}$.

$G \leftarrow$ breakpoint graph of $\mathcal{G}$ and $\mathcal{H}$
$S \leftarrow [\,]$
while G has less than c_{max} good cycles **do**
    if there is a non-trivial cycle in G **then**
        Apply one DCJ and up to one indel to create a good trivial cycle
    else if there is a bad trivial cycle in G **then**
        Apply up to two indels to balance and clean the cycle
    end if
    Update G
    Add the applied operations to S
end while
return S

Lemma 6. *Algorithm 1 can turn the genome $\mathcal{G}$ into the genome $\mathcal{H}$ with at most $2(c_{max} - c_g)$ DCJs and indels.*

Proof. If we have a non-trivial cycle, the algorithm applies one DCJ in the origin edges around a target edge of this cycle and connects the vertices of this target edge, which splits the cycle, creating one trivial cycle and another cycle. We can choose the point where the intergenic regions are cut to ensure that the trivial cycle is non-negative and has no label on its origin edge. Afterward, up to one indel can clean and balance this trivial cycle (Lemma 5). If we have a bad trivial cycle, the algorithm balances and cleans it with up to two indels.

Note that, for every balanced cycle, this process will eventually result in one indel creating one trivial cycle to be balanced and clean and one already good trivial cycle. Consequently, we need to create $c_{max} - c_g$ good cycles, each requiring up to two operations. The additional cycles required are created by insertions and do not affect the number of operations needed. $\qquad\square$

Theorem 1. *Algorithm 1 is a 2-approximation for the DCJ and indel distance problem.*

Proof. Directly from lemmas 4 and 6. $\qquad\square$

4 Assumptions and Implications

This section discusses the consequences of our two main assumptions: (i) each gene occurs in a single copy in each genome, and (ii) the input genomes are co-tailed. Although these constraints limit direct applicability to real datasets, both assumptions can be relaxed at additional computational cost.

Relaxing the Co-tailed Assumption. The co-tailed requirement can be removed by introducing sentinel (capping) genes at the extremities of each linear chromosome so that both genomes share matching extremities. If the sentinel genes are treated as fixed markers, any sequence of DCJs and indels transforming the capped version of $\mathcal{G}$ into the capped version of $\mathcal{H}$ can be mapped to a sequence of the same length transforming the original genomes. The sentinel genes must all be equal, consequently, we must assume the presence of multiple copies of a gene.

Relaxing the Single-Copy Assumption. When genes may have multiple copies, the main additional challenge is to determine a correspondence between copies (orthology assignment) before applying a single-copy method. Breakpoint-graph adaptations have been proposed to guide this correspondence under indels [13] and under intergenic-region information [12]. Combining these ideas can be used to induce a single-copy instance (via a selected correspondence), after which Algorithm 1 can be applied.

Improving Correspondences via String Partition. A complementary approach is to use the String Partition problem to refine correspondences. Rubert *et al.* [8] relate String Partition to DCJ distance, and Siqueira *et al.* [11] extend it to incorporate intergenic regions and indels, providing both heuristic and fixed-parameter algorithms. An exact solution for the partition algorithm, as is the case of the fixed-parameter algorithm, yields an 2α-approximation for the distance problem, where α is the approximation for the case with a single copy of each gene; in particular, with our algorithm we have $\alpha = 2$, which gives a factor of 4. This combined approach is no longer polynomial due to the fixed-parameter subroutine.

5   Conclusion

In this work, we integrate the presence of intergenic regions and genes exclusive to each genome in the DCJ and Indel Distance problem for co-tailed genomes with a single copy of each gene. Our main result is an approximation algorithm with factor 2 for the problem. We also described the initial steps to apply this version of the problem in a more general scenario considering gene repetition and genomes that are not co-tailed. In future works, these generalizations can be further explored as well as alternative representations for the intergenic region information.

Besides the study of the generalization of the problem, another important step for future works includes the implementation of the proposed algorithm to evaluate its performance in practice. A simulated database where the target genome is produced with a known number of operations can be used to verify how close the algorithm gets to the actual distance. A test in real genomes can also help to verify if the algorithm is suitable to estimate the evolutionary distance. A study about the complexity of the problem is also a relevant next step. This consists of either the development of a polynomial-time algorithm or NP-hardness proof for the problem.

Acknowledgement. This work was supported by the São Paulo Research Foundation, FAPESP (grant 2021/13824-8).

References

1. Alexandrino, A.O., Brito, K.L., Oliveira, A.R., Dias, U., Dias, Z.: Reversal distance on genomes with different gene content and intergenic regions information. In: Martín-Vide, C., Vega-Rodríguez, M.A., Wheeler, T. (eds.) AlCoB 2021. LNCS, vol. 12715, pp. 121–133. Springer, Cham (2021). https://doi.org/10.1007/978-3-030-74432-8_9
2. Alexandrino, A.O., Brito, K.L., Oliveira, A.R., Dias, U., Dias, Z.: Reversal and indel distance with intergenic region information. IEEE/ACM Trans. Comput. Biol. Bioinf. **20**(3), 1628–1640 (2023)
3. Alexandrino, A.O., Oliveira, A.R., Jean, G., Fertin, G., Dias, U., Dias, Z.: Transposition Distance Considering Intergenic Regions for Unbalanced Genomes. In: Proceedings of the 18th International Symposium on Bioinformatics Research and Applications (ISBRA'2022), vol. 13760, pp. 100–113. Springer Nature Switzerland (2022). https://doi.org/10.1007/978-3-031-23198-8_10
4. Braga, M.D., Willing, E., Stoye, J.: Double cut and join with insertions and deletions. J. Comput. Biol. **18**(9), 1167–1184 (2011)
5. Bulteau, L., Fertin, G., Tannier, E.: Genome rearrangements with indels in intergenes restrict the scenario space. BMC Bioinf. **17**(14), 426 (2016)
6. Elias, I., Hartman, T.: A 1.375-Approximation algorithm for sorting by transpositions. IEEE/ACM Trans. Comput. Biol. Bioinf. **3**(4), 369–379 (2006)
7. Fertin, G., Jean, G., Tannier, E.: Algorithms for computing the double cut and join distance on both gene order and intergenic sizes. Algorithms Mol. Biol. **12**(1) (2017)
8. Rubert, D.P., Feijão, P., Braga, M.D.V., Stoye, J., Martinez, F.H.V.: Approximating the DCJ distance of balanced genomes in linear time. Algorithms Mol. Biol. **12**(1), 1–13 (2017)
9. Shao, M., Lin, Y.: Approximating the edit distance for genomes with duplicate genes under DCJ, insertion and deletion. BMC Bioinformatics **13**(Suppl 19) (2012)
10. Simonaitis, P., Chateau, A., Swenson, K.M.: A general framework for genome rearrangement with biological constraints **14**(15) (2019)
11. Siqueira, G., Alexandrino, A.O., Dias, Z.: Signed rearrangement distances considering repeated genes, intergenic regions, and indels. J. Comb. Optim. **46**, 16 (2023). https://doi.org/10.1007/s10878-023-01083-w
12. Siqueira, G., Alexandrino, A., Oliveira, A., Dias, Z.: Heuristics based on adjacency graph packing for DCJ distance considering intergenic regions. In: Anais do XVII Simpósio Brasileiro de Bioinformática, pp. 71–82. SBC, Porto Alegre, RS, Brasil (2024). https://doi.org/10.5753/bsb.2024.245554
13. Siqueira, G., Oliveira, A.R., Alexandrino, A.O., Jean, G., Fertin, G., Dias, Z.: Assignment of orthologous genes in unbalanced genomes using cycle packing of adjacency graphs. J. Heurist. **30**, 269–289 (2024). https://doi.org/10.1007/s10732-024-09528-z
14. Yancopoulos, S., Attie, O., Friedberg, R.: Efficient sorting of genomic permutations by translocation inversion block interchange. Bioinformatics **21**(16), 3340–3346 (2005)

ABUS: A Framework for Evaluating Protein Language Models with Multimodal Capabilities

Fatemeh Ensafitakaldani$^{(\boxtimes)}$, Nurit Haspel, and Daniel Haehn

University of Massachusetts Boston, Boston, MA, USA
`k.ensafitakaldani001@umb.edu`

Abstract. Protein language models (PLMs) are increasingly used for tasks such as structure prediction, function annotation, and protein–protein interaction modeling. However, the rapid expansion of available models including sequence-only, structure-informed, and multimodal architectures makes it challenging for researchers to determine which model best fits a specific biological task, computational budget, or deployment setting. Existing surveys often emphasize architectural or benchmark comparisons but provide limited practical guidance for real-world usability and integration. To address this gap, we introduce the **Adaptability Bioinformatics Usability Score (ABUS)**, a structured evaluation framework that characterizes PLMs across five key dimensions: adaptability, bioinformatics relevance, usability, computational efficiency, and output suitability. Each dimension is decomposed into specific rubric-based sub-features scored on a 0/1/2 scale based on objective evidence. These sub-category values are then aggregated and transformed into a normalized overall ABUS score ranging from 0 to 100, providing both granular qualitative insights and a comparative quantitative metric. ABUS is implemented as an evidence-driven scoring engine supported by human-in-the-loop validation. The framework includes a curated database of over 20 models, an interactive web interface for exploring detailed model profiles, and a prototype AI agent designed to extract evidence and recommend scores directly from research papers and repositories.

Keywords: Protein language models · Evaluation frameworks · Structural bioinformatics · Model profiling

1 Introduction

1.1 Proteins and Protein Language Models

Proteins carry out essential biological functions through the coordinated interplay of their sequence, structure, and dynamics. Accurately modeling these relationships is central to understanding protein function, predicting interactions,

K. L. Kabir et al. (Eds.): BICOB 2026, CCIS 2977, pp. 140–155, 2026.
https://doi.org/10.1007/978-3-032-26028-4_11

and supporting therapeutic design [1, 9]. Recent advances in deep learning, particularly transformer-based architectures have enabled the development of *protein language models* (PLMs), which learn rich latent representations directly from large-scale protein sequence corpora [27, 35]. These representations capture biochemical constraints and long-range dependencies, providing a foundation for tasks such as structure prediction, function annotation, variant effect estimation, and protein–protein interaction (PPI) modeling [2, 22, 28].

1.2   PLM Modalities and Model Families

Modern protein language models differ widely in their architectural design and input modalities. Sequence-only models, such as ESM-2 [25], rely on masked language modeling over massive protein databases like BFD [37]. Structure-aware models incorporate contact maps, distance matrices, or full 3D geometry to enhance spatial reasoning and capture long-range structural constraints [17, 20]. Multimodal architectures extend these approaches by integrating additional biological signals, such as evolutionary profiles, Gene Ontology annotations, literature-derived text, or small-molecule features [4, 13]. As a result, PLMs vary substantially in their bioinformatics relevance, adaptability, usability, computational requirements, and the types of outputs they produce.

1.3   Challenges in Selecting a PLM

Despite rapid progress, selecting an appropriate PLM for a given biological or computational workflow remains difficult. Benchmark studies often rely on heterogeneous datasets and evaluation protocols, limiting direct comparability across models [32, 47]. Practical factors, such as documentation quality, installation complexity, licensing constraints, and hardware requirements are frequently under-reported or inconsistently documented [46]. Moreover, existing surveys typically focus on architectural descriptions rather than providing a systematic approach for assessing trade-offs between biological capability and practical usability.

1.4   The ABUS Framework

To address these limitations, we introduce the **Adaptability bioinformatics Usability Score (ABUS)**, a structured evaluation framework that profiles PLMs across five key dimensions: adaptability, bioinformatics relevance, usability, computational efficiency, and output suitability. Together, these dimensions capture both the scientific capabilities of a model and the practical considerations that influence its deployability in real-world bioinformatics workflows.

1.5   Contributions

ABUS applies a standardized 0/1/2 rubric to model attributes and aggregates these scores into interpretable category-level and overall assessments, with each

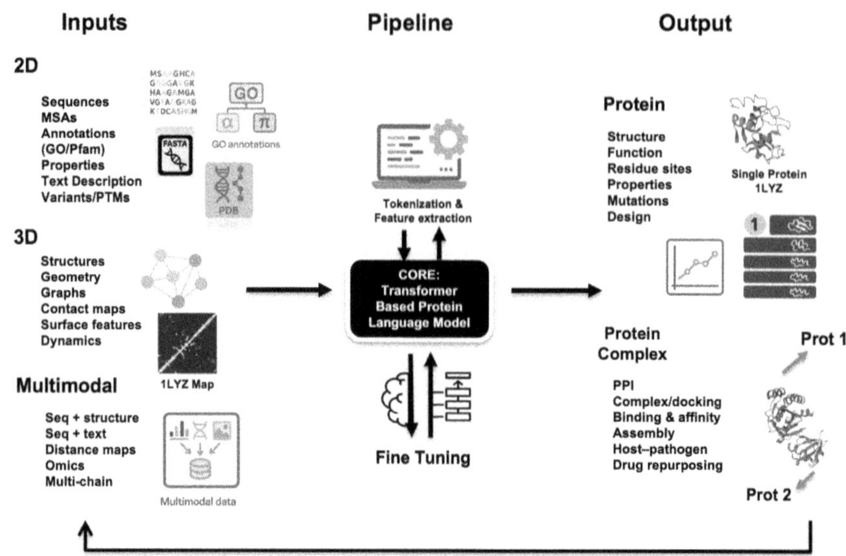

Fig. 1. Protein language models support a wide range of applications across structure prediction, function annotation, interaction analysis, and multimodal reasoning. The figure illustrates the diverse inputs these models consume such as sequences, evolutionary profiles, and structural data. The transformer-based core integrates these inputs, and creates the resulting outputs, including structures, functions, interactions, and complex biological predictions.

decision grounded in explicit evidence drawn from publications, repositories, and model cards. In this work, we make three primary contributions. First, we define the ABUS scoring formula, its five dimensions, and the associated scoring methodology. Second, we construct a curated, machine-readable database and an accompanying interactive web interface that summarize more than twenty protein language models. Third, we analyze the resulting landscape to illustrate how models vary in adaptability, biological alignment, usability, computational efficiency, and output suitability. Finally, we outline a constraint-based PLM recommender as a natural extension of the ABUS framework (Fig. 1).

2   Materials and Methods

2.1   Related Work

Protein language models (PLMs) have advanced rapidly, evolving from early sequence-only transformers to multimodal architectures that integrate structural information, evolutionary context, and literature-derived text. Fan et al. [11] survey this progression and highlight both the advantages of richer multimodal inputs and the challenges posed by sparse annotations and uneven generalization across protein families. Recent systems such as MULAN, PoET-2, and DPLM-2 [13, 21, 40] demonstrate the effectiveness of combining structural geometry, evolutionary features, and textual descriptions to produce more bioinformaticsly informed PLMs.

Beyond protein modeling, the machine learning and visualization communities have developed structured evaluation methodologies that extend beyond raw performance metrics. The Scientific Visualization Future Readiness Score (FRS) [12] illustrates the value of rubric-based, evidence-driven assessment, showing that qualitative dimensions such as usability, openness, documentation quality, and composability can be evaluated systematically. Similarly, model reporting frameworks including Model Cards [29] and Datasheets for Datasets [14] emphasize transparent, evidence-grounded characterization of model capabilities and data provenance.

Benchmarking platforms such as TAPE [32], FLIP [8], and ProteinBench [47] provide standardized downstream tasks but focus primarily on performance rather than usability or practical deployment factors. In contrast, ABUS evaluates PLMs through a broader multidimensional lens, capturing model capabilities, bioinformatics relevance, usability, computational efficiency, and output suitability. To our knowledge, ABUS is the first framework to define a structured scoring scheme for PLMs, implement explicit evidence logging, and provide downstream tools including a curated database, visualization interface, and AI-assisted scoring workflow.

2.2 ABUS Framework

The Adaptability Bioinformatics Usability Score (ABUS) converts qualitative properties of PLMs into quantitative, interpretable scores. The framework has three primary components: (1) a set of categories and subfeatures capturing biological and practical dimensions of a model, (2) a standardized scoring rubric and aggregation formula, and (3) an implementation that stores both final scores and the evidence supporting them.

Categories and Subfeatures. Each PLM is evaluated across five categories designed to capture core dimensions of bioinformatics relevance and practical usability. These categories were developed collaboratively by our team of computer science and bioinformatics researchers through an iterative process of discussion and consensus, in which the final scoring categories and their corresponding weights were jointly selected to reflect their relative importance in practical research settings.

1. Adaptability reflects a model's flexibility for fine-tuning, transfer learning, and reuse, including checkpoint availability, licensing, and support for parameter-efficient techniques such as LoRA [16] and prefix tuning [26].

2. Bioinformatics Relevance measures alignment with key biological data modalities (sequence, structure, evolutionary information) and suitability for tasks such as structure prediction, PPI modeling, and function annotation [13, 22, 28].

3. Usability captures practical barriers to adoption, including documentation quality, code maturity, installation complexity, and community engagement. This aligns with reproducibility literature emphasizing transparency and ease of use [30].

4. Computational Efficiency assesses parameter count, memory footprint, inference cost, and deployability on commonly available research hardware.

5. Output Suitability evaluates interpretability, residue-level resolution, compatibility with downstream workflows, and adherence to explainability principles in machine learning [10].

Each category consists of multiple subfeatures scored on a three-point rubric: 0 (unsupported), 1 (partially supported or ambiguous), and 2 (fully supported with clear evidence). ABUS also includes a dedicated subfeature for benchmark clarity under the *Usability* category, which captures whether a model's evaluation setup, datasets, and reported metrics are clearly described and reproducible. However, unlike benchmarking frameworks such as TAPE [32] or FLIP [8], ABUS does not incorporate or aggregate heterogeneous downstream performance values into its scores.

Scoring Formula. Let $s_f \in \{0, 1, 2\}$ denote the score assigned to subfeature f within category c, and let w_f be its weight. For each category c with subfeature set F_c, the category score is computed as a weighted average:

$$S_c = \frac{\sum_{f \in F_c} w_f s_f}{\sum_{f \in F_c} w_f}.$$ (1)

Let C denote the set of ABUS categories and w_c the weight assigned to category c. The overall and normalized ABUS scores are then:

$$\text{ABUS Score} = \sum_{c \in C} w_c S_c, \qquad \text{ABUS}_{\text{norm}} = \frac{\text{ABUS Score}}{\sum_{c \in C} w_c} \times 100.$$ (2)

By default, ABUS weights emphasize bioinformatics relevance (30%), followed by adaptability and output suitability (20% each), and usability and computational efficiency (15% each), consistent with usage patterns reported in PLM surveys [46]. These weights are configurable to support task- or application-specific evaluation profiles (Fig. 2).

Evidence-Grounded Manual Scoring. ABUS scores are assigned through a manual, evidence-driven review performed by our research team. For each model, reviewers extract supporting evidence from papers, repositories, configuration files, and model cards [29]. AI-assisted tools (e.g., ChatGPT and Gemini) are used to surface relevant passages, cross-reference findings with other AI agent systems, check consistency, and verify claims, but all scoring decisions are ultimately made by human evaluators. Following a conservative rubric [12], fully documented capabilities receive a score of 2, partially supported or ambiguous

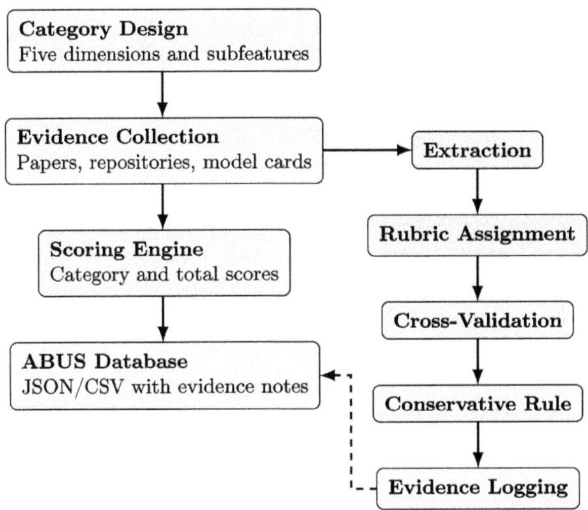

Fig. 2. Overview of the ABUS evaluation workflow. The process begins with defining category dimensions, followed by systematic evidence collection and extraction. Human reviewers assign rubric-based scores, cross-validate findings using conservative rules, and log all evidence into a structured ABUS database.

claims receive 1, and undocumented features receive 0. All evidence notes are stored in structured, machine-readable form, and we also release the protocols and prompts used for AI-assisted evidence collection on GitHub to ensure transparency and reproducibility.

2.3   Data Collection and Implementation

We curated over twenty representative PLMs, including general-purpose sequence models, multimodal systems, structure-aware architectures, and specialized predictors. For each model, we extracted metadata covering modalities, supported tasks, licensing, checkpoint availability, dependencies, implementation, and benchmark documentation. Information sources included peer-reviewed publications, preprints, GitHub repositories, model cards, and aggregation platforms such as Hugging Face [42] and the Language Model Zoo [5] [6]. This curation process follows established practices for documenting biological foundation models [49].

2.4   Preprocessing and Database Construction

Metadata and evidence were processed using a standardized pipeline inspired by dataset documentation frameworks such as Datasheets [14]. The pipeline harmonized task names, input modalities, and metric terminology, and validated inconsistencies according to ABUS's conservative evidence rules [12]. Final structured records were exported as version-controlled JSON and CSV files forming the ABUS database. A web interface renders radar plots, PCA projections, score distributions, and model summaries, following visualization conventions used in PLM evaluation [33].

2.5 Evaluation Setup

All models in the curated set were evaluated using the complete ABUS rubric. Reviewers inspected at least one primary publication and one implementation artifact per model, documented evidence for every sub-feature, and computed category-level and normalized ABUS scores using the formulas in Sect. 2.2. To analyze global patterns and variability across models, we computed descriptive statistics, generated radar charts, and produced PCA embeddings, similar to approaches used in prior PLM comparison studies [4, 13].

3 Results

3.1 Category-Level Patterns

Table 1 summarizes the aggregated ABUS category scores for all evaluated models. Multimodal and structure-aware systems such as MULAN [13], PoET-2 [21], DPLM-2 [40], HelixProtX [7], and EquiBind [36] achieve high bioinformatics relevance by explicitly modeling structure and interactions. General-purpose PLMs including ProCyon [31], Evola [51], ProtLLM [52], and ProteinChat [38] tend to perform well in adaptability and output suitability, but display more variability in bioinformatics relevance. Open-source pipelines and ensembles such as MetaDegron [50], FAPM [44], and DeepDrug [48] tend to show stronger usability signals in our rubric, reflecting relatively clearer implementation artifacts, while others such as Prot2Chat [41] remain more mixed on usability despite strong capability scores. Task-specialized architectures like TUnA [23] and VisionTransformer-PPI [18] attain strong output suitability by aligning directly with PPI or docking scenarios.

Notable models such as EquiBind [36], DeepDrug [48], PoET-2 [21], and AlphaFold 2 [22] combine strong bioinformatics relevance with good output suitability and at least moderate usability. Systems like MULAN [13], DPLM-2 [40], and Prot2Chat exhibit relatively balanced profiles across ABUS dimensions are generally strong across categories, with some dips in usability/efficiency depending on the model. In contrast, models such as ProCyon and Evola highlight trade-offs in usability or computational efficiency despite strong adaptability and bioinformatics relevance. HyboWaveNet achieves moderate overall performance driven primarily by task alignment, but shows relatively low biological input richness and limited efficiency, reflecting its specialization for a network-based PPI setting rather than broad multimodal bioinformatics modeling. Because model repositories, checkpoints, and capabilities are frequently updated, ABUS scores are maintained as a living resource, and values may be revised as new evidence, versions, or implementations become available.

Table 1. ABUS category and total scores for evaluated protein language models across Adaptability, bioinformatics Relevance, Usability, Computational Efficiency, and Output Suitability. Category scores are on a 0–2 scale, and the reported $ABUS_{norm}$ score is the normalized weighted sum on a 0–100 scale.

Model	Adaptability	Bio. relevance	Usability	Computational efficiency	Output suitability	$ABUS_{norm}$
MULAN	2.00	2.00	1.00	1.50	1.50	83.8
ProCyon	2.00	2.00	0.67	0.50	2.00	78.8
EvoLa	2.00	1.67	0.67	1.00	1.67	74.2
PoET-2	2.00	2.00	1.33	2.00	1.67	91.7
DPLM-2	2.00	2.00	1.33	1.00	1.50	82.5
Prot2Chat	2.00	2.00	0.67	1.50	1.50	81.3
MetaDegron	1.50	2.00	1.00	0.50	1.50	71.3
FAPM	1.50	1.00	1.33	0.00	1.50	55.0
ProtST	1.50	2.00	0.67	1.00	1.50	72.5
HelixProtX	2.00	1.50	1.00	0.50	1.50	68.8
S2F	1.50	2.00	0.33	1.50	1.00	68.8
AFTGAN	1.50	1.50	1.33	1.00	1.50	70.0
CAT-CPI	1.50	1.50	0.67	1.00	1.50	65.0
DeepDrug	2.00	2.00	1.33	1.00	1.50	82.5
TUnA	1.50	1.00	1.33	1.50	1.50	66.3
HyboWaveNet	1.50	1.00	1.33	0.50	1.50	58.8
MM-StackEns	1.50	1.40	0.50	1.00	1.33	58.8
EquiBind	1.50	2.00	1.33	1.50	1.50	81.3
VisionTransformer-PPI	1.50	2.00	0.67	1.00	1.50	72.5
RF-Ensemble	1.00	1.00	1.00	2.00	2.00	67.5
AlphaFold 2	1.50	2.00	1.33	0.50	1.50	73.8

3.2 Trade-off Analysis

ABUS profiles make cross-model trade-offs explicit. Table 2 summarizes normalized category scores for three representative multimodal PLMs, illustrating how similar biological capabilities can be paired with different usability and efficiency characteristics.

Table 2. Category-level ABUS scores (0–100) for three representative models, illustrating different trade-off profiles.

Model	Adaptability	Bio. Relevance	Usability	Comp. Efficiency	Output Suit.
PoET-2	100	100	66.5	100	83.5
MULAN	100	100	50.0	75.0	75.0
DPLM-2	100	100	66.5	50.0	75.0

MULAN [13] emphasizes multimodal bioinformatics relevance but offers only moderate usability and computational efficiency. PoET-2 [21] combines strong bioinformatics relevance with high computational efficiency and relatively high usability, making it attractive in resource-limited settings. DPLM-2 [40] provides a balanced profile with high adaptability and bioinformatics relevance but lower computational efficiency, which may be acceptable when hardware constraints are less stringent. ABUS makes these trade-offs explicit and supports model choice based on task and resource requirements.

3.3   PCA Analysis of ABUS Profiles

To explore relationships between models within the five-dimensional ABUS space, we applied Principal Component Analysis (PCA) to the vector of category scores (adaptability, bioinformatics relevance, usability, computational efficiency, and output suitability). PCA is widely used in PLM evaluation studies [4,33] and provides a compact two-dimensional embedding that highlights the global structure across systems (Fig. 3).

The first two principal components explain the majority of the variance in the five-dimensional ABUS space. PC1 reflects overall capability strength, with the highest positive loadings arising from bioinformatics relevance, output suitability, and adaptability. In contrast, PC2 serves as an axis that separates models

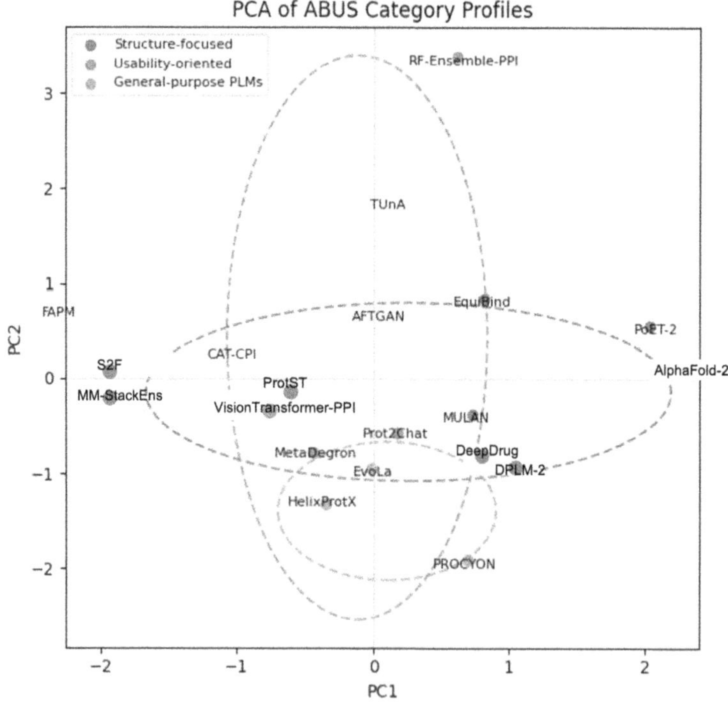

Fig. 3. Models are positioned based on their ABUS scores, revealing three natural groups: structure-focused models, pipelines with relatively strong usability or accessibility characteristics, and general-purpose PLMs. Ellipses highlight how models cluster according to their capability profiles.

emphasizing usability and computational efficiency from those more strongly focused on structural and interaction modeling, due to opposing feature contributions.

Three qualitative groups emerge in the PCA projection. First, a cluster of structure-focused and interaction-focused models appear, comprising MULAN [13], DPLM-2 [40], HelixProtX [7], EquiBind [36], and AlphaFold 2 [22]. A second group contains pipelines with relatively stronger usability or accessibility characteristics, including MetaDegron [50], ProtST [45], DeepDrug [48], and Prot2Chat [41]. Finally, a general-purpose PLM cluster is formed by ProCyon [31], Evola [51], ProtLLM [52], and ProteinChat [38].

Taken together, the PCA reveals a continuous landscape ranging from multimodal, structure-aware PLMs to more usability-focused pipelines and general-purpose sequence models. The projection also highlights that relatively few systems combine very high usability with deep multimodal bioinformatics relevance, pointing to opportunities for future PLM development (Fig. 4).

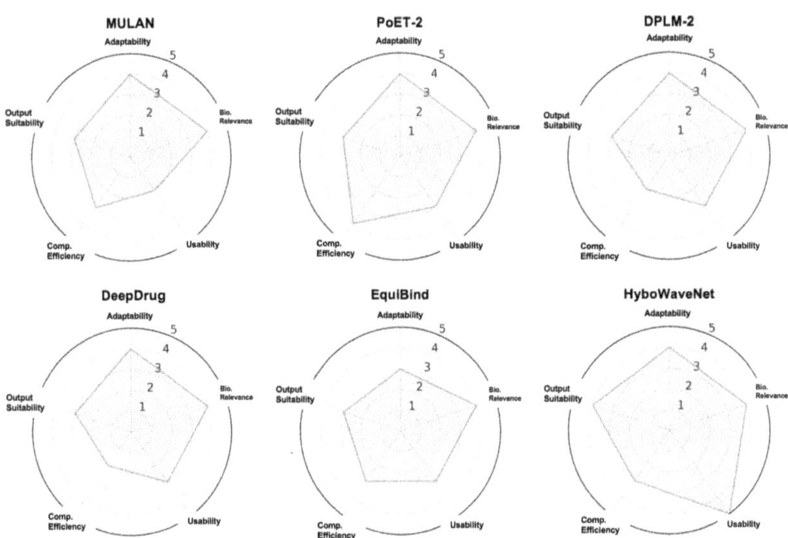

Fig. 4. Radar plots of ABUS category scores for a subset of models. Each polygon highlights the strengths and weaknesses of an individual model across the five ABUS dimensions.

4   Conclusion

We presented the **Adaptability Bioinformatics Usability Score (ABUS)**, a structured and evidence-based framework for evaluating protein language models. ABUS defines five meaningful categories, applies a standardized $0/1/2$ rubric, and aggregates scores into interpretable category-level and overall assessments.

In addition, we developed a curated ABUS database, a lightweight web interface for interactive exploration, and a prototype AI agent that assists with evidence extraction and preliminary scoring, an approach supported by recent work showing that large language models can reliably aid annotation and information extraction [15,39]. Applying ABUS to more than twenty modern PLMs shows that models occupy distinct regions of the evaluation space: some emphasize multimodal bioinformatics relevance, others prioritize usability or output suitability, and others achieve strong computational efficiency. Rather than identifying a single best model, ABUS provides a principled way to reason about trade-offs and select models aligned with specific tasks or resource constraints.

The ABUS framework, database, and visualization code are available at https:// github.com/kattens/ABUS.

4.1   Discussion

ABUS encourages viewing protein language models as *multidimensional capability profiles* rather than as single scalar metrics, aligning with principles of structured model documentation and evaluation [14,29]. By decomposing evaluation into adaptability, bioinformatics relevance, usability, computational efficiency, and output suitability, the framework helps practitioners determine which models fit structure-aware workflows, which are best suited for rapid prototyping or teaching, and where gaps remain in the current model landscape.

The ABUS database and accompanying visual interface make these profiles accessible, enabling comparison across systems and exploration of broader model clusters. While the current scores are verified through a human-in-the-loop process to ensure high fidelity, the framework is designed as a scalable, evidence-driven engine. By leveraging an AI agent to extract technical evidence and recommend scores directly from research repositories, ABUS minimizes subjective bias and addresses the challenge of keeping up with the rapid evolution of the field. This hybrid approach—combining automated extraction with expert validation—complements benchmark-driven evaluations by incorporating usability, efficiency, and multimodality into a unified, transparent framework, aligning with broader trends in human-centered system assessment [3].

4.2   Future Work and Research Vision

Recommender Systems. A natural next step is a constraint-aware recommender built on the ABUS database. Recommender-system paradigms [34] are well-suited for filtering models, ranking candidates using ABUS or custom weights, and visualizing trade-offs through interactive components. Developing and evaluating such a recommender, including user studies, is left for future work.

Automated scoring with LLMs. Our prototype scoring agent suggests that large language models can assist in evidence extraction and rubric assignment

[15, 39]. Future extensions include formal agreement studies with expert labels, domain-specific fine-tuning, and a standalone agent enabling external groups to maintain ABUS-style databases.

Database Expansion and Community Contributions. The ABUS database will be extended as new models are released and opened to community submissions via version-controlled pull requests. In addition, researchers will be able to use the scoring system and submit their own papers through our website portal to receive preliminary ABUS evaluations. Approved submissions and their corresponding scores will be incorporated into the database, further supporting continuous expansion. Variants tailored to specific tasks (e.g., ABUS-PPI or ABUS-Structure) can be supported through configurable weights and expanded subfeatures.

Validation and Extension. We plan to conduct several follow-up studies, including a user agreement study to assess inter-rater reliability, an evaluation of practitioner feedback, and a sensitivity analysis of category weights. These experiments aim to further validate the robustness and practical relevance of the ABUS framework.

Beyond Protein Modeling. The ABUS category-based scoring framework with explicit evidence logging and customizable weights is broadly applicable to other scientific foundation models, including genomics [19], chemistry [43], and climate modeling [24]. In these domains, issues of usability, computational cost, and multimodal inputs are similarly central, positioning ABUS as a general template for structured, evidence-grounded model evaluation.

Acknowledgements. The authors thank colleagues in the Computer Science Department for helpful discussions and feedback on the ABUS rubric, database design, and visualization interface.

Disclosure of Interests. The authors have no competing interests to declare that are relevant to the content of this article.

References

1. Alberts, B., Johnson, A., Lewis, J., Raff, M., Roberts, K., Walter, P.: Molecular Biology of the Cell. Garland Science, New York, 6 edn. (2014). alberts B, Johnson A, Lewis J, et al. Molecular Biology of the Cell. 4th edition. New York: Garland Science; 2002. Available from: https://www.ncbi.nlm.nih.gov/books/NBK21054/
2. Alley, E.C., Khimulya, G., Biswas, S., AlQuraishi, M., Church, G.M.: Unified rational protein engineering with sequence-based deep representation learning. Nat. Methods **16**(12) (2019). https://doi.org/10.1038/s41592-019-0598-1

3. Amershi, S., Cakmak, M., Knox, W.B., Kulesza, T.: Power to the people: the role of humans in interactive machine learning. AI Mag. **35**(4), 105–120 (2014). https://doi.org/10.1609/aimag.v35i4.2513, https://ojs.aaai.org/aimagazine/index.php/aimagazine/article/view/2513

4. Brandes, N., Ofer, D., Peleg, Y., Rappoport, N., Linial, M.: ProteinBERT: a universal deep-learning model of protein sequence and function. Bioinformatics **38** (2022). https://doi.org/10.1093/bioinformatics/btac020

5. Chen, Z., et al.: LLM ZOO: democratizing ChatGPT. https://github.com/FreedomIntelligence/LLMZoo (2023)

6. Chen, Z., et al.: Phoenix: Democratizing ChatGPT across languages. arXiv preprint arXiv:2304.10453 (2023)

7. Chen, Z., et al.: Unifying sequences, structures, and descriptions for any-to-any protein generation with the large multimodal model helixprotx (2024). https://arxiv.org/abs/2407.09274

8. Dallago, C., et al.: FLIP: benchmark tasks in fitness landscape inference for proteins. bioRxiv **1** (2021). https://doi.org/10.1101/2021.11.09.467890, https://datasets-benchmarks-proceedings.neurips.cc/paper_files/paper/2021/file/2b44928ae11fb9384c4cf38708677c48-Paper-round2.pdf

9. Dill, K., Maccallum, J.: The protein-folding problem, 50 years on. Science (New York, N.Y.) **338**, 1042–6 (2012). https://doi.org/10.1126/science.1219021

10. Doshi-Velez, F., Kim, B.: Towards a rigorous science of interpretable machine learning. arXiv: Machine Learning (2017). https://api.semanticscholar.org/CorpusID:11319376

11. Fan, W., et al.: Computational protein science in the era of large language models (LLMs) (2025). https://doi.org/10.48550/arXiv.2501.10282

12. Franke, L., Haehn, D.: Modern scientific visualizations on the web. Informatics **7**, 37 (2020). https://api.semanticscholar.org/CorpusID:224899153

13. Frolova, D., Pak, M., Litvin, A., Sharov, I., Ivankov, D., Oseledets, I.: MULAN: multimodal protein language model for sequence and structure encoding. Bioinform. Adv. **5**(1), vbaf117 (2025). https://doi.org/10.1093/bioadv/vbaf117

14. Gebru, T., et al.: Datasheets for datasets. Commun. ACM **64**, 86–92 (2018). https://api.semanticscholar.org/CorpusID:4421027

15. Gilardi, F., Alizadeh, M., Kubli, M.: ChatGPT outperforms crowd workers for text-annotation tasks. Proc. Nat. Acad. Sci. US Am. **120**(40), e2305016120 (2023). https://doi.org/10.1073/pnas.2305016120, https://pmc.ncbi.nlm.nih.gov/articles/PMC10372638/

16. Hu, E.J., et al.: LoRA: low-rank adaptation of large language models. In: International Conference on Learning Representations (2022). https://openreview.net/forum?id=nZeVKeeFYf9

17. Ingraham, J., Garg, V., Barzilay, R., Jaakkola, T.: Generative models for graph-based protein design. In: Wallach, H., Larochelle, H., Beygelzimer, A., d'Alché-Buc, F., Fox, E., Garnett, R. (eds.) Advances in Neural Information Processing Systems. vol. 32. Curran Associates, Inc. (2019). https://proceedings.neurips.cc/paper_files/paper/2019/file/f3a4ff4839c56a5f460c88cce3666a2b-Paper.pdf

18. Jha, K., Saha, S., Karmakar, S.: Prediction of protein-protein interactions using vision transformer and language model. IEEE/ACM Trans. Comput. Biol. Bioinform. **20**(5), 3215–3225 (2023). https://doi.org/10.1109/TCBB.2023.3248797

19. Ji, Y., Zhou, Z., Liu, H., Davuluri, R.V.: DNABERT: pre-trained bidirectional encoder representations from transformers for DNA sequence analysis. Bioinformatics **37**(15), 2112–2120 (2021). https://doi.org/10.1093/bioinformatics/btab083

20. Jing, B., Eismann, S., Soni, P.N., Dror, R.O.: Equivariant graph neural networks for 3d macromolecular structure. ArXiv abs/2106.03843 (2021). https://api.semanticscholar.org/CorpusID:235364032

21. Jr, T.F.T., Bepler, T.: PoET: a generative model of protein families as sequences-of-sequences. In: Thirty-seventh Conference on Neural Information Processing Systems (2023). https://openreview.net/forum?id=1CJ8D7P8RZ

22. Jumper, J., et al.: Highly accurate protein structure prediction with AlphaFold. Nature **596**(7873), 583–589 (2021). https://doi.org/10.1038/s41586-021-03819-2, publisher Copyright: 2021, The Author(s)

23. Ko, Y.S., Parkinson, J., Liu, C., Wang, W.: . Briefings TUnA: an uncertainty-aware transformer model for sequence-based protein–protein interaction predictionBioinform. **25**(5), bbae359 (2024). https://doi.org/10.1093/bib/bbae359

24. Lam, R., et al.: GraphCast: learning skillful medium-range global weather forecasting. Science **382**(6674), 1309–1314 (2023). https://doi.org/10.1126/science.adi2336

25. Lin, Z., et al.: Evolutionary-scale prediction of atomic-level protein structure with a language model. Science **379**(6637), 1123–1130 (2023). https://doi.org/10.1126/science.ade2574, https://www.science.org/doi/abs/10.1126/science.ade2574

26. Liu, X.L., Zheng, Y., Du, J., Hildebrandt, M., Neubig, G.: Prefix-tuning: optimizing continuous prompts for generation. In: Proceedings of the 60th Annual Meeting of the Association for Computational Linguistics (ACL) (2021). https://doi.org/10.48550/arXiv.2101.00190

27. Madani, A., et al.: ProGen: Language modeling for protein generation. bioRxiv (2020). https://api.semanticscholar.org/CorpusID:214725226

28. Meier, J., Rao, R., Verkuil, R., Liu, J., Sercu, T., Rives, A.: Language models enable zero-shot prediction of the effects of mutations on protein function. In: Ranzato, M., Beygelzimer, A., Dauphin, Y., Liang, P., Vaughan, J.W. (eds.) Advances in Neural Information Processing Systems. vol. 34, pp. 29287–29303. Curran Associates, Inc. (2021). https://proceedings.neurips.cc/paper_files/paper/2021/file/f51338d736f95dd42427296047067694-Paper.pdf

29. Mitchell, M., et al.: Model cards for model reporting. In: Proceedings of the Conference on Fairness, Accountability, and Transparency, pp. 220–229. FAT* '19, Association for Computing Machinery, New York, NY, USA (2019). https://doi.org/10.1145/3287560.3287596

30. Pineau, J., et al.: Improving reproducibility in machine learning research (a report from the NeurIPS 2019 reproducibility program). J. Mach. Learn. Res. **22**(164), 1–20 (2021). https://doi.org/10.48550/arXiv.2003.12206, http://jmlr.org/papers/v22/20-303.html

31. Queen, O., et al.: ProCyon: A multimodal foundation model for protein phenotypes. bioRxiv (2025). https://doi.org/10.1101/2024.12.10.627665, [Preprint]

32. Rao, R., et al.: Evaluating protein transfer learning with TAPE. bioRxiv (2019). https://doi.org/10.1101/676825

33. Rao, R.M., et al.: MSA transformer. In: Meila, M., Zhang, T. (eds.) Proceedings of the 38th International Conference on Machine Learning. Proceedings of Machine Learning Research, vol. 139, pp. 8844–8856. PMLR (2021). https://proceedings.mlr.press/v139/rao21a.html

34. Ricci, F., Rokach, L., Shapira, B. (eds.): Recommender Systems Handbook. Springer, New York, 3 edn. (2022)

35. Rives, A., et al.: Biological structure and function emerge from scaling unsupervised learning to 250 million protein sequences. Proc. Nat. Acad. Sci. **118**(15), e2016 (2021)

36. Stärk, H., Ganea, O., Pattanaik, L., Barzilay, D., Jaakkola, T.: EquiBind: Geometric deep learning for drug binding structure prediction. In: Chaudhuri, K., Jegelka, S., Song, L., Szepesvari, C., Niu, G., Sabato, S. (eds.) Proceedings of the 39th International Conference on Machine Learning. Proceedings of Machine Learning Research, vol. 162, pp. 20503–20521. PMLR (2022). https://proceedings.mlr.press/v162/stark22b.html
37. Steinegger, M., Söding, J.: Clustering huge protein sequence sets in linear time. Nature Commun. **9**(1) (2018). https://doi.org/10.1038/s41467-018-04964-5, publisher Copyright: 2018 The Author(s)
38. Wang, C., Fan, H., Quan, R., Yao, L., Yang, Y.: ProtChatGPT: towards understanding proteins with hybrid representation and large language models. In: Proceedings of the 48th International ACM SIGIR Conference on Research and Development in Information Retrieval, pp. 1076–1086. Association for Computing Machinery (2025). https://doi.org/10.1145/3726302.3730064
39. Wang, L., et al.: A survey on large language model based autonomous agents. Front. Comput. Sci. **18**(6), 186345 (2024). https://doi.org/10.1007/s11704-024-40231-1
40. Wang, X., Zheng, Z., Ye, F., Xue, D., Huang, S., Gu, Q.: DPLM-2: A multimodal diffusion protein language model. ArXiv abs/2410.13782 (2024). https://api.semanticscholar.org/CorpusID:273403705
41. Wang, Z., Ma, Z., Cao, Z., Zhou, C., Zhang, J., Gao, Y.Q.: Prot2Chat: protein large language model with early fusion of text, sequence, and structure. Bioinformatics **41**(8), btaf396 (2025). https://doi.org/10.1093/bioinformatics/btaf396
42. Wolf, T., et al.: Transformers: state-of-the-art natural language processing. In: Liu, Q., Schlangen, D. (eds.) Proceedings of the 2020 Conference on Empirical Methods in Natural Language Processing: System Demonstrations, pp. 38–45. Association for Computational Linguistics, Online (2020). https://doi.org/10.18653/v1/2020.emnlp-demos.6, https://aclanthology.org/2020.emnlp-demos.6/
43. Wu, Z., et al.: MoleculeNet: a benchmark for molecular machine learning. Chem. Sci. 513–530 (2018). https://doi.org/10.1039/C7SC02664A
44. Xiang, W., Xiong, Z., Chen, H., Xiong, J., et al.: FAPM: functional annotation of proteins using multimodal models beyond structural modeling. Bioinformatics **40**(12), btae680 (2024). https://doi.org/10.1093/bioinformatics/btae680
45. Xu, M., Yuan, X., Miret, S., Tang, J.: ProtST: multi-modality learning of protein sequences and biomedical texts. In: Proceedings of the 40th International Conference on Machine Learning (ICML). Proceedings of Machine Learning Research, vol. 202, pp. 38714–38731. PMLR (2023). https://proceedings.mlr.press/v202/xu23t.html
46. Yang, Q., Yu, J., Zheng, J.: A survey of downstream applications of evolutionary scale modeling protein language models. Quant. Biol. **14** (2025). https://doi.org/10.1002/qub2.70013
47. Ye, F., et al.: ProteinBench: A holistic evaluation of protein foundation models (2024). https://arxiv.org/abs/2409.06744
48. Yin, Q., Fan, R., Cao, X., Liu, Q., Jiang, R., Zeng, W.: DeepDrug: a general graph-based deep learning framework for drug–drug interactions and drug–target interactions prediction. Quant. Biol. **11**(3), 260–274 (2023). https://doi.org/10.15302/J-QB-022-0320
49. Zhang, Z., et al.: BioCAP: Exploiting synthetic captions beyond labels in biological foundation models (2025). https://arxiv.org/abs/2510.20095
50. Zheng, M., et al.: MetaDegron: multimodal feature-integrated protein language model for predicting E3 ligase targeted degrons. Briefings Bioinform. **25**(6), bbae519 (2024). https://doi.org/10.1093/bib/bbae519

51. Zhou, X., et al.: Decoding the molecular language of proteins with Evola. bioRxiv (2025). https://doi.org/10.1101/2025.01.05.630192, https://www.biorxiv.org/content/10.1101/2025.01.05.630192v2, [Preprint]
52. Zhuo, L., et al.: ProtLLM: an interleaved protein-language LLM with protein-as-word pre-training. In: Ku, L.W., Martins, A., Srikumar, V. (eds.) Proceedings of the 62nd Annual Meeting of the Association for Computational Linguistics (Volume 1: Long Papers), pp. 8950–8963. Association for Computational Linguistics, Bangkok, Thailand (2024). https://doi.org/10.18653/v1/2024.acl-long.484, https://aclanthology.org/2024.acl-long.484/

Modernizing Open-TGGATEs Through Data and AI Methods

Guojing Cong[1(✉)], Frank Chao[2], Parker Combs[2], Jeremy Erickson[2], and Scott S. Auerbach[2]

[1] Data and AI Section, Oak Ridge National Laboratory, Oak Ridge, TN, USA
`congg@ornl.gov`
[2] Division of Translational Toxicology, National Institute of Environmental Health Sciences, RTP, NC, Durham, USA

Abstract. Open-TGGATEs is an important toxicogenomics database that has been used and referenced in many studies. Unfortunately, Open-TGGATEs were curated decades ago on the Affymetrix platform. This creates barriers for current studies that often use profiles from modern platforms such as TempO-Seq. In our study we present an AI-based methodology for modernizing Open-TGGATEs. We demonstrate the effectiveness of our approach using the measured liver profiles. Our results show that the modernized profiles maintain important relationships among treatment profiles in the original measurements.

1 Introduction

Transcriptomics captures genome-wide patterns and enables systematic investigation of cellular states, disease mechanisms, toxicological responses, and therapeutic effects. Transcriptomics data underpin large-scale efforts in research and development, and the extensive data repositories generated over the past two decades collectively represent an invaluable resource for analysis and hypothesis generation. Currently there are two main toxicogenomics databases, Open-TGGATEs and DrugMatrix, that contain data measured on historical platforms developed decades ago, including CodeLink [1] and Affymetrix [2]. In later studies these data are complemented and partially supplanted by new technologies such as RNA sequencing [3], Templated Oligo Sequencing (TempO-Seq) [4], targeted panels such as S1500+ [5], and high-throughput panels such as BioSpyder Whole Transcriptome (BioSpyderWT) [6]. Each technological shift has brought improvements in sensitivity, dynamic range, and coverage, but also introduced platform-specific characteristics. Without modernization, it is almost impossible to combine these rich historical resources with current data sources. It is desirable that historical data on prior platforms be translated onto modern ones.

As an example of practical challenges associated with un-harmonized data across platforms, we consider machine learning approaches for identifying biomarkers. Since the transcriptomic profiles on all three platforms are high-dimensional vectors as shown in Table 1, these vectors cannot be directly combined into a single dataset. In Table 1, the smallest dimension among the three is 8,565 for CodeLink. The largest is 31,042 for Affymetrix.

K. L. Kabir et al. (Eds.): BICOB 2026, CCIS 2977, pp. 156–169, 2026.
https://doi.org/10.1007/978-3-032-26028-4_12

Table 1. Dimensions of transcriptomic profiles on three platforms. In our experiments we use 22,794 identified genes from a total of 31,042 probsets on Affymetrix as there are many repeats among the probesets

CodeLink	Affymetrix	BioSpyderWT
8,565	31,042	22,794

In this study we present a methodology to modernize one of the most popular toxicogenomics databases, Open-TGGATEs, by leveraging data and AI methods. Open-TGGATEs is a comprehensive public toxicogenomics database established by the Japanese toxicogenomics project consortium with the goal of advancing mechanistic understanding of chemical and drug-induced toxicity and supporting improved predictive toxicology [7]. Open-TGGATEs was developed over the course of a decade (20022012) through collaboration among government agencies, academic institutions, and pharmaceutical companies, and data from approximately 170 compounds, predominantly therapeutic agents and known toxicants, are included in the released dataset. The database was designed to enable integrative analyses linking gene expression changes to phenotypic and pathological outcomes across species and experimental contexts [8].

In Open-TGGATEs, gene expression profiles are generated primarily using Affymetrix GeneChip microarrays from both in vivo and in vitro experimental systems. In vivo data comprise rat liver and kidney gene expressions following single-dose and repeat-dose exposures at multiple time points across different treatment levels, while in vitro measurements capture transcriptomic responses in primary cultured hepatocytes from rats and humans exposed to the same set of compounds. The raw expression measurements are provided as CEL files, allowing users to employ standard Affymetrix preprocessing and normalization methods to derive expression matrices for analysis [8,9].

Open-TGGATEs does not contain any measurements on modern platforms, for example, TempO-Seq. To modernize Open-TGGATEs, we consider leveraging cross-platform translation schemes built for another database, DrugMatrix. The original DrugMatrix contains data from two different microarray platforms: CodeLink and Affymetrix [10]. The extended DrugMatrix contains data from the third-generation TempO-Seq platform (BioSpyderWT), a targeted transcriptomics technology developed by BioSpyder and derived from a subgenomic platform processed through GeniE [11]. The measured data from all three platforms form a matrix G with approximately 375,000 rows and 2,700 columns. The rows are the genes (or probesets) and the columns are the treatments. Approximately 12% of the endpoints in the matrix are present. The endpoint values are log10 ratio of treatment gene expression level versus control gene expression level for each gene for each sample.

Comparing the two toxicogenomics databases, Open-TGGATEs has data for two tissues, liver and kidney, while DrugMatrix contains data for eight tissues. However, tissue coverage is not even with the three platforms in DrugMatrix, as shown in Table 2. The CodeLink and TempO-Seq platforms cover all eight tissues including liver (LI), kidney (KI), heart (HE), skeletal muscle (SM), spleen (SP), bone marrow (BM), intestine (IN), and brain (BR). In contrast, Affymetrix covers only LI, KI, HE, and SM.

Table 2. Amount (in Million) of measured data for different tissues on three Platforms in DrugMatrix

Platform	LI	KI	HE	SM	BM	BR	IN	SP
CodeLink	14.2	7.7	5.3	0.2	2.7	0.55	0.17	1.5
Affymetrix	20.3	11.3	6.5	1.3	0	0	0	0
BioSpyderWT	17.6	15.9	10.7	6.5	0.09	0.07	0.5	1.6

As DrugMatrix contains data on both Affymetrix and BioSpyderWT, it is possible to build an AI translation model using these data. In fact we have developed *TransPlatformer*, a transformer-based deep learning framework designed to translate gene expression profiles across generations of transcriptomic platforms [12]. *TransPlatformer* produces a model that is trained on DrugMatrix to predict BioSpyderWT profiles from measured Affymetrix inputs.

Since Open-TGGATEs does not contain any BioSpyderWT measurements and there are no overlaps in treatments between Open-TGGATEs and DrugMatrix, it is not immediately clear whether and how *TransPlatformer* might be applied in modernizing Affymetrix measurements onto the BioSpyderWT platform. In addition, even if *TransPlatformer* can be trained on DrugMatrix data and applied to translating Affymetrix profiles in Open-TGGATEs, there might not be enough training data to produce a high quality translation. We explore solutions to these open questions and challenges, and present our study of leveraging data and AI methods for modernizing Open-TGGATEs. We focus on in vivo LI data, and astute readers will see the same methodology can be applied to modernizing in vivo KI profiles.

2   Inspection of the Two Datasets

Since both Open-TGGATEs and DrugMatrix contain Affymetrix profiles, in the ideal scenario, these data samples can be merged into one set and a model trained with DrugMatrix can be used to translate the BioSpyderWT profiles for measurements in Open-TGGATEs. However, systematic biases and batch effects inherent in different studies, in this case, one Japanese and one American, may prevent combining the two datasets. In general, differences in probe design, gene coverage, dynamic range, and normalization methods create batch effects that complicate direct comparisons and data integration.

As a first step to alleviate the adverse impact of batch effects in our translation, we adopt $\log_{10}$ fold-change values for both DrugMatrix and Open-TGGATEs. The LI Affymetrix values in DrugMatrix fall approximately within the range of $(-3, 3)$. Depending on the expression level, the data are divided into five categories: extremely under-expressed $(-3, -1]$, under-expressed $(-1, -0.3]$, normal $(-0.3, 0.3)$, over-expressed $[0.3, 1)$, and extremely over-expressed $[1, 3)$. Table 3 shows the data distribution in the 5 categories.

Table 3. Data distribution in DrugMatrix and Open-TGGATEs for five categories

Dataset	extremely-under	under	normal	over	extremely-over
DrugMatrix	0.05%	5.38%	89.21%	5.28%	0.05%
TGGATEs	0.005%	0.26%	99.3%	0.32%	0.01%

In both datasets, the normal category dominates the other categories, with about 89% and 99% total samples. The over and under expressed categories represent a small percentage of the data, but they carry important biological and toxicological meaning. It is imperative that they are preserved as rare signals in translation and not lost to noise. The statistics in Table 3 also agrees with our general understanding of the experiments in the two studies. The chemicals in DrugMatrix tend to be more potent than those studied in Open-TGGATEs. The fact that DrugMatrix covers a wider range of gene expression values is good news that a model trained on DrugMatrix data may be applied to Open-TGGATEs. The converse may not be true.

Table 4. Statistics in DrugMatrix and Open-TGGATEs. Minimum (min), maximum (max), mean, median, and standard deviation (STD) are shown

Dataset	min	max	mean	median	STD
DrugMatrix	−3.16	3.54	−0.0002	0.000	0.068
Open-TGGATEs	−2.98	2.88	0.0001	0.000	0.192

Table 4 shows the statistics of the two datasets. The min, max, mean, and median values for the two datasets are similar. Open-TGGATEs has a standard deviation (STD) of 0.192 while DrugMatrix has a STD of 0.068. Table 4 shows DrugMatrix covers a wider range of values than Open-TGGATEs.

We further inspect the two datasets and compute two widely used metrics, the *Adjusted Rand Index* (ARI) and the *Silhouette Coefficient*, for evaluating the separability of the data samples. These metrics allow us to quantify how effectively unsupervised clustering captures known partitions of the data. In our study, they provide a way to assess whether Open-TGGATEs samples form distinguishable groups from DrugMatrix samples, and thereby whether a model trained on one may generalize to the other.

ARI measures agreement between two partitions by counting pairwise decisions: whether pairs of elements are placed in the same cluster or in different clusters [13]. ARI with value of 1 indicates perfect correspondence between the clustering labels and the ground-truth labels. Given two partitions U and V of the same set of n samples, the ARI is defined as

$$\text{ARI} = \frac{\sum_{ij} \binom{n_{ij}}{2} - \left[\sum_i \binom{a_i}{2} \sum_j \binom{b_j}{2} \right] / \binom{n}{2}}{\frac{1}{2} \left[\sum_i \binom{a_i}{2} + \sum_j \binom{b_j}{2} \right] - \left[\sum_i \binom{a_i}{2} \sum_j \binom{b_j}{2} \right] / \binom{n}{2}},$$

where n_{ij} denotes the number of samples assigned to cluster i in partition U and cluster j in partition V, while a_i and b_j denote the marginal sums. Moderate positive values of ARI indicate partial but incomplete separability.

The Silhouette Coefficient introduced by Rousseeuw [14] evaluates clustering quality using only intra-cluster and inter-cluster distances. For a given sample i, let $a(i)$ denote the average distance between i and all other points within the same cluster, and let $b(i)$ denote the minimum average distance between i and all points in any other cluster. The silhouette coefficient for sample i is then defined as

$$s(i) = \frac{b(i) - a(i)}{\max\{a(i), b(i)\}},$$

which takes values in the interval $[-1, 1]$. A value close to 1 indicates that the sample is well matched to its own cluster and poorly matched to neighboring clusters, whereas values near 0 suggest overlapping clusters. Negative values indicate potential mis-assignment of samples to clusters. The overall silhouette score for a clustering solution is computed as the average of $s(i)$ over all samples. Values near 1 indicate strongly separated clusters, while values between 0.25 and 0.50 typically reflect weak-to-moderate cluster structure.

Applying these metrics to the column-wise representation of our dataset, we obtain ARI ≈ 0.27 and Silhouette ≈ 0.37, both suggesting there are significant overlaps in the samples from a clustering perspective between the two datasets. Furthermore, the UMAP visualization of Open-TGGATEs and DrugMatrix profiles are shown in Fig. 1. In the figure the samples from the two datasets show significant overlap.

3   Translation Methodology

As Open-TGGATEs and DrugMatrix data are not clearly separable, our strategy to modernize Open-TGGATEs is to train a translation model using the Affymetrix data with the corresponding BioSpyderWT profiles in DrugMatrix. Although DrugMatrix contains about 3000 treatments, as the matrix is not complete, there are a total of 340 matched (Affymetrix, BioSpyderWT) measured sample pairs. We use these pairs to train a cross-platform translation model.

The AI model, *TransPlatformer*, is a deep learning model for cross-platform translation of transcriptomic profiles that leverages a novel attention mechanism tailored to toxicogenomics data. For completeness, an overview of the model architecture is provided in Fig. 2. While *TransPlatformer* is inspired by sequence-to-sequence transformer models, it departs from the standard formulation in

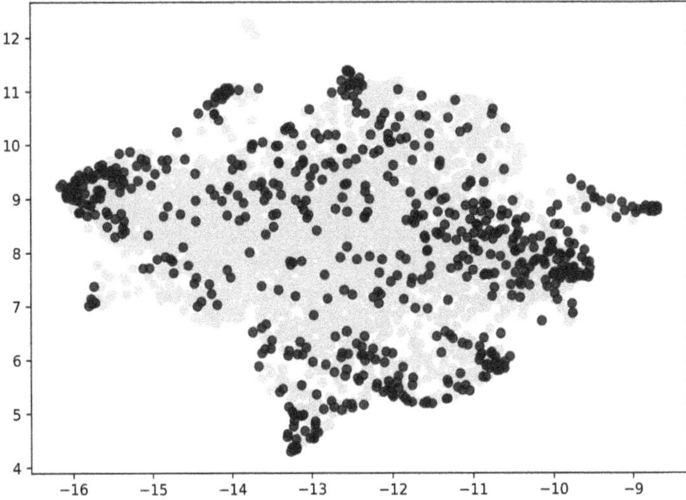

Fig. 1. UMAP visualization of Open-TGGATEs and DrugMatrix profiles on Affymetrix. The yellow color is for Open-TGGATEs and the purple color is for Drug-Matrix (Color figure online)

several key aspects to better accommodate high-dimensional gene expression inputs and outputs.

The model takes as input a batch of expression vectors, where each sample in a batch of size B is represented as a vector of dimensionality n. Optionally, the input vectors are first passed through a fully connected bottleneck layer that maps the original feature space from n to s dimensions. The resulting representations are then projected through an embedding layer of dimension r, yielding an intermediate tensor of shape $B \times s \times r$. This tensor serves as the input to a stack of TransPlatformer attention blocks, arranged analogously to the encoder component of conventional transformer architectures.

The output of the encoder captures a latent representation of the transcriptomic profiles that encodes platform-invariant structure learned from the data. To generate predictions in the target feature space of dimensionality m, the encoded tensor is fed into $\lceil m/s \rceil$ parallel decoding TransPlatformer modules. Each decoder produces an output tensor of size $B \times s \times r$, corresponding to a disjoint subset of the target dimensions. These decoded representations are subsequently passed through a linear projection layer mapping from r to 1 and concatenated to form a final prediction tensor of shape $B \times m$.

The *TransPlatformer* model we use consists of 32 stacked TransPlatformer layers with an embedding dimensionality of $r = 16$, and dropout is applied with a rate of 0.1 throughout the network. For platform-specific translations, in the case of Affymetrix-to-BioSpyderWT translation, the corresponding dimensions are $n = 31042$ and $m = 22794$. The segment size parameter is fixed at $s = 512$, resulting in $\lceil m/s \rceil = 45$ parallel output segments in the *TransPlatformer*

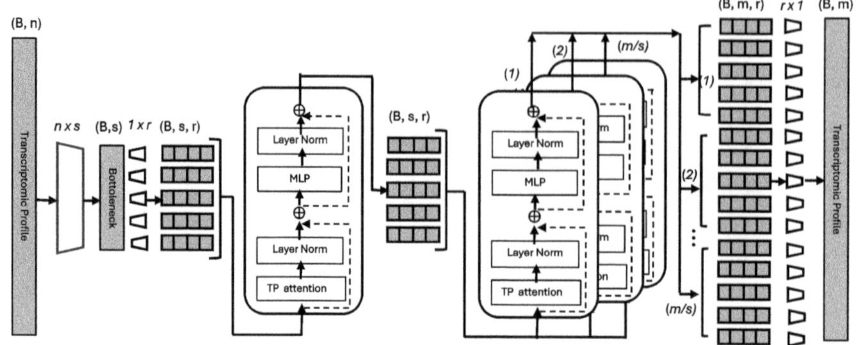

Fig. 2. Architecture of *TransPlatformer*. The input is a vector of size n, and the output is a translated vector of size m. Colored blocks represent the data (tensors) with the middle green block as the encoded states of the inputs. The rest are neural network constructs, and on the right, there are $\lceil m/s \rceil$ concurrent decoding structures. The attention mechanism is *TransPlatformer* attention (Color figure online)

decoding stage for both translation tasks. Model training is performed using the Adam optimizer with default hyperparameters, including a learning rate of 10^{-3} and weight decay set to 5×10^{-4}. A batch size of 16 is used consistently across all experiments.

Unless noted otherwise, we use 9:1 train-test split of the data in our study. We perform five random splits and report the average of five runs in terms of mean squared error (MSE), mean absolute error (MAE) for all genes and those over- and under-expressed genes (including extremely over- and extremely under-expressed ones) and Pearson correlation coefficient (PCC) between prediction and ground truth.

Table 5. Predictive performance of *TransPlatformer* with measured DrugMatrix pairs

test MSE	test MAE	rare MAE	PCC
0.0041	0.0361	0.131	0.799

Table 5 shows the test performance for the measured dataset of 340 pairs of samples. In our experiments, 306 pairs are used in training and 34 pairs are used in test. Of these 34 pairs, the MSE and MAE between prediction and ground-truth are 0.0041 and 0.0361, respectively. For rare signals, that is, those over (including extremely over) expressed and under (including extremely under) expressed genes, the MAE is 0.131. The PCC is 0.799. Considering the input is in the fold-change format and there are inherent noises in the measurement, these metrics indicate that the translation performance is quite good.

As DrugMatrix contains data from tissues beyond LI, we seek to further improve the model performance by increasing the number of training samples.

If DrugMatrix contained data for all endpoints, we would find in each column an (Affymetrix, BioSpyderWT) pair of samples. In total there would be 2711 samples, almost an order of magnitude more than 340 measured samples.

We apply matrix completion on DrugMatrix to impute all the missing endpoints. We adopt the Funk-SVD approach [15] for completion where the matrix $G_{n_1 \times n_2}$ is factored into two factors, $P \times Q$. Intuitively, P maps the genes/probes into a d dimensional latent space, and Q maps the treatments into the same d dimensional latent space. P is of size $n_1 \times d$, and Q is of size $n_2 \times d$. This formulation is common in corroborative filtering applications such as the Netflix challenge [16]. P and Q are randomly initialized and then learned from the observed entries in the matrix, using the following objective function

$$ min \sum \left(G_{u,i} - \left(bu_u + bi_i + \sum_{f=1}^{r} P_{u,f} \times Q_{i,f} \right) \right)^2 $$

with $M_{u,i}$ being the observed entries, bu_u and bi_i being the biases for the probes and treatments, respectively. A solution to the problem can be found with stochastic optimization. We use the Adam [17] optimizer with the default hyper-parameters.

With the completed matrix, we still use the 9:1 train:test split on the 340 pairs. We augment the training set with new (Affymetrix,BioSpyderWT) pairs produced by matrix completion.

Table 6. Predictive performance with augmented dataset. The training set includes pairs with imputed Affymetrix and BioSpyderWT profiles

test MSE	test MAE	rare MAE	PCC
0.0006	0.016	0.060	0.956

Table 6 shows the performance with the augmented training set. In comparison to Table 5, all metrics improve.

Figure 3 compares in four plots with bar charts the performance metrics of training with measured data pairs with that with augmented data pairs. In the figure test MSE is reduced by 4 times, test MAE and rare MAE by more than 2 times. PCC increases from 0.79 to 0.95. It is clear that data augmentation greatly improves the performance of *TransPlatformer*.

4   Evaluation of Translated Profiles

We use the model trained on augmented DrugMatrix dataset to translate Open-TGGATEs profiles onto the BioSpyderWT platform. As there is no ground-truth for the predicted Open-TGGATEs profiles, we evaluate the prediction performance by inspecting whether biological relationships present in the original transcriptomic profiles are preserved after cross-platform translation.

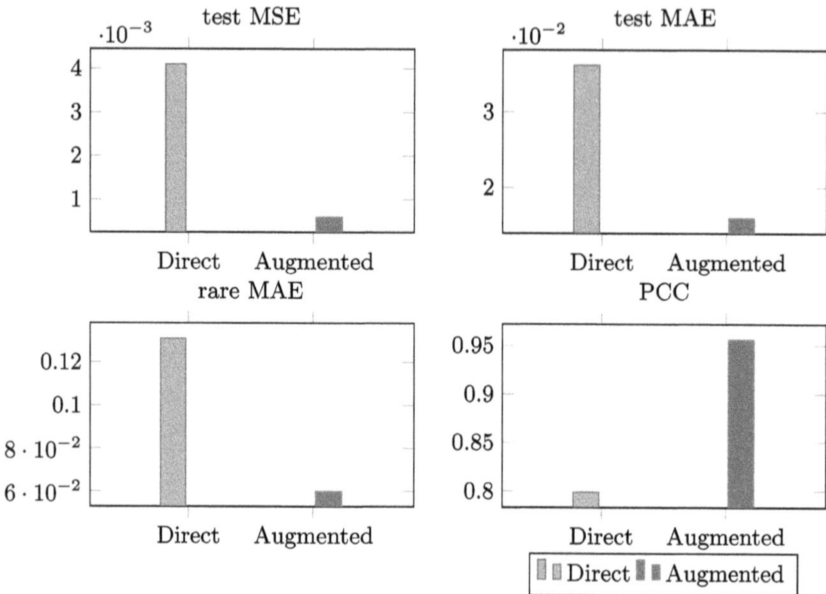

Fig. 3. Performance comparison of two methods. Direct is for the model trained with measured data pairs, and Augmented is for the model trained with data augmented from matrix completion. The smaller the test MSE, test MAE, and rare MAE, the better the performance; The larger the PCC, the better the performance

4.1 ARI and Silhouette Score

Recall that before translation, the ARI and silhouette score for the Open-TGGATEs profiles and the measured DrugMatrix Affymetrix profiles are 0.27 and 0.37, respectively, indicating that there are no clear clusters forming with the data. Post translation, we compute the ARI and silhouette score with the translated Open-TGGATEs profiles (specifically, on the BioSpyderWT platform) together with the measured BioSpyderWT profiles in DrugMatrix. The ARI and silhouette score are 0.29 and 0.47, respectively. This shows the translated Open-TGGATEs data are not separable from DrugMatrix data on the modern BioSpyderWT platform and one can combine them as a harmonized dataset for downstream tasks.

4.2 Median Rank Preservation of Nearest-Neighbor Relationships

To assess whether local neighborhood structure among treatments is preserved after cross-platform translation, we perform a median rank preservation test based on nearest-neighbor relationships in transcriptomic space.

For each treatment column x in the original dataset (Open-TGGATEs), we identify its nearest neighboring treatment y based on pairwise cosine distances computed between all treatment profiles. Cosine distances are applied consistently to both datasets. Let (x, y) denote this nearest-neighbor pair in Open-TGGATEs.

We then identify the corresponding translated treatments (x', y') in the translated dataset, Open-TGGATEs translated to BioSpyderWT. For each translated treatment x', we compute its distances to all other treatments and rank these distances in ascending order. The rank of y' within this ordered list was recorded as the neighbor rank for x'.

This procedure is repeated for all treatments, producing a distribution of neighbor ranks. The median of this distribution is used as a summary statistic, reflecting the typical preservation of nearest-neighbor relationships after translation. Lower median rank values indicate stronger preservation of local treatment relationships.

To assess statistical significance, we construct a null distribution by randomly permuting the correspondence between treatments in Open-TGGATEs and Open-TGGATEs translated to BioSpyderWT. Specifically, treatment labels in the translated dataset are randomly permuted while preserving the overall distance structure. For each permutation, the median neighbor rank is recomputed following the same procedure described above.

This process is repeated for 1,000 permutations, yielding an empirical null distribution of median ranks. A one-sided empirical p-value is computed as the proportion of permutations in which the null median rank is less than or equal to the observed median rank.

Our experiments show nearest-neighbor relationships among treatments is strongly preserved following cross-platform translation. Across all treatments, the observed median neighbor rank in the translated dataset is $\tilde{r} = \mathbf{2}$, indicating that, for a typical treatment, its closest counterpart in the original dataset remains among the top-ranked neighbors after translation.

To evaluate statistical significance, we compare the observed median rank to a null distribution generated from 1,000 random permutations of treatment correspondence. Under the null model, the median rank is $\mathbf{1729.17 \pm 28.33}$ (mean $\pm$ standard deviation), reflecting the expectation under random association between datasets.

The observed median rank is substantially lower than the null expectation, yielding an empirical p-value of $\boldsymbol{p < 10^{-4}}$. This result demonstrates that the preservation of nearest-neighbor relationships is highly unlikely to occur by chance and provides strong evidence that local transcriptomic structure among treatments is conserved after translation.

4.3   Preservation of Dosage–Response Relationships

Beyond ARI and silhouette score, we evaluate whether *dose–response relationships* for individual drug treatments were preserved after cross-platform translation. For each drug and exposure duration, treatments are administered at three dosage levels (low, medium, and high) in Open-TGGATEs, defined by the reported dosage in mg/kg.

For each drug–duration group, we focus on the relative distances between transcriptomic profiles corresponding to increasing dosage levels. Specifically, using the original dataset (Open-TGGATEs), we compute pairwise distances between the low-dose profile and the corresponding medium-dose and high-dose profiles:

$$d^{(1)}_{\text{low, med}}, \quad d^{(1)}_{\text{low, high}}.$$

Distances are computed using a cosine-based metric to reduce sensitivity to platform-dependent scaling effects.

The same distances are computed for the translated dataset, that is, Open-TGGATEs on BioSpyderWT:

$$d^{(2)}_{\text{low, med}}, \quad d^{(2)}_{\text{low, high}}.$$

Dosage–response distance preservation is assessed by testing whether the relative ordering of these distances is conserved after translation, that is,

$$d^{(1)}_{\text{low, med}} < d^{(1)}_{\text{low, high}} \quad \Longleftrightarrow \quad d^{(2)}_{\text{low, med}} < d^{(2)}_{\text{low, high}}.$$

This criterion evaluates whether increasing dosage continues to induce progressively larger transcriptomic deviations following translation, independent of absolute distance magnitudes.

Across all evaluated drug–duration groups with complete low, medium, and high dosage measurements, the relative ordering of low–medium and low–high distances is perfectly preserved after translation. For all drugs and durations, the distance between low- and high-dose profiles exceed the distance between low- and medium-dose profiles in both the original and translated datasets, corresponding to a preservation rate of **100%**.

These results indicate that the translated transcriptomic profiles retain coherent dose–response structure, with higher dosages consistently producing larger deviations in gene expression space. The complete preservation of relative dosage distances provides strong evidence that biologically meaningful dose-dependent transcriptional effects are conserved across platforms.

4.4   Preservation of Duration–Response Relationships

In addition to dosage effects, we evaluate whether temporal response structure is preserved after cross-platform translation. For each drug and fixed dosage level, we identify treatments corresponding to multiple durations (e.g., 1d, 4d, and 28d) that are present in both the original dataset Open-TGGATEs and the translated dataset. For each such group, we compute pairwise distances between transcriptomic profiles corresponding to different durations within each dataset. Cosine distances are used as a consistent metric across datasets.

To assess preservation of temporal structure, we compare the relative ordering of distances between duration pairs in Open-TGGATEs and Open-TGGATEs translated to BioSpyderWT. Specifically, for each drug–dosage combination, we evaluate whether the relative distances between short, intermediate, and long durations (e.g., 1d–4d and 1d–28d) observed in Open-TGGATEs were preserved in the translated data. A duration group is considered preserved if the ordering of distances between duration pairs is identical in the translated data.

Across all drug–dosage combinations with multiple annotated durations, the relative distances between transcriptomic profiles corresponding to different exposure durations are fully preserved following translation. In all evaluated cases, the ordering of distances between duration pairs observed in the original dataset is identical in the translated dataset, resulting in a preservation rate of **100%**.

This result indicates that the translation procedure maintains not only static transcriptomic similarity but also the temporal progression of transcriptional responses to chemical exposure. The full preservation of duration–response structure provides strong evidence that biologically meaningful temporal relationships are retained across platforms.

5   Conclusion and Future Work

We have presented a methodology for modernizing the Open-TGGATEs dataset onto the BioSpyderWT platform. We demonstrate with LI profiles. Since Open-TGGATEs does not contain any measurement on BioSpyderWT, we leverage measured (Affymetrix,BioSpyderWT) pairs from DrugMatrix. We establish that data in fold-change in these two datasets have significant overlap and do not separate into clear clusters by the database they belong to. The Adjusted Rand Index is 0.27 for the combined measured Affymetrix profiles. We then train a model on DrugMatrix achieving 0.0041 test MSE, 0.0361 test MAE, 0.131 rare MAE, and 0.799 PCC. The performance is further improved by augmenting the training dataset with imputed profiles using matrix completion. The test MSE, MAE, rare MAE, and PCC become 0.0006, 0.016, 0.060, and 0.956, respectively.

We evaluate the modernized Open-TGGATEs profiles and inspect the ARI of predicted profiles with measured DrugMatrix profiles. The ARI is 0.29. This shows the predicted Open-TGGATEs profiles and measured DrugMatrix profiles on the BioSpyderWT platform appear to have significant overlap. We further consider the preservation of median rank preservation of nearest neighbors, the preservation of dosage-response relationship, and the preservation of duration-response relationship. The test results show these relationships in the Affymetrix measurements are well preserved in the modernized prediction profiles.

We are currently in the process of open-sourcing these predictions and in future work we will consider modernizing the KI profiles.

Acknowledgments. This research used resources of the Compute and Data Environment for Science (CADES) at the Oak Ridge National Laboratory, which is supported by the Office of Science of the U.S. Department of Energy under Contract No. DE-AC05-00OR22725. This research was also supported by the NIH, National Institute of Environmental Health Sciences, through Intramural Research Project ZIAES103385 and Interagency Agreement No. AES24003001-24-001 between NIEHS and Oak Ridge

National Labs. The authors would like to acknowledge Andrew Rooney for his support and management of the interagency agreement between ORNL and NIEHS. The contributions of the NIH/NIEHS author(s) are considered Works of the United States Government. The findings and conclusions presented in this paper are those of the author(s) and do not necessarily reflect the views of the NIEHS, NIH or the U.S. Department of Health and Human Services.

References

1. Lockhart, D.J., et al.: Expression monitoring by hybridization to high-density oligonucleotide arrays. Nat. Biotechnol. **14**(13), 1675–1680 (2000)
2. Affymetrix rat genome 230 2.0 array. https://www.ncbi.nlm.nih.gov/geo/query/acc.cgi?acc=GPL1355
3. Wang, Z., Gerstein, M., Snyder, M.: RNA-Seq: a revolutionary tool for transcriptomics. Nat. Rev. Genet. **10**(1), 57–63 (2009)
4. Krämer, A., Green, J., Pollard Jr, J., Tugendreich, S.: Causal analysis approaches in ingenuity pathway analysis. Bioinformatics **30**(4), 523–530 (2014)
5. Mav, D., et al.: A hybrid gene selection approach to create the S1500+ targeted gene sets for use in high-throughput transcriptomics. PloS One **13**(2), e0191105 (2018)
6. Yeakley, J.M., et al.: A trichostatin a expression signature identified by tempo-seq targeted whole transcriptome profiling. PloS One **12**(5), e0178302 (2017)
7. Igarashi, Y., et al.: Open TG-GATEs: a large-scale toxicogenomics database. Nucleic Acids Res. **43**(D1), D921–D927 (2015)
8. Igarashi, Y., et al.: Open TG-GATEs: a large-scale toxicogenomics database. Nucleic Acids Res. **43**(Database issue), D921–D927 (2015)
9. Database description: Open TG-GATEs. https://dbarchive.biosciencedbc.jp/en/open-tggates/desc.html. Accessed 12 2025
10. Svoboda, D.L., Saddler, T., Auerbach, S.S.: An overview of national toxicology program's toxicogenomic applications: DrugMatrix and ToxFX. In: Advances in Computational Toxicology: Methodologies and Applications in Regulatory Science, pp. 141–157 (2019)
11. Genie. https://www.sciome.com/genie/
12. Transplatformer – translating toxicogenomic profiles between generations of platforms. under review
13. Hubert, L., Arabie, P.: Comparing partitions. J. Classif. **2**(1), 193–218 (1985)
14. Rousseeuw, P.J.: Silhouettes: a graphical aid to the interpretation and validation of cluster analysis. J. Comput. Appl. Math. **20**, 53–65 (1987)
15. Piatetsky, G.: Interview with Simon Funk. ACM SIGKDD Explorations Newsletter **9**(1), 38–40 (2007)

16. Goel, V.: Netflix challenge–improving movie recommendations. Recommender System with Machine Learning and Artificial Intelligence: Practical Tools and Applications in Medical, Agricultural and Other Industries, pp. 251–267 (2020)
17. Kingma, D.P., Ba, J.: Adam: A method for stochastic optimization. arXiv preprint arXiv:1412.6980 (2014)

SpaHMS: Dynamic Multi-scale Integration of Multi-batch Spatial Multi-omics

Yong Zhao, Canqun Yang, Tao Tang, and Yingbo Cui[✉]

College of Computer Science and Technology, National University of Defense Technology, Changsha, China
yingbocui@nudt.edu.cn

Abstract. Spatial multi-omics technologies enable joint profiling of molecular modalities within native tissue contexts, but integrating data across heterogeneous tissue sections and experimental batches remains a fundamental challenge. A key limitation of existing methods is that spatial graphs are constructed using manually defined radii or fixed neighbor sizes prior to training, preventing models from adapting spatial interaction scales across tissues with variable architectures and cell densities. To address this limitation, we propose SpaHMS, a fully dynamic framework for spatial multi-omics integration that learns hierarchical, multi-scale spatial receptive fields in an end-to-end manner. SpaHMS introduces a Dynamic Hierarchical Multi-Scale Graph Neural Network that jointly optimizes spatial graph structure and multimodal representations, enabling adaptive spatial modeling across heterogeneous tissues. Each modality is encoded using a dedicated network with cross-section parameter sharing, preserving modality-specific structure while mitigating batch effects through explicit mutual nearest neighbor (MNN) alignment. Across datasets involving RNA, ADT, and ATAC, SpaHMS produces robustly aligned, batch-corrected, and biologically coherent embeddings that preserve meaningful spatial organization.

Keywords: Spatial Multi-omics Integration · Multi-batch Integration · Mosaic Integration · Dynamic Learning

1 Introduction

In recent years, spatial multi-omics technologies have advanced rapidly, enabling researchers to simultaneously acquire multi-layer molecular information together with precise spatial localization. Although high-performance sequencing alignment and processing tools have greatly accelerated large-scale omics data analysis [1–4], spatial multi-omics technologies overcome a fundamental limitation of conventional sequencing, which requires tissue dissociation and consequently loses spatial context. Spatially resolved single-cell profiling technologies now allow high-resolution mapping of molecular and structural tissue organization [5–7].

As these technologies continue to expand, large-scale studies increasingly generate multi-modal data across multiple tissue sections, experimental batches, and technological platforms [8,9]. This rapid growth of spatial multi-omics datasets further accelerates the need for scalable and robust computational frameworks. Integrating these highly heterogeneous datasets into a unified latent representation is essential for delineating spatial domains, identifying cellular states, and comparing tissue organization across biological samples. However, multi-batch spatial multi-omics integration remains challenging due to substantial batch effects, modality-specific noise characteristics, variability in sequencing depth, and spatial-scale differences arising from tissue microenvironmental heterogeneity and variable cell density [10,11].

Current computational frameworks for spatial multi-omics integration typically rely on pre-defined spatial radii or fixed neighbor sizes to construct spatial graphs before model training. Such static graph definitions cannot accommodate the heterogeneous spatial organization observed in real tissues, where densely packed regions and sparse anatomical zones require different effective receptive fields. Moreover, graph construction is often decoupled from model optimization, resulting in mismatches between the neighborhoods used for message passing and the learning objectives used for alignment or reconstruction. Several existing frameworks, including MOFA+ [12], scMoMaT [13], GLUE [14], MultiVI [15], SpatialGlue [16], SpaBalance [17], SpaMosaic [18], and StabMap [19], have advanced multimodal or spatial integration. However, these methods generally rely on fixed or heuristically defined neighborhood structures, which limits generalization across tissue sections with variable cell densities and distinct anatomical architectures.

Recent advances in graph neural networks have underscored the importance of flexible and learnable neighborhood aggregation mechanisms. Models such as GraphSAGE [20] enable scalable and inductive neighborhood sampling, while hierarchical graph pooling approaches, including Hierarchical Graph Pooling and Readout (HGP-SL) [21] and DiffPool [22], demonstrate that explicitly modeling hierarchical graph structures can substantially enhance representation learning by capturing both local and global topological patterns. In parallel, deep message passing frameworks such as DeeperGCN [23] show that increasing network depth facilitates the progressive integration of broader contextual information. Moreover, recent studies on dynamic structure learning, such as Learning Discrete Structures (LDS) [24], further advance this direction by learning graph connectivity through gradient-based optimization rather than relying on predefined graph constructions.

Despite these advances, a key unresolved challenge in spatial single-cell multi-omics integration is that existing methods still rely on static, manually defined spatial graphs constructed prior to training. Such fixed neighborhood definitions prevent models from adapting spatial interaction scales across heterogeneous tissue architectures and experimental batches, and decouple graph structure learning from multimodal representation learning.

To address these challenges, we propose **SpaHMS**, a dynamic hierarchical multi-scale framework for multi-batch spatial single-cell multi-omics integration. SpaHMS introduces a Dynamic Hierarchical Multi-Scale Graph Neural Network (Dynamic HMS-GNN) that learns spatial receptive fields in an end-to-end manner, allowing the underlying graph structure to adapt to tissue-specific spatial organization through backpropagation. Each modality is encoded using a modality-specific Dynamic HMS-GNN with parameters shared across tissue sections to ensure cross-batch consistency while preserving modality-specific characteristics. The framework jointly optimizes spatially informed representations through reconstruction-based objectives, spatial smoothness regularization, and explicit mutual nearest neighbor (MNN) alignment across batches. By tightly coupling spatial graph structure learning with multimodal representation learning, SpaHMS preserves biologically meaningful spatial organization and enables robust integration across heterogeneous tissue architectures, experimental batches, and modality combinations.

2   Materials and Methods

2.1   Dataset

In this study, we evaluate SpaHMS using two complementary datasets: a biologically informed spatial tri-omics simulation dataset and a real human lymph node dataset derived from the SpaMosaic framework. The simulation dataset comprises RNA, ADT, and ATAC modalities generated through a hierarchical procedure. Gene expression is simulated with a GMM-based model following the design principles of Splatter [25]. Protein abundance patterns are generated using marker-guided rules consistent with CITE-seq measurements [26,27]. Chromatin accessibility is simulated in a manner that preserves cis-regulatory coupling with transcriptional output [28,29]. Spatial domains are constructed using Voronoi partitioning and perturbed with batch-specific spatial shifts to mimic realistic spatial variation across tissue sections. The human lymph node dataset provides spatially resolved RNA and ADT measurements obtained using the 10x Genomics Visium platform, including pixel-level coordinates and expert-curated tissue annotations. Together, these datasets serve as robust benchmarks for evaluating multimodal integration accuracy, spatial structure preservation, and cross-batch correction performance.

2.2   Overview

SpaHMS is a dynamic hierarchical multi-scale framework designed for integrating spatially resolved single-cell multi-omics data across multiple batches with heterogeneous modality compositions. The framework addresses two major challenges in existing spatial integration methods: (1) reliance on fixed or heuristic spatial neighborhood definitions, and (2) decoupling between graph construction and representation learning.

SpaHMS performs modality-wise global spatial coordinate normalization to unify batch-specific spatial scales, followed by hybrid intra-batch spatial and inter-batch mutual nearest neighbor (MNN) graph construction. It then applies a Dynamic HMS-GNN that learns adaptive spatial receptive fields through differentiable, learnable radii and soft graph masks, enabling multi-scale message passing and context-aware scale selection for each node. The model is trained end-to-end using a unified objective that integrates reconstruction loss, spatial smoothness regularization, and explicit MNN-based batch alignment. The overall training procedure is summarized in Algorithm 1.

Algorithm 1. End-to-end training pipeline of SpaHMS

Require: Multi-batch multi-modal datasets $\mathcal{D}$, number of scales S, epochs T
1: Perform modality-wise global spatial normalization (Section 2.3, Eq. 1)
2: Construct hybrid intra/inter-batch graph (Section 2.4, Eq. 2)
3: **for** epoch $= 1$ to T **do**
4:    **for** each modality **do**
5:       Predict learnable multi-scale radii (Section 2.5, Eq. 3)
6:       Compute differentiable soft graph masks (Section 2.5, Eqs. 4, 5)
7:       Perform multi-scale message passing (Section 2.5, Eq. 6)
8:       Fuse multi-scale representations via scale-selection gate (Section 2.5, Eqs. 7, 8)
9:    **end for**
10:    Compute reconstruction, spatial smoothness, and MNN alignment losses (Section 2.6, Eqs. 9–11)
11:    Update model parameters using the total objective (Section 2.6, Eq. 12)
12: **end for**

2.3 Spatial Coordinate Normalization

Spatial coordinates from different batches often exhibit significant variations in scale due to differences in imaging resolution, tissue size, and spot density. To ensure cross-batch comparability, SpaHMS performs modality-wise global spatial normalization.

For modality m, let (x_i, y_i) denote raw coordinates. The global minimum and maximum across all batches are computed and coordinates are normalized to $[0, 1]$:

$$\tilde{x}_i = \frac{x_i - x_{\min}^{(m)}}{x_{\max}^{(m)} - x_{\min}^{(m)} + \epsilon}, \qquad \tilde{y}_i = \frac{y_i - y_{\min}^{(m)}}{y_{\max}^{(m)} - y_{\min}^{(m)} + \epsilon}, \tag{1}$$

where ϵ is a small constant to avoid numerical instability.

In addition, SpaHMS augments normalized coordinates with distance to tissue center and local density estimated via Gaussian kernel density estimation. The final spatial feature vector is:

$$\mathbf{s}_i = [\tilde{x}_i, \tilde{y}_i, d_{\text{center},i}, \rho_i].$$

2.4   Hybrid Intra/Inter-batch Graph Construction

SpaHMS constructs a hybrid graph that simultaneously captures local spatial continuity within batches and alignment relationships across batches. Intra-batch edges are formed using a spatial kNN graph based on normalized coordinates, while inter-batch edges are constructed using Mutual Nearest Neighbors identified in a low-dimensional feature space.

The final graph is defined as:

$$\mathcal{E} = \mathcal{E}_{\text{intra}} \cup \mathcal{E}_{\text{inter}}. \tag{2}$$

2.5   Dynamic HMS-GNN

Spatial tissue organization exhibits hierarchical patterns at multiple spatial scales, ranging from local cellular neighborhoods to global tissue regions. To capture such heterogeneity, SpaHMS employs a **Dynamic Hierarchical Multi-Scale Graph Neural Network (HMS-GNN)** that adaptively learns spatial receptive fields and selects the most relevant scale for each node in a data-driven manner.

Learnable Multi-scale Radii. Instead of relying on predefined spatial radii, SpaHMS learns S spatial scales end-to-end from data. A global summary vector $\mathbf{g}$ is first obtained by pooling node features across the graph (e.g., mean pooling). An MLP then predicts S latent variables $\{z_s\}_{s=1}^{S}$, which are mapped to valid radii:

$$r_s = r_{\min} + (r_{\max} - r_{\min}) \cdot \sigma(z_s), \tag{3}$$

where $\sigma(\cdot)$ denotes the sigmoid function, constraining each radius within a biologically plausible range $[r_{\min}, r_{\max}]$. This mechanism enables the model to automatically adjust receptive fields according to tissue scale and data characteristics.

Differentiable Soft Graph Mask. Hard neighborhood selection based on fixed radii is non-differentiable and prevents joint optimization. To address this, SpaHMS introduces a differentiable soft graph mask. For any pair of nodes i and j with Euclidean distance d_{ij}, the soft mask at scale s is defined as:

$$\text{mask}_{ij}^{(s)} = \sigma\left(\frac{r_s - d_{ij}}{\tau}\right), \tag{4}$$

where τ is a temperature parameter controlling the sharpness of the transition. To further model spatial decay, the mask is combined with a Gaussian kernel:

$$w_{ij}^{(s)} = \text{mask}_{ij}^{(s)} \cdot \exp\left(-\frac{d_{ij}^2}{2\sigma_s^2}\right), \tag{5}$$

where σ_s controls the effective smoothing strength at scale s. This formulation enables gradients to flow through r_s, allowing spatial scales to be optimized jointly with model parameters.

Multi-scale Message Passing. For each scale s, message passing is performed independently to extract scale-specific spatial representations. The update rule for node i is:

$$\tilde{\mathbf{h}}_i^{(s)} = \text{ReLU}\left(\mathbf{W}^{(s)}\mathbf{h}_i + \sum_j w_{ij}^{(s)}\mathbf{W}^{(s)}\mathbf{h}_j\right), \tag{6}$$

where $\mathbf{h}_i$ is the input node feature, $\mathbf{W}^{(s)}$ is a learnable weight matrix specific to scale s, and $w_{ij}^{(s)}$ is the adaptive edge weight defined in Eq. 5. This operation aggregates information from a soft, learnable spatial neighborhood.

Scale-Selection Gate. Different spatial contexts may require different effective scales. SpaHMS therefore incorporates a scale-selection gate to adaptively fuse multi-scale representations. For node i, scale-specific representations are concatenated:

$$\mathbf{H}_i = [\tilde{\mathbf{h}}_i^{(1)}, \ldots, \tilde{\mathbf{h}}_i^{(S)}].$$

Attention weights over scales are computed as:

$$a_{i,s} = \frac{\exp(\mathbf{H}_i^{(s)}\mathbf{q}_i/\sqrt{d})}{\sum_{s'=1}^{S}\exp(\mathbf{H}_i^{(s')}\mathbf{q}_i/\sqrt{d})}, \tag{7}$$

where $\mathbf{q}_i$ is a learnable query vector and d denotes feature dimensionality. The final node representation is obtained via weighted aggregation:

$$\mathbf{h}_i = \sum_{s=1}^{S} a_{i,s}\,\tilde{\mathbf{h}}_i^{(s)}. \tag{8}$$

This gating mechanism enables node-wise, context-aware scale selection.

2.6   Reconstruction and Spatial Smoothness

To preserve biological information, SpaHMS employs a reconstruction loss that encourages decoded features to match original inputs:

$$\mathcal{L}_{\text{recon}} = \frac{1}{N}\sum_i \|\mathbf{x}_i - \hat{\mathbf{x}}_i\|_2^2. \tag{9}$$

To enforce spatial coherence, a spatial smoothness regularization penalizes embedding discrepancies between neighboring spots:

$$\mathcal{L}_{\text{spatial}} = \frac{1}{|\mathcal{E}_{\text{spatial}}|}\sum_{(i,j)\in\mathcal{E}_{\text{spatial}}} \frac{\|\mathbf{z}_i - \mathbf{z}_j\|_2^2}{d_{ij} + \epsilon}. \tag{10}$$

2.7 MNN-Based Batch Alignment

To explicitly enforce cross-batch consistency, SpaHMS minimizes distances between Mutual Nearest Neighbor pairs:

$$\mathcal{L}_{\mathrm{MNN}} = \frac{1}{|\mathcal{M}|} \sum_{(i,j)\in\mathcal{M}} \|\mathbf{z}_i - \mathbf{z}_j\|_2^2. \tag{11}$$

2.8 Overall Training Objective

The final objective function is a weighted sum of all loss components:

$$\mathcal{L}_{\mathrm{total}} = \lambda_1 \mathcal{L}_{\mathrm{recon}} + \lambda_2 \mathcal{L}_{\mathrm{spatial}} + \lambda_3 \mathcal{L}_{\mathrm{MNN}}, \tag{12}$$

where λ_1, λ_2, and λ_3 control the contribution of each term. By jointly optimizing this objective, SpaHMS learns spatially coherent, biologically meaningful, and batch-aligned representations across heterogeneous spatial multi-omics datasets.

3 Results

3.1 Evaluation Metrics

We evaluated multimodal integration using six metrics assessing batch mixing, biological structure preservation, clustering fidelity, and embedding quality. **iLISI** measures batch mixing (higher is better), while **cLISI** evaluates cell-type separation (lower is better). Although these two metrics may seem contradictory, using both together provides a balanced assessment of integration: iLISI captures the ability to remove batch effects, while cLISI ensures that biologically meaningful cell-type distinctions are preserved.**ARI** and **NMI** quantify clustering accuracy relative to ground-truth labels. **Silhouette** assesses within-cluster compactness and between-cluster separation. **Spatial Conservation** evaluates preservation of spatial neighborhood relationships. Together, these metrics provide a comprehensive framework for benchmarking spatial multi-omics integration.

3.2 Integration Performance on Simulated Spatial Tri-omics Dataset

We first benchmarked SpaHMS on a simulated spatial tri-omics mosaic dataset composed of three batches. Specifically, batch1 contained RNA, ADT, and ATAC modalities; batch2 contained RNA only; and batch3 contained RNA, ADT, and ATAC modalities. Six representative integration tools–MOFA+, MultiVI, scMo-MaT, StabMap, SpaMosaic, and SpaHMS–were applied to the mosaic dataset to evaluate their ability to integrate multiple omics profiles with heterogeneous modality compositions.

As shown in Fig. 1, SpaHMS achieves substantially improved batch mixing while preserving well-separated and modality-consistent biological structures.

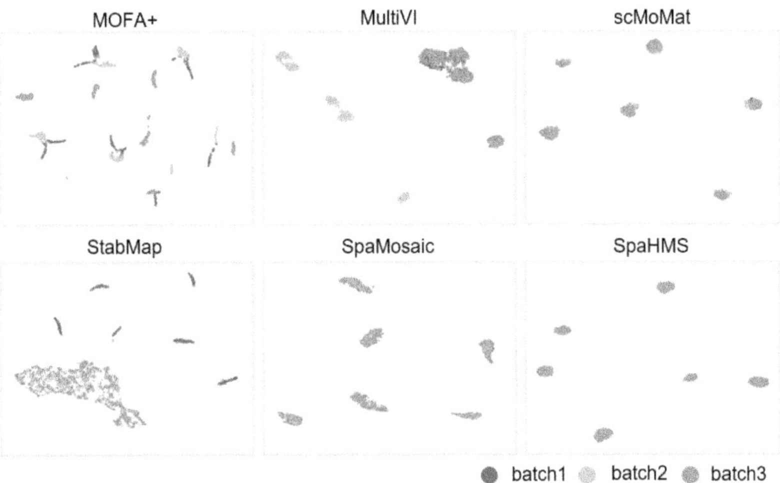

Fig. 1. UMAP visualizations of simulated tri-omics mosaic dataset integration. Batch 1 contains RNA+ADT+ATAC, batch 2 contains RNA only, and batch 3 contains RNA+ADT+ATAC. SpaHMS achieves superior batch mixing while preserving modality-specific biological structures.

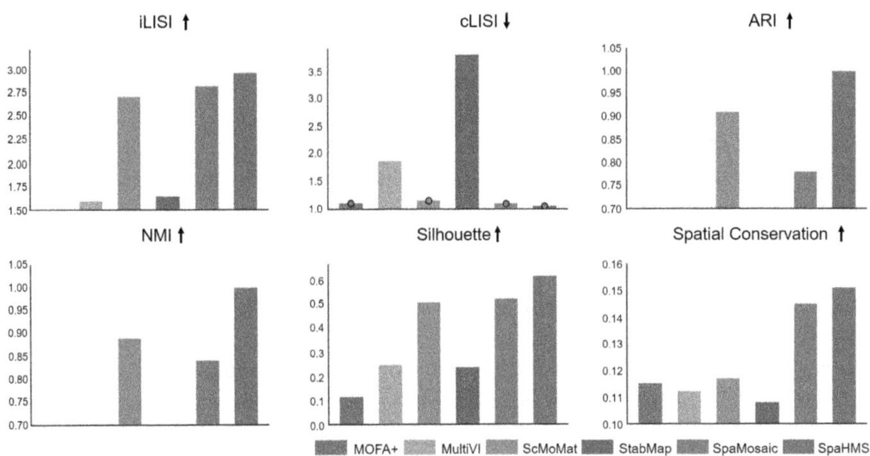

Fig. 2. Quantitative benchmarking on simulated tri-omics dataset. Six multimodal integration methods (MOFA+, MultiVI, scMoMaT, StabMap, SpaMosaic, and SpaHMS) were evaluated using six metrics: iLISI, cLISI, ARI, NMI, Silhouette, and Spatial Conservation. SpaHMS consistently achieves the best performance across all metrics. To better visualize the differences among methods, the bar plots were scaled and do not start from zero, resulting in some bars not displaying the full range for all methods.

In contrast, most baseline methods exhibit either incomplete batch alignment or distorted cluster geometry when integrating heterogeneous modality compositions. Quantitative results (Fig. 2) further confirm this advantage. SpaHMS attains the highest iLISI score (2.95), indicating near-optimal batch mixing, and achieves perfect cluster recovery with ARI and NMI both reaching 1.00. It also yields the best Silhouette score (0.603), reflecting compact and well-separated embeddings, together with the strongest spatial conservation (0.151). These results demonstrate that SpaHMS consistently balances batch alignment, biological fidelity, and spatial structure preservation, enabling robust integration of complex multimodal spatial mosaics.

3.3   Integration Performance on Human Lymph Node Spatial Multi-omics Dataset

We further evaluated SpaHMS on a human lymph node spatial multi-omics dataset comprising RNA and ADT modalities. In this dataset, batch 1 contained paired RNA and ADT measurements, whereas batch 2 and batch 3 provided RNA profiles only, resulting in a realistic multimodal mosaic setting with incomplete modality pairing across batches. This scenario poses substantial challenges for integration, as models must simultaneously align heterogeneous batches, transfer protein-informed cell identity signals from multimodal to RNA-only batches, and preserve the highly structured spatial organization characteristic of lymphoid tissue.

We compared SpaHMS with MOFA+, MultiVI, scMoMaT, StabMap, SpaMosaic, and SpaHMS across multiple complementary evaluation metrics capturing batch mixing, biological separation, and spatial structure preservation. While several methods achieved partial improvements along individual dimensions, many exhibited clear trade-offs, such as improved batch alignment at the expense of spatial coherence, or preserved spatial patterns accompanied by insufficient batch mixing.

In contrast, SpaHMS demonstrated consistently strong performance across all evaluation criteria. It achieved the highest batch mixing score (iLISI $= 2.745$) while maintaining clear cell-type separation (cLISI $= 2.264$), indicating effective removal of batch effects without over-correction. SpaHMS also attained the best agreement with ground-truth cell-type annotations, as reflected by ARI (0.158) and NMI (0.299). Importantly, SpaHMS more faithfully preserved spatial neighborhood relationships than competing approaches, achieving the highest spatial conservation score (0.166). Together, these results demonstrate that SpaHMS can robustly integrate heterogeneous spatial multi-omics data under realistic mosaic conditions while simultaneously maintaining biologically meaningful cellular identities and spatial organization Fig. 3 and 4.

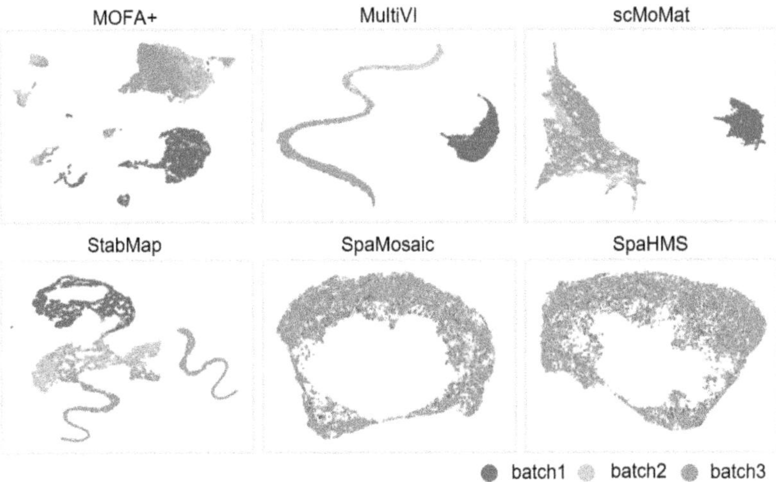

Fig. 3. UMAP visualizations of human lymph node dataset integration. Batch 1 contains RNA+ADT, while batches 2 and 3 contain RNA only. SpaHMS exhibits superior mixing of the three batches while preserving tissue structural organization.

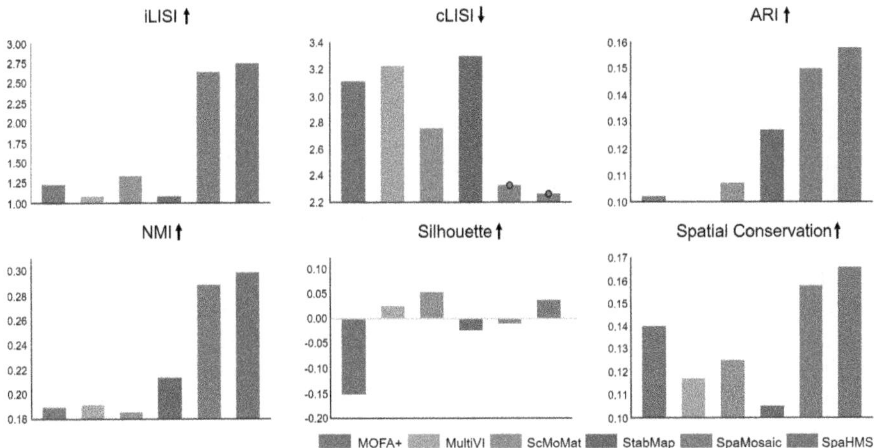

Fig. 4. Quantitative benchmarking on human lymph node dataset. Comparison of six multi-omics integration methods across six metrics: iLISI, cLISI, ARI, NMI, Silhouette, and Spatial Conservation. SpaHMS achieves consistently superior or comparable performance. To better visualize differences among methods, the bar plots were scaled and do not start from zero, resulting in some bars not displaying the full range for all methods.

4   Conclusion

Existing spatial multi-omics integration methods often rely on fixed, manually defined spatial graphs, which limits their ability to adapt to heterogeneous tissue architectures and multi-batch settings. To address this limitation, we propose SpaHMS, a dynamic hierarchical multi-scale framework for integrating spatially resolved single-cell multi-omics data across batches. SpaHMS jointly learns spatial receptive fields and constructs hybrid intra- and inter-batch graphs, enabling end-to-end spatial structure learning without manual parameter tuning. Through a dynamic HMS-GNN backbone and alignment-driven optimization, the framework preserves biologically meaningful spatial organization while achieving robust batch alignment across heterogeneous modalities, including RNA, ADT, and ATAC.

Acknowledgments. This work was supported by the National Natural Science Foundation of China No. 62102427, the Science and Technology Innovation Program of Hunan Province No. 2024RC3115, the Innovative Talent Program of the National University of Defense Technology.

References

1. Langmead, B., Salzberg, S.L.: Fast gapped-read alignment with bowtie 2. Nat. Methods **9**(4), 357–359 (2012)
2. Kim, D., Paggi, J.M., Park, C., Bennett, C., Salzberg, S.L.: Graph-based genome alignment and genotyping with hisat2 and hisat-genotype. Nat. Biotechnol. **37**(8), 907–915 (2019)
3. Wang, Z., Cui, Y., Peng, S., Liao, X., Yangbo, Yu.: MinimapR: a parallel alignment tool for the analysis of large-scale third generation sequencing data. Comput. Biol. Chem. **99**, 107735 (2022)
4. Xia, Z., et al.: A review of parallel implementations for the Smith-Waterman algorithm. Interdiscipl. Sci. Comput. Life Sci. **14**(1), 1–14 (2022)
5. Ståhl, P.L., et al.: Visualization and analysis of gene expression in tissue sections by spatial transcriptomics. Science **353**(6294), 78–82 (2016)
6. Goltsev, Y., et al.: Deep profiling of mouse splenic architecture with codex multiplexed imaging. Cell **174**(4), 968–981 (2018)
7. Liu, Y., et al.: High-spatial-resolution multi-omics sequencing via deterministic barcoding in tissue. Cell **183**(6), 1665–1681 (2020)
8. Moffitt, J.R., et al.: Molecular, spatial, and functional single-cell profiling of the hypothalamic preoptic region. Science **362**(6416), eaau5324 (2018)
9. Stickels, R.R., et al.: Highly sensitive spatial transcriptomics at near-cellular resolution with slide-seqv2. Nat. Biotechnol. **39**(3), 313–319 (2021)
10. Liu, X., et al.: Spatial multi-omics: deciphering technological landscape of integration of multi-omics and its applications. J. Hematol. Oncol. **17**(1), 72 (2024)
11. Yingcheng, W., Cheng, Y., Wang, X., Fan, J., Gao, Q.: Spatial omics: navigating to the golden era of cancer research. Clin. Transl. Med. **12**(1), e696 (2022)
12. Argelaguet, R., et al.: Mofa+: a statistical framework for comprehensive integration of multi-modal single-cell data. Genome Biol. **21**(1), 111 (2020)

13. Zhang, Z., et al.: scMoMaT jointly performs single cell mosaic integration and multi-modal bio-marker detection. Nat. Commun. **14**(1), 384 (2023)
14. Cao, Z.-J., Gao, G.: Multi-omics single-cell data integration and regulatory inference with graph-linked embedding. Nat. Biotechnol. **40**(10), 1458–1466 (2022)
15. Ashuach, T., Gabitto, M.I., Koodli, R.V., Saldi, G.A., Jordan, M.I., Yosef, N.: Multivi: deep generative model for the integration of multimodal data. Nat. Methods **20**(8), 1222–1231 (2023)
16. Long, Y., et al.: Deciphering spatial domains from spatial multi-omics with spatialglue. Nat. Methods **21**(9), 1658–1667 (2024)
17. Cui, Y., et al.: Spabalance: balanced learning for efficient spatial multi-omics decoding. Adv. Sci. e12973 (2025)
18. Yan, X., et al.: Mosaic integration of spatial multi-omics with spamosaic. BioRxiv, p. 2024–10 (2024)
19. Ghazanfar, S., Guibentif, C., Marioni, J.C.: Stabilized mosaic single-cell data integration using unshared features. Nat. Biotechnol. **42**(2), 284–292 (2024)
20. Liu, J., Ong, G.P., Chen, X.: Graphsage-based traffic speed forecasting for segment network with sparse data. IEEE Trans. Intell. Transp. Syst. **23**(3), 1755–1766 (2020)
21. Zhang, Z., et al.: Hierarchical graph pooling with structure learning. arXiv preprint arXiv:1911.05954 (2019)
22. Ying, Z., You, J., Morris, C., Ren, X., Hamilton, W., Leskovec, J.: Hierarchical graph representation learning with differentiable pooling. In: Advances in Neural Information Processing Systems, vol. 31 (2018)
23. Li, G., Xiong, C., Thabet, A., Ghanem, B.: DeeperGCN: all you need to train deeper gcns. arXiv preprint arXiv:2006.07739 (2020)
24. Franceschi, L., Niepert, M., Pontil, M., He, X.: Learning discrete structures for graph neural networks. In: International Conference on Machine Learning, pp. 1972–1982. PMLR (2019)
25. Zappia, L., Phipson, B., Oshlack, A.: Splatter: simulation of single-cell RNA sequencing data. Genome Biol. **18**(1), 174 (2017)
26. Stoeckius, M., et al.: Simultaneous epitope and transcriptome measurement in single cells. Nat. Methods **14**(9), 865–868 (2017)
27. Hao, Y., et al.: Integrated analysis of multimodal single-cell data. Cell **184**(13), 3573–3587 (2021)
28. Grandi, F.C., Modi, H., Kampman, L., Corces, M.R.: Chromatin accessibility profiling by atac-seq. Nat. Protocols **17**(6), 1518–1552 (2022)
29. Ma, S., et al.: Chromatin potential identified by shared single-cell profiling of RNA and chromatin. Cell **183**(4), 1103–1116 (2020)

Predicting Obstetric and Non-obstetric Diagnoses Co-occurrences During Pregnancy

Akash Singh, Samuel Infante, Seungbae Kim, and Anowarul Kabir[✉]

University of South Florida, Tampa, FL 33620, USA
{akashsingh,sinfante,seungbae}@usf.edu, akabir@usf.edu

Abstract. Pregnancy care often involves simultaneous obstetric and other medical conditions, but their co-occurrence patterns are rarely modeled explicitly in a systematic, network-based approach. In this work, we formulate obstetric and non-obstetric diagnoses co-occurrences as a link prediction problem on a diagnosis-level homogeneous graph constructed from pregnancy encounters. Diagnoses are represented as nodes connected by co-occurrence edges, with node features capturing graph structure and demographic statistics.

We address this challenge by leveraging collected electronic health records data and study several standalone and hybrid graph neural network (GNN) architectures, including GCN, GAT, GraphSAGE, and three hybrid encoders that combine complementary aggregation mechanisms, namely GCN+GraphSAGE, GCN+GAT, and GAT+GraphSAGE. All models used consistent train-validation-test splits and are evaluated on 5-fold cross-validation sets. Among standalone models, GraphSAGE achieved the strongest performance, whereas hybrid GraphSAGE-based models (GCN+GraphSAGE and GAT+GraphSAGE) are best performers. The GCN+GraphSAGE hybrid, reaching an AUROC and AUPRC of approximately 0.90, consistently outperformed all other architectures. Further analysis of top-ranked predicted links revealed clinically plausible associations between pregnancy stage and risk-related diagnoses and common endocrine, metabolic, and hematological conditions. These findings indicate that graph-based link prediction may effectively prioritize obstetric and non-obstetric diagnosis pairs, providing a scalable framework for identifying clinically meaningful comorbidity patterns. They may further support hypothesis generation and downstream obstetric risk stratification efforts.

Keywords: Obstetric and Non-obstetric Diagnoses · Graph Neural Networks · Pregnancy

1 Introduction

Pregnancy is a high-stakes period where preexisting chronic conditions, acute complications, and social risk factors interact in a complex way. Maternal morbidity and mortality are often diagnosed by conditions related to pregnancy or postpartum period, known as obstetric diagnoses. However, combinations of preexisting or developing medical or surgical non-obstetric conditions, such as cardiovascular, metabolic, hypertension and/or anxiety disorders, may also contribute to the ill cause simultaneously. Since clinicians often focus on obstetric diagnoses, automatic diagnosis of non-obstetric and obstetric co-occurrences during pregnancy is of great interest for risk assessment, pregnancy care and fetal outcomes.

There is a growing body of interest in developing machine and deep learning based methods for risk prediction [18] and clinical diagnosis [6,25]. This is due to the availability of high dimensional patients' electronic health records (EHRs) that document patients' symptoms, prescriptions, clinical notes, and medical images which is heterogeneous and longitudinal in nature [14,20,22]. Despite this high-volume and rich information present in many EHRs, almost no research has been conducted concentrating both obstetric and non-obstetric conditions co-occurrences focusing on pregnant morbidity and mortality. Recent systematic review of ML models focuses on obstetric outcomes (e.g., preeclampsia) [7,12,17]. Meredith et al. [11] highlight that including preexisting co-occurring medical conditions (multimorbidity) improves predictive performance for severe preeclampsia. Several works have concentrated on maternal and adverse birth outcomes for obstetric diagnoses, broadly summarized in [1,2,13]. Therefore, to address this gap in pregnancy care, we propose graph neural network (GNN) based frameworks and evaluate multiple architectures to predict obstetric and non-obstetric diagnosis co-occurrences during pregnancy, essentially providing a benchmark for future development.

We formulate the problem as a network of co-occurring diagnoses rather than isolated events, since EHRs store these obstetric and non-obstetric conditions as coded diagnoses across temporal encounters of the same patient. We ask *which obstetric and non-obstetric diagnosis pairs are more likely to co-occur in future pregnancy encounters?* We construct a diagnosis-level graph from pregnancy-related encounters. Each node represents a diagnosis concept, edges represent co-occurrence of two diagnoses within the same pregnancy visit. Then, the nodes are enriched with structural features from the graph plus demographic statistics, such as average patient age at which the diagnosis appears in pregnancy encounters. We then formulate a link prediction problem, given the current pregnancy diagnosis graph and node features. To solve this problem, we systematically applied and studied various existing GNNs and hybrid architectures for robust and explainable predictive modeling towards co-occurrence diagnoses.

Several studies have applied GNNs to EHR-like data. Zhu et al. [26] proposed a variationally regularized graph-based representation learning framework that uses GCN layers over a medical concept graph to learn patient embeddings for downstream prediction tasks. Cho et al. [3] built a heterogeneous graph of patients and medical entities and used HinSAGE, a variant of GraphSAGE for heterogeneous graphs, to learn patient representations that improved prognostic prediction compared with non-graph baselines. More recent heterogeneous GNN frameworks, such as GraphEHR, also use neighborhood-sampling-style message passing to predict multiple diseases from hospital EHR data [9]. Graph attention networks (GATs) have been explored for several clinical tasks. GRAM introduced an attention mechanism over a medical ontology tree to compute concept representations for disease prediction, highlighting that attention over graph neighbors can improve interpretability [4]. HealthGAT proposed a hierarchical GAT architecture that learns embeddings for medical codes and visits from EHR graphs and showed gains on node classification and readmission prediction tasks [15]. GAT variants have also been used together with ontology trees for medication recommendation, again showing that attention over clinically meaningful graph structure can enhance performance [24].

There is growing interest in graph transformers for EHR. Hypergraph Transformers for EHR-based clinical predictions use hyperedge-based self-attention over sets of medical codes within visits, and achieve strong results on readmission and mortality prediction tasks [23]. Poulain et al. [16] proposed a graph-based time-aware transformer that builds visit embeddings with a graph module over medical codes and then applies a BERT-style model to patient timelines, improving prediction of future diagnoses. These works suggest that transformer-style attention over graph-structured EHR data can capture complex dependencies beyond what local message-passing GNNs see. The Graph Convolutional Transformer (GCT) combines GCN layers over a diagnosis code graph with a transformer over visit sequences, and shows that explicitly learning the graphical structure of EHR codes improves the prediction of future clinical events [5].

Although the application of GNNs to EHR data is not new, recent works [14, 18] mentioned that translating EHRs into meaningful insights is challenging due to the heterogeneity, high-dimensionality, poor-quality, sparse and multimodality present in the data. Unlike temporal or heterogeneous EHR graph models that learn representations at the patient or visit level across longitudinal data, our approach deliberately adopts a diagnosis-centric homogeneous graph abstraction. Pregnancy encounters are collapsed into a co-occurrence network to capture structural comorbidity patterns that are often diluted in patient sequence models. This design complements time-aware EHR graph approaches by offering an interpretable view of interactions between obstetric and non-obstetric diagnoses. In this work, we create and propose a novel benchmark of standalone and hybrid GNN architectures that focuses on predicting the likelihood of obstetric and non-obstetric diagnoses co-occurrence in pregnancy encounters, rather than patient-level outcome prediction or general diagnosis forecasting. We leverage a diagnosis co-occurrence graph, systematically evaluate multiple GNN families, including GCN [10], GraphSAGE [8], GAT [19], and hybrid approaches, in order to identify which architecture choice affects the ability to predict clinically meaningful comorbidity links in pregnancy care. Our proposed model can enable non-obstetric conditions to appear together with pregnancy-related diagnoses and may act as comorbidity hubs. With our reliable and rigorously evaluated approach, we aim to support earlier identification of clinically relevant comorbidity patterns that may inform downstream risk stratification, targeted monitoring, and counseling for patients at risk of multi-morbid pregnancy courses. However, this work is designed to model diagnosis-level co-occurrence patterns within pregnancy-related encounters, rather than to perform patient-level outcome prediction or real-time clinical decision support. Accordingly, the proposed framework is most applicable to retrospective EHR analysis for hypothesis generation, cohort characterization, and risk pattern discovery. It is not intended to replace clinician judgment or establish causal relationships.

2   Problem Setup

To identify the co-occurring obstetric and non-obstetric diagnoses, we define a homogeneous undirected diagnosis graph, $G = (V, E)$ where V defines the set of diagnoses concepts as graph nodes that often appear during pregnancy encounters in addition with preexisting conditions, and E denotes the set of edges or relations between various diagnoses concepts. Next, we define the edges' weight and compute feature vector for each node, described as follows.

2.1   Defining Relation Among Diagnosis Concepts as Edges

Edges encode the co-occurrence of diagnoses within pregnancy encounters. For each pregnancy visit r, we collect the set of unique diagnoses

$$D_r \subseteq V \tag{1}$$

For each unordered pair of diagnosis, $(i, j) \subseteq D_r$, we compute a visit level co-occurrence counter by incrementing by one each time.

$$w_{ij} \leftarrow w_{ij} + 1 \tag{2}$$

After processing all pregnancy visits, we define an undirected edge $(i, j) \in E$ if and only if $w_{ij} \geq 2$. Each edge therefore has an integer weight (w_{ij}) equal to the number of pregnancy encounters in which diagnoses i and j co-occur. Additionally, we define an auxiliary edge distance for path-based measures and community detection

$$d_{ij} = \frac{1}{w_{ij}} \tag{3}$$

such that paths through strongly co-occurring pairs are shorter.

2.2   Defining Diagnosis Concepts as Nodes

Each node $v \in V$ corresponds to a unique diagnosis concept that appears in the dataset by at least ten pregnancy encounters. We define the node features $\mathbf{x}_v \in \mathbb{R}^9$ by considering both graph structural and patient demographic information, discussed as follows:

- **Pregnancy degree.** The unweighted degree of v in the pregnancy graph,

$$k_v = \big|\{u \in V : (v, u) \in E\}\big| \tag{4}$$

- **Pregnancy co-occurrence strength.** Let w_{vu} be the edge weight between v and u. The co-occurrence strength is defined by

$$s_v = \sum_{u \in N(v)} w_{vu} \tag{5}$$

where $N(v)$ is the neighboring nodes of v. This counts how often v co-occurs with any other retained diagnosis in pregnancy encounters.
- **Betweenness centrality in the pregnancy graph.** Using the edge distance definition from Eq. 3, we compute standard betweenness centrality for each concept

$$BC(v) = \sum_{s \neq v \neq t} \frac{\sigma_{st}(v)}{\sigma_{st}} \tag{6}$$

where σ_{st} is the number of shortest paths between s and t and $\sigma_{st}(v)$ counts those that pass through v. This measures how often v lies on shortest co-occurrence paths.
- **Community participation coefficient.** We run Louvain method [13] to extract non-overlapping communities on our defined concept network G. Let $\mathcal{C}$ denotes the set of communities and k_v^c the number of neighbors of v that lie in community $c \in \mathcal{C}$. The participation coefficient is computed as

$$P_v = 1 - \sum_{c \in \mathcal{C}} \left(\frac{k_v^c}{k_v}\right)^2 \tag{7}$$

which is high when v distributes its edges across many communities and low when it is embedded inside a single community.
- **Cross community neighbor share.** Let $\ell(v) \in \{\mathrm{OB}, \mathrm{nonOB}\}$ be the obstetric label of node v. The cross-community neighbor share is

$$\mathrm{CrossShare}_v = \frac{\big|\{u \in N(v) : \ell(u) \neq \ell(v)\}\big|}{|N(v)|} \tag{8}$$

which captures the proportion of neighbors that belong to the opposite class (obstetric vs non-obstetric).
- **Mean age at diagnosis in pregnancy.** The mean age of patients at which diagnosis v appears in pregnancy encounters.
- **Age variability at diagnosis in pregnancy.** Standard deviation of patient age when the diagnosis is recorded.
- **Obstetric concept indicator.** Binary feature indicator of whether the diagnosis belongs to the obstetric vocabulary.
- **Bridge neighborhood indicator.** Binary feature equal to 1 if the diagnosis has at least one obstetric neighbor and at least one non-obstetric neighbor.

All continuous features (items 1–7) are standardized globally to have a mean of zero and unit variance. The last two remain binary indicators. This feature design allows the models to combine global graph structure, community connectivity, and demographic information when predicting which potential edges are likely to exist.

2.3   Obstetric and Non-obstetric Diagnoses Co-occurrences

We define the prediction task of obstetric (OB) and non-obstetric (nonOB) diagnoses co-occurrences as the link prediction problem over the set of possible obstetric–non-obstetric pairs. Positive edges are the set of observed cross-community edges, defined as

$$E_{\text{OB-nonOB}} = \{(i,j) \in E : \ell(i) \neq \ell(j)\} \tag{9}$$

And, the negative edges are between concept pair (i,j) such that

$$\ell(i) \neq \ell(j) \qquad \text{and} \qquad (i,j) \notin E. \tag{10}$$

The full graph G, including obstetric–obstetric and non-obstetric–non-obstetric edges, provides the structural context used by the GNN encoders.

3   Dataset and Preprocessing Steps

We utilize the EHRShot sampled EHR dataset [21] $(2,000$ patients in OMOP format) and focused on conditions recorded in the `sampled_condition_occurrence` file table. We kept only rows with non-null `person_id`, `visit_occurrence_id`, and `condition_concept_id`, then created string identifiers for patient, visit, and diagnosis. Empty diagnosis codes were removed. Then we apply several global and local filters.

As a first global filter, we required that a diagnosis concept appear at least 20 times in the full dataset. This removed extremely rare codes that would produce unstable statistics. We also capped the vocabulary to the $2,000$ most frequent diagnoses by global count, although the pregnancy graph became much smaller after subsequent restrictions.

Pregnancy-related concepts were identified in `concept.csv` by searching condition names for a curated list of substrings such as "pregnan", "obstetric", "prenatal", "antepartum", "postpartum", "gestational", "preeclampsia", and "labor". The corresponding concept IDs define the set of obstetric diagnoses. A visit was considered a pregnancy encounter if at least one of its diagnoses belonged to this obstetric set. All subsequent graph construction is restricted to these pregnancy encounters.

Within the pregnancy cohort, we applied a second level of filtering. A diagnosis had to appear in at least 10 pregnancy encounters to be retained as a node, and a pair of diagnoses had to co-occur in at least two distinct pregnancy encounters to be kept as an edge. These explicit thresholds balance two goals: removing noisy one-off co-occurrences and keeping enough connectivity for meaningful community structure and centrality measures.

Demographic information was added by joining the filtered condition records with `sampled_person.csv` on `person_id`. We used `year_of_birth` and the condition date (when available) to compute age at each diagnosis event. Implausible ages (negative or greater than 120) were set to missing. For each diagnosis, we then computed the mean and standard deviation of age across all pregnancy encounters where it appeared. Missing age statistics were imputed with the global mean age (for the mean) and zero (for the standard deviation).

All continuous node features were standardized with a global z-score transform before being saved. Every model in the project, including GCN, Graph-SAGE, GAT, and the hybrid architectures, loads exactly this preprocessed graph and uses the same train, validation, and test splits for link prediction.

After all filtering and preprocessing, the pregnancy diagnosis co-occurrence graph used for training contains 279 unique diagnosis concepts as nodes, $5,285$ undirected co-occurrence edges and 9 features per node. Based on the obstetric label of each diagnosis, 124 nodes (44.4 %) are obstetric, and 155 nodes (55.6 %) are non-obstetric. The edge set partitions as: (i) Obstetric–non-obstetric edges: $2,717$ edges (51.4 %), (ii) Obstetric–obstetric edges: $1,628$ edges (30.8 %), and (iii) Non-obstetric–non-obstetric edges: 940 edges (17.8 %).

Data Splits. We focus on edges between obstetric and non-obstetric diagnoses:

- Collect all observed cross-community edges and randomly split them into 70% train, 15% validation and 15% test positives.
- For each split, sample an equal number of negative pairs between obstetric and non-obstetric nodes that are not directly connected in the graph.
- Validation and test positive edges are removed from the adjacency matrix used for message passing to avoid label leakage.

The same splits are reused across all architectures.

4   Methodology

We perform binary link prediction on obstetric and non-obstetric diagnosis pairs, using observed cross-community edges as positives and randomly sampled unconnected pairs as negatives, with fixed splits reused across all architectures. We consider three GNNs and three hybrid architectures in model development process that follow the same encoder-decoder structure. First, we briefly discuss the GNN architectures that computes the nodes representation from the graph, followed by the decoder as the prediction module.

4.1   Graph Convolutional Network (GCN)

The GCN uses three graph convolutional layers with hidden dimension 128, batch normalization, and dropout. At layer ℓ, node representations $\mathbf{H}^{(\ell)}$ are updated by

$$\mathbf{H}^{(\ell+1)} = \sigma\left(\tilde{\mathbf{D}}^{-\frac{1}{2}}\tilde{\mathbf{A}}\tilde{\mathbf{D}}^{-\frac{1}{2}}\mathbf{H}^{(\ell)}\mathbf{W}^{(\ell)}\right) \tag{11}$$

where $\tilde{\mathbf{A}}$ is the adjacency matrix with self-loops, $\tilde{\mathbf{D}}$ is its degree matrix, $\mathbf{W}^{(\ell)}$ is a learnable weight matrix and σ is a ReLU nonlinearity. This corresponds to a degree-normalized neighborhood average that smooths features over local pregnancy co-occurrence structure.

4.2   Graph Attention Network (GAT)

The GAT model uses two attention layers with four heads, exponential linear unit (ELU) activations, batch normalization, and dropout. It replaces uniform aggregation with learned attention weights. For one attention head, coefficients between node v, neighbor u and the update are

$$\alpha_{vu} = \text{softmax}_u\left(\mathbf{a}^\top[\mathbf{W}\mathbf{h}_v \,\|\, \mathbf{W}\mathbf{h}_u]\right) \tag{12}$$

$$\mathbf{h}'_v = \sigma\left(\sum_{u\in N(v)} \alpha_{vu}\,\mathbf{W}\mathbf{h}_u\right) \tag{13}$$

Multiple heads are combined (first by concatenation, then by averaging in the last layer) before the decoder. This allows the model to focus on specific comorbid diagnoses that are more informative for predicting cross-obstetric vs non-obstetric links.

4.3   GraphSAGE

The GraphSAGE model also uses three layers with a hidden size of 128, but replaces spectral smoothing with an explicit aggregation of neighbors features. For a node v at layer ℓ,

$$\mathbf{m}_v^{(\ell)} = \frac{1}{|N(v)|}\sum_{u\in N(v)} \mathbf{h}_u^{(\ell)} \tag{14}$$

$$\mathbf{h}_v^{(\ell+1)} = \sigma\left(\mathbf{W}^{(\ell)}[\mathbf{h}_v^{(\ell)} \,\|\, \mathbf{m}_v^{(\ell)}]\right). \tag{15}$$

where $N(v)$ is the set of neighbors and $\|$ denotes concatenation. This preserves the node's own representation while injecting a learned summary of its neighborhood, which can capture heterogeneous local patterns between different diagnosis communities.

4.4   Hybrid GCN + GAT

The first hybrid model runs a GCN branch and a GAT branch in parallel on the same graph. The GCN branch applies a stack of GCNConv layers with ReLU and dropout between layers. In contrast, the GAT branch applies a stack of GATConv layers with multi-head attention using a nonlinear transformation through ELU and dropout between layers. Importantly, both branches are configured to produce the same final embedding dimension (`out_channels`), after which layer normalization is applied separately to the GCN and GAT outputs to stabilize their scales before fusion. For each node, the two normalized branch embeddings are then fused either by (i) concatenation followed by a linear projection back to `out_channels` (`projected_concatenation`), or (ii) a learnable gated weighted sum (`gated_sum`) that adaptively balances the contribution of the GCN and GAT representations per node. Conceptually, the GCN part captures stable, smoothed structure, whereas the GAT part introduces attention-driven reweighting of neighbors, and the updated fusion ensures both signals are combined in a controlled way rather than relying on raw concatenation alone.

4.5   Hybrid GAT + GraphSAGE

The third hybrid model combines a GAT-based encoder and a GraphSAGE-based encoder applied to the same diagnosis co-occurrence graph, with both components operating directly on the original node features. The GAT component models attention-weighted interactions among neighboring diagnoses, allowing the model to emphasize more informative connections, whereas the GraphSAGE component applies mean aggregation to capture robust local neighborhood statistics inductively. The outputs of the two encoders are normalized to control their relative scales and then fused into a single node representation through a learned combination mechanism, enabling the model to balance attention-driven selectivity with stable neighborhood aggregation. The resulting fused embedding is used for link prediction via the same dot-product (Hadamard-based) decoder as in the other architectures.

4.6   Hybrid GCN + GraphSAGE

The second hybrid model applies a GCN-based encoder and a GraphSAGE-based encoder to the same diagnosis co-occurrence graph, with both components operating on the original node features. The GCN component captures globally smoothed structural context through multiple convolution layers, while the GraphSAGE component uses mean aggregation to model localized neighborhood information. The outputs of the two encoders are normalized and then fused into a single node representation using a learned combination mechanism, ensuring that neither global smoothing nor local aggregation dominates the final embedding. The resulting representation is used for link prediction with the same dot-product (Hadamard-based decoder) as above. This design combines complementary inductive biases, allowing the model to exploit both broad graph structure and flexible neighborhood aggregation.

4.7   Decoder as the Prediction Module

Given node embeddings $\mathbf{z}_i, \mathbf{z}_j \in \mathbb{R}^d$, the decoder computes a logit, followed by a sigmoid function, and predicts co-occurrence likelihood of obstetric and non-obstetric diagnoses.

$$s_{ij} = \sigma(\mathbf{w}^\top (\mathbf{z}_i \odot \mathbf{z}_j)) \tag{16}$$

$$\hat{y}_{ij} = \mathbf{1}\big[s_{ij} \geq 0.5\big]. \tag{17}$$

If $\hat{y}_{ij} = 1$, the model considers that obstetric and non-obstetric diagnosis to form a clinically meaningful bridge within pregnancy care. We train the models using binary cross-entropy loss over logits given the ground-truth binary labels $y_{ij} \in \{0, 1\}$.

$$L = -\sum_{(i,j)} \Big(y_{ij} \log s_{ij} + (1 - y_{ij}) \log(1 - s_{ij})\Big) \tag{18}$$

We report area under the receiver operating characteristic curve (AUROC) and area under the precision-recall curve (AUPRC) as primary means of co-occurrence prediction metrics, with F1-score, Matthews correlation coefficient (MCC), accuracy, sensitivity and specificity.

5   Results

5.1   Overall Performance Analysis

Table 1 summarizes the overall results of all considered models in terms of four well-known performance metric: AUROC, AUPRC, F1, and MCC. We report mean $\pm$ standard deviation over randomly selected 5-fold cross validation sets.

We first attempt to understand the efficacy of the standalone architectures, such as GCN, GAT and GraphSAGE. Among these, GraphSAGE outperforms GCN and GAT in all metrics, particularly by an average of 5% and 2%, respectively, when comparing AUROC. Later hybrid models demonstrate that GraphSAGE is the core component of developing an effective model for predicting obstetric and non-obstetric diagnoses co-occurrences. Additionally, GCN results poor performance among these set of models, especially GCN drops to 0.134 ± 0.051 of MCC compared with GraphSAGE's 0.621 ± 0.025.

Next, we studied the hybrid architectures. GCN+GraphSAGE outperforms all hybrid and standalone methods across all performance metrics. Hybrid GAT +GraphSAGE model performs closely compared with our best performer. We want to highlight that hybrid GCN+GAT performs worst among all considered methods. The strong influence of GraphSAGE on the top-performing hybrid models, as well as the standalone GraphSAGE model, suggests that on this moderately sized and relatively dense diagnosis graph, neighborhood aggregation is particularly effective at capturing the underlying predictive structure. While GAT achieves competitive results, it lags behind GraphSAGE, indicating that attention-based mechanisms alone are somewhat less effective than

Table 1. Obstetric and non-obstetric diagnoses co-occurrences prediction performance across models (mean±SD over 5-fold cross-validation).

Model	AUROC	AUPRC	F1	MCC
GCN	0.835 ± 0.023	0.836 ± 0.027	0.676 ± 0.006	0.134 ± 0.051
GAT	0.860 ± 0.015	0.853 ± 0.019	0.790 ± 0.017	0.560 ± 0.039
GraphSAGE	0.889 ± 0.013	0.880 ± 0.018	0.812 ± 0.010	0.621 ± 0.025
GCN+GAT	0.824 ± 0.011	0.810 ± 0.015	0.760 ± 0.006	0.465 ± 0.017
GAT+GraphSAGE	0.889 ± 0.013	0.883 ± 0.015	0.811 ± 0.009	0.614 ± 0.020
GCN+GraphSAGE	**0.897** ± 0.012	**0.894** ± 0.016	**0.817** ± 0.012	**0.636** ± 0.025

aggregation-based inductive biases in this setting. In contrast, the standalone GCN model performs the worst among the single-GNN family, suggesting that spectral smoothing alone is not enough to fully utilize the graph structure. Overall, the results support the idea that combining complementary inductive biases-such as GCN-style smoothing with GraphSAGE-style aggregation gives the most reliable prediction of obstetric and non-obstetric diagnoses co-occurrences.

5.2 Precision–Recall and Receiver-operating Curve Analysis

The precision–recall curves and ROC curves in Fig. 1(A,B) provide a more detailed view of ranking quality across thresholds. All models achieve AUPRC values well above the expected baseline for a balanced test set, confirming that they rank true cross-community links above negatives most of the time. Graph-SAGE shows the strongest precision-recall curve among the standalone baselines, maintaining precision above 0.8 across a wide range of recall, followed by GAT, while GCN performs the worst. The hybrid GCN+GraphSAGE achieves the best AUPRC, with a curve that stays close to the upper left corner, meaning it can recover a large fraction of true obstetric and non-obstetric co-occurrence links while retaining high precision. The hybrid GAT+GraphSAGE performs similarly, with only a slight drop in precision at very high recall. In contrast, the standalone GAT curve drops earlier as recall increases, which reflects its difficulty in distinguishing borderline positive edges from negatives.

5.3 Error Patterns from Confusion Bars

Figure 1(C) presents averaged confusion bars on the held out test set across 5-fold cross validation. Most models, except GCN, achieved high true negative counts, confirming that they are effective at rejecting non-existent obstetric and non–obstetric diagnoses co-occurrence links. The standalone GCN did not perform well, as it rendered very few true negatives and significantly higher false positives as compared to other models, which aligns with its lower F1 and MCC. The Hybrid GCN+GAT, although not as bad as the standalone GCN model, did render fewer true negatives and more false positives compared to its other

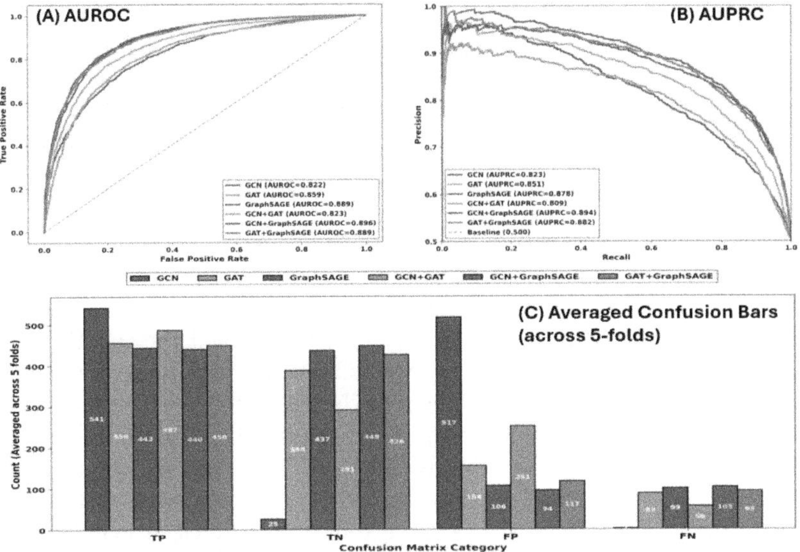

Fig. 1. Overall performance analysis of the GNN models. (A) Area under the received operating curve (AUROC) and (B) area under the precision-recall curve (AUPRC) on the test set. (C) Confusion bars averaged over 5-fold cross-validation test sets. TP: true positive, TN: true negative, FP: false positive, FN: false negative.

better-performing counterparts, which corresponds to its lower F1 and MCC as well. The hybrid GCN+GraphSAGE had the overall best confusion bar pattern as it identified the most true negatives and the fewest false positives, although having slightly higher false negatives as compared to other counterparts. Hybrid GAT+GraphSAGE followed behind, reducing false negatives at the cost of a modest increase in false positives. The pure GAT model displays more false positives but slightly fewer false negatives than the Hybrid GCN+GraphSAGE model. In qualitative terms, the top two hybrid models are better at identifying clinically meaningful bridges without dramatically inflating the number of spurious links, which is important if the predictions are to be used as candidates for risk stratification or hypothesis generation in pregnancy care.

The ROC curves tell a consistent story. All models dominate the diagonal random baseline, but the top two hybrid architectures show the largest area under the curve and stay closest to the upper left corner, indicating a better trade-off between sensitivity and specificity across thresholds. Taken together, the table and plots demonstrate that while GraphSAGE is already a strong single-architecture baseline, GCN performs comparatively worse, and hybrid models that combine aggregation with attention or stacked aggregation deliver the best link prediction performance on the pregnancy diagnosis co-occurrence graph. This curve trend is further supported by the validation AUROC trajectories, where GCN shows an early plateau and limited improvement with continued training, suggesting over-smoothing and reduced discriminative capacity in dense neighborhoods. In contrast, GraphSAGE and GraphSAGE-based hybrids continue to improve and converge to higher AUROC, indicating that explicit neighborhood aggregation better preserves local structural signals, while hybridization mitigates the limitations of spectral smoothing alone.

5.4 Top Predicted Obstetric and Non-obstetric Diagnoses Co-occurrences

To characterize the most salient co-occurrences between obstetric and non-obstetric diagnoses, we applied the best-performing hybrid GCN+GraphSAGE model to all non-existing cross-community pairs, retained the top 10 links per fold according to the predicted edge probabilities, and then aggregated predictions across the 5-folds using a ranking score based on the mean predicted probability multiplied by a recurrence term, mean_probability $\times$ log(num_folds + 1), which favors edges that both achieve high confidence and consistently appear across multiple folds. Table 2 reports the 10 co-occurrence edges with the highest mean scores. All pairs in the table have mean predicted probabilities above 0.999, which indicates that the model is extremely confident that these diagnosis pairs behave like true co-occurring links in the pregnancy graph, despite not being observed in the original data.

The top-ranked connections highlight clinically plausible bridges involving obstetric complications and systemic medical conditions. Many of the highest scoring pairs link endocrine and metabolic disorders, particularly "hypothyroidism" and "type 2 diabetes mellitus without complication", to trimester-specific pregnancy diagnoses and broader obstetric risk labels such as "disorder of pregnancy" and "high-risk pregnancy". These links reflect the fact that chronic conditions are frequently documented along pregnancy-related stage or risk stratification codes, rather than being encoded as a single combined diagnosis.

Several additional high-scoring connections associate pregnancy stage codes with "anemia", "maternal and/or fetal conditions affecting labor and delivery", and other pregnancy-related findings. From a clinical perspective, these predictions suggest that the model is capturing recurring documentation patterns in which systemic conditions and obstetric status are recorded in parallel across encounters, even when their interaction is implied rather than explicitly specified. In the context of our research question, the table provides concrete candidate bridges where non-obstetric diagnoses may coincide with pregnancy progression or delivery-related risk, and it highlights specific diagnosis pairs that could warrant targeted chart review and longitudinal analysis in future obstetric risk stratification studies.

Table 2. Top 10 Predicted Cross-Community Diagnosis Associations Between Obstetric and Non-Obstetric Conditions Using hybrid GCN+GraphSAGE Model with 5-fold cross-validation.

Rank	node 1 name	node 2 name	Mean score
1	Hypothyroidism	Disorder of pregnancy	0.9999
2	High risk pregnancy	Elderly primigravida	0.9999
3	Suspected fetal abnormality affecting management of mother	High risk pregnancy	0.9999
4	Single live birth	Endocrine, nutritional and metabolic disease complicating pregnancy, childbirth and puerperium	0.9999
5	Hypothyroidism	Third trimester pregnancy	0.9998
6	Hypothyroidism	Second trimester pregnancy	0.9997
7	Uterine leiomyoma	Disease of the digestive system complicating pregnancy, childbirth and/or the puerperium	0.9997
8	Third trimester pregnancy	Anemia	0.9995
9	Abnormal findings on antenatal screening of mother	Disorder of pregnancy	0.9995
10	Abnormal findings on antenatal screening of mother	Finding related to pregnancy	0.9995

6   Conclusion

In this work, we formulate the obstetric and non-obstetric diagnosis co-occurrence in pregnancy as a link prediction problem on a diagnosis graph and show that GNNs can accurately rank clinically meaningful comorbidity links from noise. Using a homogeneous diagnosis co-occurrence graph enriched with structural and demographic node features, most models achieved strong discrimination between true and negative cross-community pairs except the standalone GCN model. Particularly, hybrid GCN+GraphSAGE and GAT+GraphSAGE encoders reach the highest performance and identify top-ranked links that align with endocrine, metabolic, and hematologic conditions co-occurring with pregnancy stage and risk-related diagnoses, catching clinically plausible patterns that are often under-recorded within single visits. These results suggest that graph-based link prediction is a useful tool to generate prioritized lists of obstetric and non-obstetric diagnosis pairs that may influence pregnancy trajectories and postpartum recovery.

In downstream clinical workflows, the proposed model could serve as a retrospective screening and prioritization layer rather than a diagnostic tool. Predicted high-confidence pairs of obstetric-non-obstetric diagnosis may be used to identify comorbidity patterns for targeted chart review, guide risk-stratified monitoring during prenatal care, or inform the design of rule-based alerts within existing EHR systems. Importantly, such integration would occur offline or asynchronously, supporting clinicians by detecting plausible diagnosis associations rather than generating automated clinical decisions.

This study has several limitations that open up directions for future work. Our experiments rely on a single sampled EHR dataset, focus exclusively on diagnosis codes within pregnancy-related encounters, and treat co-occurrence within a visit as the only notion of relatedness, without modeling temporal order, severity, treatment, or patient-level outcomes. The link predictions are observational and should be interpreted as hypotheses rather than evidence of causal relationships, fairness across patient subgroups, or prospective utility at the point of care. Future work should validate these findings in larger and more diverse cohorts, incorporate medications, procedures, and laboratory results into heterogeneous or temporal graphs, explore graph transformers and uncertainty-aware models, and connect diagnosis-level link prediction to patient-level risk scores that could be evaluated in collaboration with clinicians in real obstetric surveillance workflows.

References

1. Aoyama, K., D'Souza, R., Pinto, R., et al.: Risk prediction models for maternal mortality: a systematic review and meta-analysis. PLoS ONE **13**(12), e0208563 (2018)
2. Bertini, A., Salas, R., Chabert, S., Sobrevia, L., Pardo, F.: Using machine learning to predict complications in pregnancy: a systematic review. Front. Bioeng. Biotechnol. **9** (2022)
3. Cho, H.N., Ahn, I., Gwon, H., et al.: Heterogeneous graph construction and hinsage learning from electronic medical records. Sci. Reports **12**(1) (2022)
4. Choi, E., Bahadori, M.T., Song, L., et al.: Gram: graph-based attention model for healthcare representation learning. In: Proceedings of the 23rd ACM SIGKDD International Conference on Knowledge Discovery and Data Mining.,pp. 787–795. KDD '17, Association for Computing Machinery, New York, NY, USA (2017)
5. Choi, E., Xu, Z., Li, Y., et al.: Learning the graphical structure of electronic health records with graph convolutional transformer (2020)
6. Fu, S., Leung, L.Y., Raulli, A.O., et al.: Assessment of the impact of EHR heterogeneity for clinical research through a case study of silent brain infarction. BMC Med. Inf. Decision Making **20**(1) (2020)
7. Giaxi, P., Vivilaki, V., Sarella, A., et al.: Artificial intelligence and machine learning: an updated systematic review of their role in obstetrics and midwifery. Cureus (2025)

8. Hamilton, W.L., Ying, R., Leskovec, J.: Inductive representation learning on large graphs (2018)
9. Jung, J., Choe, M., Agarwal, K., Dhanasekaran, N.: GraphEHR: heterogeneous graph neural network for electronic health records. In: AI for Critical Infrastructure Workshop @ IJCAI-24 (2024)
10. Kipf, T.N., Welling, M.: Semi-supervised classification with graph convolutional networks (2017)
11. Meredith, M.E., Steimle, L.N., Stanhope, K.K., et al.: Racial/ethnic differences in pre-pregnancy conditions and adverse maternal outcomes in the numom2b cohort: a population-based cohort study. PLoS ONE **19**(8), e0306206 (2024)
12. Mohamed Dkeen, N.O., Dawelbait Radwan, M.E., Alnaw Zumam, I.A., et al.: Artificial intelligence applications in obstetric risk prediction: a systematic review of machine learning models for preeclampsia. Cureus (2025)
13. Muche, A.A., Baruda, L.L., Pons-Duran, C., et al.: Prognostic prediction models for adverse birth outcomes: a systematic review. J. Global Health **14** (2024)
14. Oss Boll, H., Amirahmadi, A., Ghazani, M.M., et al.: Graph neural networks for clinical risk prediction based on electronic health records: a survey. J. Biomed. Inform. **151**, 104616 (2024)
15. Piya, F.L., Gupta, M., Beheshti, R.: Healthgat: node classifications in electronic health records using graph attention networks. In: 2024 IEEE/ACM Conference on Connected Health: Applications, Systems and Engineering Technologies (CHASE), pp. 132–141 (2024)
16. Poulain, R., Beheshti, R.: Graph transformers on EHRS: better representation improves downstream performance. In: Kim, B., Yue, Y., Chaudhuri, S., et al. (eds.) International Conference on Representation Learning. vol. 2024, pp. 38492–38508 (2024)
17. Ranjbar, A., Montazeri, F., Ghamsari, S.R., et al.: Machine learning models for predicting preeclampsia: a systematic review. BMC Pregnancy and Childbirth **24**(1) (2024)
18. Si, Y., Du, J., Li, Z., et al.: Deep representation learning of patient data from electronic health records (EHR): a systematic review. J. Biomed. Inform. **115**, 103671 (2021)
19. Veličković, P., Cucurull, G., Casanova, A., et al.: Graph attention networks (2018)
20. Wells, B.J., Nowacki, A.S., Chagin, K., Kattan, M.W.: Strategies for handling missing data in electronic health record derived data. eGEMs (Generating Evidence & Methods to improve patient outcomes) **1**(3), 7 (2013)
21. Wornow, M., Thapa, R., Steinberg, E., Fries, J.A., Shah, N.H.: Ehrshot: an EHR benchmark for few-shot evaluation of foundation models (2023)
22. Xie, F., Yuan, H., Ning, Y., et al.: Deep learning for temporal data representation in electronic health records: a systematic review of challenges and methodologies. J. Biomed. Inform. **126**, 103980 (2022)
23. Xu, R., Ali, M.K., Ho, J.C., Yang, C.: Hypergraph transformers for EHR-based clinical predictions. AMIA Summits Transl. Sci. Proc. **2023**, 582–591 (2023)
24. Yue, W., Zhang, L., Zhang, L., et al.: Med-tree: a medical ontology tree combined with the graph attention networks for medication recommendation. Electronics **11**(21) (2022)
25. Zheng, T., Xie, W., Xu, L., et al.: A machine learning-based framework to identify type 2 diabetes through electronic health records. Int. J. Med. Informatics **97**, 120–127 (2017)

26. Zhu, W., Razavian, N.: Variationally regularized graph-based representation learning for electronic health records. In: Proceedings of the Conference on Health, Inference, and Learning, pp. 1–13. CHIL '21, Association for Computing Machinery, New York, NY, USA (2021)

Classification of a Post-traumatic Stress Disorder-Related Brain Condition Using Dynamic Functional Connectivity Statistics in Functional MRI

Unal Sakoglu[(⊠)] and Amaresh Mishra

University of Houston – Clear Lake, Houston, TX 77058, USA
sakoglu@uhcl.edu

Abstract. We present a dynamic functional connectivity (DFC)-based classification analysis of functional magnetic resonance imaging (fMRI) data from veterans with a type of post-traumatic stress disorder (PTSD) and from matched normal control (NC) veterans. Whole-brain resting-state fMRI (rsfMRI) data which were scanned from 23 PTSD (mean age 49) and 30 NC (mean age 50) veterans were used for analyses. A computational method using statistics of DFC and support-vector machine (SVM) classifier were used to correctly classify PTSD vs NC with up to 98% accuracy. Results show that, SVM-based machine learning technique, combined with simple t-test method for feature extraction utilizing the standard deviation of DFC, performed significantly better than a convolutional neural network (CNN) based deep learning method in terms of classification accuracy, with around 98% accuracy for the former vs. around 60% accuracy for the latter; it also outperformed static functional connectivity-based classification, which resulted in only up to 72% accuracy.

Keywords: MRI · fMRI · dynamic functional connectivity · classification · traumatic brain injury

1 Introduction

Human brain is highly complex, and while aging and brain disorders are known to affect brain structure and function, specific underlying changes remain not so well understood. A range of neuroimaging technologies enable direct or indirect assessment of the nervous system's structure, function, and physiology, with MRI playing a central role to advance our understanding of brain function and disease [1]. Neuroimaging is broadly classified into structural imaging, which examines nervous system anatomy and aids in diagnosing conditions such as brain tumors, and functional imaging, which investigates brain activity and dysfunction by tracking neural or neurovascular dynamics [2]. Functional magnetic resonance imaging (fMRI) is a non-invasive, blood-oxygenation level-dependent (BOLD) functional neuroimaging technique, known simply as BOLD fMRI or fMRI, which has been employed for over three decades to study brain function, evaluate disease effects, and guide treatments [3].

© The Author(s), under exclusive license to Springer Nature Switzerland AG 2026
K. L. Kabir et al. (Eds.): BICOB 2026, CCIS 2977, pp. 199–220, 2026.
https://doi.org/10.1007/978-3-032-26028-4_15

Extensive fMRI datasets from human and animal studies have been collected under a variety of task and stimulus conditions, alongside the development of advanced analytical methods. This work applies some of these methodologies to previously collected, fully de-identified fMRI data from Gulf War Illness (GWI)-type post-traumatic stress disorder (PTSD) subjects and matched normal control (NC) veterans. The fMRI data were obtained as part of a larger study which was done at UT Southwestern Medical Center in Dallas, TX.

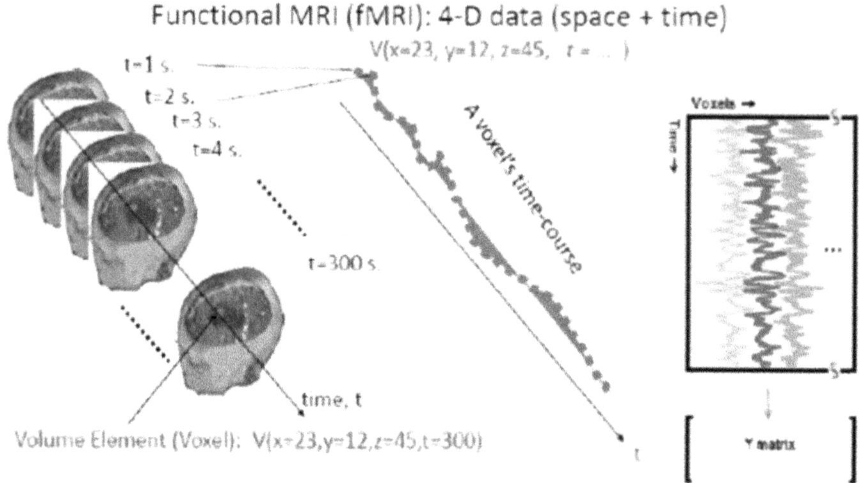

Fig. 1. Visualization of functional MRI (fMRI), 4-D data (3-D space and time) [4].

1.1 Functional Connectivity (FC)

Functional connectivity (FC) describes the functional relationships among brain regions and is defined as the temporal correlation (i.e., Pearson's correlation) between two different brain regions of interest. Two brain regions are considered functionally connected when their activity signals show a statistically significant correlation [5]. FC is studied in both resting-state (i.e. task-free) and task-based fMRI, though rsfMRI-based FC specifically examines correlations across individual BOLD fMRI signals during rest [6].

FC analyses have been widely used to investigate mental health disorders such as bipolar disorder, schizophrenia, addiction, and PTSD, as well as to assess treatment effects [7, 8]. Unlike structural connectivity, which examines physical connections, FC reflects interactions among brain networks that support higher-level cognitive functions [8]. Moreover, functional connectivity is dynamic and can vary over timescales of seconds in fMRI data [9].

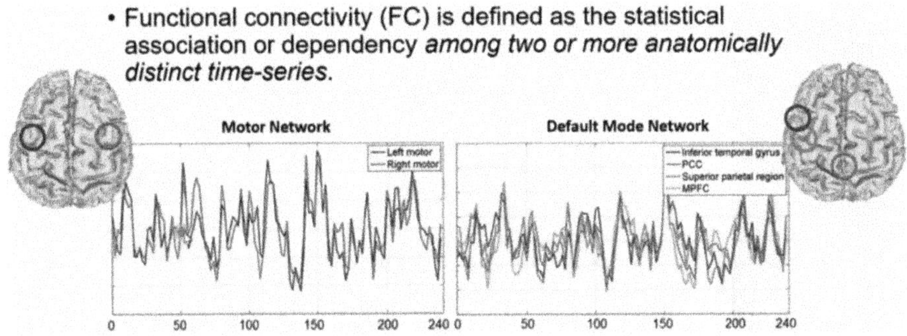

Fig. 2. Pictorial representation of functional connectivity of a 'region-pair' [10].

1.2 Dynamic Functional Connectivity (DFC)

Dynamic functional connectivity (DFC) refers to functional connectivity measured over short time intervals relative to the full fMRI scan, allowing changes in connectivity to be captured across time windows [11–13]. Unlike traditional FC analyses that assume static brain networks, DFC accounts for the temporal variability of functional interactions and has been applied to a wide range of neurological disorders, offering a more realistic representation of brain network behavior [14, 15]. Although DFC is most commonly used with fMRI data, it can also be applied to other time-varying neuroimaging modalities such as EEG and perfusion MRI [16, 17]. The development of DFC was motivated by observed temporal fluctuations in connectivity within steady-state analyses [14, 18, 19]. When connectivity dynamics are computed using brain networks which are derived using methods such as independent component analysis, the approach is often termed dynamic network functional connectivity (DFNC), which follows the same underlying principles as DFC [14].

1.3 Types of Analysis Techniques

Sliding-window DFC analyses. Sliding-window DFC analysis is the most widely used approach for studying functional connectivity dynamics. First introduced by Sakoglu *et al.* in 2009 and applied to schizophrenia, it was termed dynamic functional network connectivity when combined with independent component analysis (ICA) to derive brain networks [15, 18, 19]. The method involves analyzing a fixed number of consecutive fMRI scans, defined as the window length, and then shifting this window forward in time to repeat the analysis. The window shift is typically characterized by the degree of overlap between successive windows [20]. A key advantage of sliding window analysis is its flexibility, as many steady-state connectivity methods can be adapted to this framework when sufficiently large windows are used. The approach is relatively straightforward and intuitive, which aids interpretation [21]. As a result, it has been extensively applied to examine DFC across different contexts, including comparisons between patient and healthy groups, individuals with varying cognitive performance, and distinct large-scale brain states.

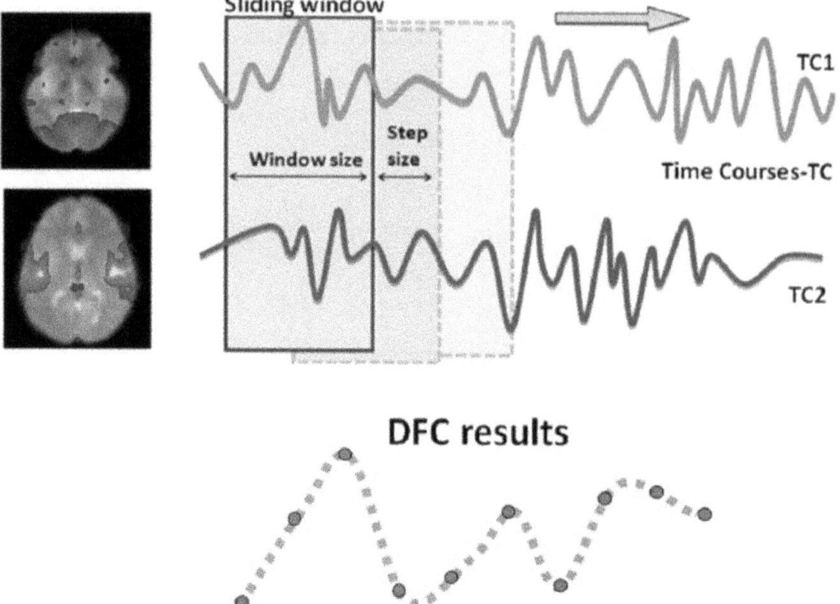

Fig. 3. Dynamic Functional Connectivity (DFC) analysis on two brain regions. The brain regions, each with their fMRI time course, constitute a 'region-pair'. For each pair, and for each time window, one DFC time point is obtained. By sliding the time window, the pair's DFC time course is obtained [20].

Activation Patterns. One of the earliest approaches to analyzing DFC involved pattern analysis of fMRI images, which identifies activation patterns in spatially separated brain regions which exhibit synchronized activity. This reveals a spatial and temporal periodicity in the brain, likely reflecting ongoing fundamental processes [22]. Repeating network patterns are estimated to account for a quarter to half of the variance in fMRI BOLD signals [23–25]. Such activity patterns have been observed as propagating waves along the cortex in rats and are linked to underlying neural activity, with similar patterns also detected in humans [24].

Point process analysis. A recent approach departs from traditional methods by analyzing rapidly changing functional activation patterns through transforming fMRI data into a point process [25, 26]. This involves selecting the fMRI signal's points of inflection (peaks) for each voxel, which capture most functional connectivity information. Despite reducing data size by over 95%, this method produces connectivity inferences comparable to standard full-signal analyses [27, 28]. This high information content aligns with Petridou *et al.* [29], who showed that "spontaneous events" significantly contribute to correlation strength and slow fluctuation power spectra. Building on this, similar strategies have been successfully applied as co-activation patterns (CAP) analyses [30, 31].

Time-frequency. Time-frequency analysis is an alternative to sliding window methods, allowing simultaneous investigation of frequency and amplitude in DFC. Using wavelet transforms, this approach has confirmed the presence of dynamic functional connectivity and revealed that features such as anticorrelation between the default mode network and task-positive networks are transient rather than constant [32]. Independent component analysis (ICA), commonly used to generate steady-state networks, separates fMRI signals into spatially independent components with similar temporal patterns. More recently, temporal ICA has been applied to fMRI data to identify statistically independent components, capturing about 25% of the variability in correlations between anatomical nodes [33].

1.4 Clinical Importance of fMRI Analyses

Dynamic functional connectivity (DFC) is motivated by the brain's highly dynamic nature, tracking changes in connectivity among networks over time. Both static FC and DFC have been valuable for understanding various disorders, including depression [33], cocaine addiction [20], schizophrenia [34], and Alzheimer's disease [35]. For instance, Alzheimer's patients show altered network connectivity and changes in time spent in certain networks [35]. While DFC changes do not imply causation, they can aid in understanding disease effects and improving diagnosis.

Exploratory ICA-based FC analysis in veterans with GWI-type PTSD (simply, GWI-PTSD) versus matched normal controls revealed impaired connectivity between functional networks, supporting the idea that GWI-type is a different brain disorder [36]. GWI PTSD affects roughly 250,000 of the 700,000 U.S. veterans deployed in the 1991 Gulf War, causing multi-symptom deficits across cognitive, affective, sensory, and motor domains [36]. Using previously collected, de-identified rsfMRI data, deep neural networks and DFC methods were applied to identify discriminative brain networks between GWI-PTSD and controls. Prior studies showed impaired FC in GWI-PTSD veterans between language, sensory, motor, and sensorimotor networks, providing a potential mechanism for central nervous system dysfunction [36]. These network features were then used for deep neural net and SVM-based classification.

2 Background Work

About 200,000-to-250,000 of the veterans deployed in the 1991 Persian Gulf War (GW) are estimated to suffer from a type of PTSD known as GW illness (GWI) [37]. Post-war, GW veterans developed certain signs or symptoms, with a higher-than-average rate, which cannot be explained by any specific medical problems. These symptoms include sustained and debilitating fatigue, sleeping difficulty, migraines, headache, problems with memory and cognition, and digestive ailments [37].

2.1 Participant Demographics and Characteristics

Participant demographics and additional clinical features are summarized in Table 1 [36]. As shown in the table, there were no statistically significant differences in age or years of education between the GWI-PTSD and NC groups. The demographic data presented correspond to the full dataset, in which GWI-PTSD participants from three symptom categories were pooled. However, the present study focused exclusively on the

"Symptom-2" (S2) subgroup (n = 23), consisting of GWI-PTSD patients reporting the highest symptom severity. In this work, now on we will simply refer to our GWI-PTSD S2 patient group as GWI-PTSD.

In this study, de-identified resting-state fMRI (rsfMRI) data from GWI-PTSD veterans were analyzed using deep neural networks, support vector machines, and DFC methods to identify brain networks that distinguish GWI-PTSD patients from matched controls. Previous research showed impaired functional connectivity in GWI-PTSD veterans across language, sensory, motor, and sensorimotor networks, consistent with self-reported symptoms [36]. These impairments suggest central nervous system dysfunction. Using fMRI data from these networks, features were constructed for neural network and SVM classification, along with DFC-based windowed correlation analyses over brain signal time courses [38].

Table 1. Demographics and clinical characteristics of 60 GWI-PTSD patients and 30 NC.

Demographics and Clinical Characteristics	GWI-PTSD (60 participants)	NC (30 participants)
Age in Years	50.1 ± 8	50.0 ± 8
Education in Years	5.2 ± 2	5.3 ± 2
Gender(F/M)	15	6
Right-handed in sample	57	29
CDC GWI-PTSD Case Definition	60	0
Modified Kansas Case Definition	60	0
Chronic Fatigue Syndrome	6	0
Fibromyalgia	34	0
PTSD	24	0
Other Mood Disorders	40	3

2.2 Preprocessing of Data

The rsfMRI time-series were shifted temporally for slice acquisition time correction, 3-D volume-registered to a base volume to correct global rigid motion, spatially smoothed with an isotropic Gaussian filter with FWHM of 6 mm, and detrended of motion-related and physiological artifacts using a technique known as ICA-AROMA [46]. These corrected rsfMRI time-series were aligned to the MNI152 template by applying the affine transformation parameters that are computed from the registration of unsmoothed EPI time-series to MNI152 template; and resampled to 3 mm × 3 mm × 3 mm voxel resolution.

2.3 DFC Analysis Between Brain Regions

We have calculated a 2-D (region-by-region) FC matrix for each participant, a 3-D (region-by-region-by-time-window) DFC matrix for each participant, and, the mean and the standard deviation of the DFC (avgDFC and stdDFC) across participants in each group, NC and GWI-PTSD, and used these as initial features in the classification models.

3 Methods

Dynamic functional connectivity (DFC) is a recently developed fMRI analysis method that captures temporally evolving interactions among brain networks [11]. DFC generates a large number of features, which can be leveraged by machine learning and deep learning algorithms. In this study, we applied DFC analysis to pre-processed (includes motion correction, noise filtering, slice-time corrections, co-registration and transformation/normalization to common MNI template brain), de-identified fMRI data from GWI-PTSD [36], extracting DFC features (mean and standard deviation) and using them in machine learning models, including deep neural networks, to classify subjects and identify the most discriminative brain networks.

Analyses were implemented primarily in MATLAB, with GWI-PTSD data converted into 4D MATLAB-readable (.mat) files, categorized into matched controls, Syndrome 1, Syndrome 2, and Syndrome 3 groups. This work focused on comparisons between controls ('group 0', NC) and GWI-PTSD Syndrome 2 ('group 2', S2), whose symptoms are more uniform. The dataset is 4D (participants × brain regions × brain regions × time windows), totaling nearly half a million features. Window size was thirty-two and step size was eight, with no tapering. The high feature-to-sample ratio poses challenges for classification, which we addressed using CNN and SVM with static FC and dynamic DFC features. Details of these methods are described in the following subsections.

3.1 Classification Using CNN

A brain region-of-interest (ROI) is a 3D shape which represents a specific brain region. Each ROI is fed through a neural network to produce output features [39]. After receiving the input data (image), training and test data were prepared using the training options available in MATLAB [40]. Table 2 lists three sets of training options ('opts') along with the training parameter values which were selected during this process using MATLAB's Machine Learning and Statistics toolboxes [40]. Many other training sets of parameters have been tried and evaluated; however, the list in Table 2 contains only three sample sets, which have been ultimately used to obtain the results presented in this work. Once training options are set, and all the training and test data sets are created, a convolutional neural network (CNN) was created. A CNN has multiple layers, where each layer defines a specific computation of weight parameters. MATLAB Machine Learning and Deep Learning toolboxes [40] were used to design the CNN layers. The details of the CNN architecture, its input, middle and final layers' details, are listed in Table 3. Once all the parameters are set then the value of those parameters is provided to the trainNetwork algorithm in MATLAB [40], classification accuracy results along with the progress and time elapsed to provide the results are generated.

Table 2. Training options ("opts") of three different sets of parameter values.

Training Options	Param. Set 1	Param. Set 2	Param. Set 3
Training Algorithm	sgdm	sgdm	sgdm
Momentum	0.9	0.95	0.95
InitialLearnRate	0.01	0.015	0.015
LearnRateSchedule	piecewise	piecewise	piecewise
LearnRateDropFactor	0.1	0.1	0.1
LearnRateDropPeriod	8	8	8
L2Regularization	0.004	0.004	0.004
ValidationFrequency	15	15	15
MaxEpochs	20	20	24
MiniBatchSize	9	9	9

3.2 Classification Using SVM

In this approach, the learning algorithm classifies data by mapping it into a high-dimensional feature space, enabling separation even when the original data are not linearly separable. A separating hyperplane is then identified for classification [41].

For the 4D GWI-PTSD DFC dataset, the standard deviation across the two groups is first calculated. A t-test (ttest2) is then applied to reduce features, retaining only those significant for classification. The selected features are input into MATLAB's fitcsvm function, using zero-mean unit standardization, a radial basis function (RBF) kernel, and automatic kernel scaling. The model outputs mean cross-validation accuracy and identifies classifying pairs [40].

Table 3. Details of the CNN architecture's input layer, middle layers and the final layer.

Layer	Details
imageInputLayer	size of the input images for the layer, vector consisting of height width and number of channels [height,width,numChannels, ~] = size(trainingImages) imageSize = [height width numChannels] inputLayer = imageInputLayer(imageSize)
middleLayer1: convolutional2dLayer	a bank of 32 7x7x3 filters; symmetric padding of 2 pixels added to ensure that image borders are included in the processing in order to avoid information at the borders being washed away too early in the network
middleLayer2: reluLayer	rectified linear unit (ReLU), performs a simple threshold operation, where any input value less than zero will be set to zero

(continued)

Table 3. (*continued*)

Layer	Details
middleLayer3: maxPooling2dLayer	layer that performs max pooling, divides the input into rectangular pooling regions, outputs the maximum of each region; poolSize, the width and height of a pooling region, is 3-by-3; stride parameter was set to 2. MaxPooling2dLayer (3, 'Stride',2)
finalLayer1: fullyConnectedLayer	creates a fully connected layer output; multiplies the input by a matrix and then adds a bias vector
finalLayer2: reluLayer	same as above, with appropriate input and output size between the preceding and subsequent layers
finalLayer3: softmaxLayer	scales the output to a probability metric and adds further nonlinearity
finalLayer4: classificationLayer	classification output layer for the neural network, applies the specified loss function that is used for training the network, using the size of the output, and the class labels

4   Results

4.1   Classification Using DFC and CNN Classifier

In this study, multiple approaches were explored to perform classification using CNN. The reported results include methods based on all available features, reduced feature sets using mean and standard deviation, and selected features derived from the AAL Brain Region Atlas [44].

DFC classification results with all available features. All features from the 4D dataset were used for classification, resulting in nearly half a million features. The control (NC) group data had dimensions $30 \times 116 \times 116 \times 37$, while GWI-PTSD group two had dimensions $23 \times 116 \times 116 \times 37$, reflecting their respective subject counts. The training algorithm was run for multiple iterations, with cross-validation (CV) accuracy recorded each time and averaged at the end. The network architecture consisted of one image input layer, nine intermediate layers, and five output layers. Full details of the architecture are deferred to a related manuscript under preparation.

The training progression for the first parameter set in Table 2 is summarized as follows. Across four iterations, validation accuracy varied widely, ranging from about 38% to 67%, with validation loss values between approximately 3.4 and 8.5. The first iteration achieved about 58% accuracy in 220 s, while subsequent iterations took about 90 s each and showed fluctuating performance. After all iterations, the mean cross-validation accuracy was 51%. The same procedure was applied to parameter sets 2 and 3, with results compared in Table 4. Parameter set 3 achieved a similar cross-validation accuracy of 50%, indicating that changes in parameter sets had minimal impact and overall model performance stabilized around 50% accuracy.

Table 4. Comparison of mean cross-validation (CV) accuracy among the three parameter sets.

Parameter set (in Table 2)	Num. Iterations	Mean CV accuracy
1	4	51.42%
2	4	48.85%
3	3	50.63%

DFC classification results with mean and standard deviation. To reduce feature dimensionality, the mean and standard deviation across DFC windows were computed, but this did not improve performance, with accuracy remaining around 50%. Three parameter sets were evaluated. For the first set, four iterations produced cross-validation accuracies of 0.42, 0.46, 0.538, and 0.538, yielding an overall mean of 0.49. Training with mean and standard deviation features required significantly more time than using all features—about 20 additional minutes—without notable accuracy gains. The remaining parameter sets were also evaluated, and their results are summarized in Table 5, showing minimal variation across settings.

Table 5. Comparison of mean cross-validation (CV) accuracy among the three parameter sets using mean and standard deviation across one dimension.

Parameter set (in Table 2)	Num. Iterations	Mean CV accuracy
1	4	51.42%
2	4	48.85%
3	4	50.63%

DFC classification results with selective features (selective brain regions of AAL ATLAS). In this case, selective features from all participants were used to compute classification accuracy. The brain regions chosen for classification are shown in Table 7. These regions were evaluated using three different parameter sets, each run for four iterations, but the resulting accuracies remained near 50%, as summarized in Table 6.

Table 6. Comparison of mean cross-validation (CV) accuracy among the three parameter sets using selective brain regions.

Parameter set (in Table 2)	Num. Iterations	Mean CV accuracy
1	4	49.33%
2	4	51.25%
3	4	48.08%

Table 7. List of selected brain regions used for calculating classification accuracy in Table 6.

AAL brain region label	AAL brain region name
19	Supp_Motor_Area_L 2401
20	Supp_Motor_Area_R 2402
47	Lingual_L 5021
48	Lingual_R 5022
81	Temporal_Sup_L 8111
82	Temporal_Sup_R 8112
85	Temporal_Mid_L 8201
86	Temporal_Mid_R 8202

4.2 Classification Results Using FC and CNN Classifier

It was necessary to carry out a comparable classification analysis on the FC dataset in order to contrast the results with those from DFC. Initially, classification accuracies were computed using all available features. Next, the mean and standard deviation across one regional dimension were calculated and applied to CNN-based methods. Finally, selective brain regions listed below were used. Throughout this process, accuracies were evaluated using the parameter sets described in Tables 2 and 3.

FC classification results utilizing all the features. The FC dataset constitutes a 3D matrix (number of object vs number of brain regions vs number of brain regions). During this computation, we utilized all the features and trained the network with the same algorithms discussed in DFC with same parameter sets. The results are listed in Table 8.

Table 8. Comparison of mean cross-validation accuracy utilizing all features for FC.

Parameter set (in Table 2)	Num. Iterations	Mean CV accuracy
1	4	45.92%
2	4	43.41%
3	4	54.08%

FC classification results utilizing mean and standard deviation. As in the prior subsection, the mean and standard deviation across the brain regions were calculated, and the classification cross-validation accuracy and results were obtained for each parameter set are shown below in Table 9.

Table 9. Comparison of mean CV accuracy utilizing mean and standard deviation, for FC.

Parameter set (in Table 2)	Num. Iterations	Mean CV accuracy
1	4	60.62%
2	3	40.02%
3	4	52.15%

FC classification results utilizing selective brain regions of AAL ATLAS. As noted earlier, this analysis used selective brain regions that were chosen based on a previous study [36]. Only a few brain regions resulted in the selection, and the corresponding results are presented in Table 10 below.

Table 10. Comparison of mean CV accuracy with selective brain regions, for FC.

Parameter set (in Table 2)	Num. Iterations	Mean CV accuracy
1	4	53.85%
2	4	52.56%
3	4	50.00%

4.3 Comparison of DFC Versus FC Using CNN

Using CNN, the accuracy results in both FC and DFC cases were around 50%, as summarized in the previous two subsections. Even when the number of iterations increased to one hundred, the classification accuracy remained around 50%. Overall, no definitive conclusions can be drawn from the CNN-based results on either FC or DFC.

4.4 Classification Results Using DFC and SVM Classifier

After obtaining the results using CNN, we implemented the same datasets this time using SVM. The results that are obtained from some of them are explained below. We used a two-sample t-test to find out the suitable pairs to be used for classification. Note that, feature selection uses the full dataset prior to CV and features correspond to certain brain region-pairs which apply to all participants; otherwise, each nested subset would have different region-pairs, albeit some overlapping, making the selection of selected common brain region-pairs intractable (it may be an empty set), in this dataset with small number of samples, where each sample is a participant.

SVM-based classification results of DFC for different p values. For the calculations, we utilized AAL atlas-based DFC of the different groups separately, using the histogram distributions of the standard deviation of the DFC (stdDFC) metrics for NC and GWI-PTSD groups separately. Histograms were obtained to check the distribution of the standard deviations of all the features. After this, we computed the unpaired t-test of the significance of the FC differences between the two groups, using the ttest2 function in MATLAB; we set the unpooled variance option to allow for different variances for the two groups [40]. p values of the t-test for each region-pair combination were computed and the region-pairs with significant p values ($p < 0.01$) of the t-test where the null hypothesis is rejected, were computed, meaning that the two groups have significantly different standard deviation of their DFC (stdDFC) between only certain pairs of brain regions. Once the t-test was performed on the data set, we obtained the 116×116 p-values. Using the previously obtained results, we first conducted classification with the default 10-fold cross-validation (CV), repeated one hundred times, and computed the average performance. The mean cross-validation loss when using all features was approximately 0.35, corresponding to about 35% loss and roughly 65% classification accuracy. Thus, although performance improved with the SVM approach, no significant differences were observed between SVM and CNN when all features were included in the analysis. After this step, we utilized the p values obtained during the t-test and checked the brain region pairs which resulted in significant differences between the two groups.

SVM-based classification for DFC for $p < 0.001$. Feature reduction was performed using a t-test, retaining only features with $p < 0.001$. Under this threshold, a single region pair was identified: Left Medial Orbital Superior Frontal Gyrus (AAL region #25) and Right Superior Occipital Gyrus (AAL region #50). Using the standard deviation of DFC (stdDFC) with default 10-fold cross-validation repeated one hundred times, the SVM achieved a mean classification accuracy of 64.57%. The stdDFC between these regions was 17% lower in the GWI-PTSD group compared to controls ($p < 0.001$).

SVM-based classification for DFC for $p < 0.05$. To include more features, the p-value threshold was relaxed, as detailed in Table 11. At $p < 0.05$, 113 region pairs were identified. Using stdDFC with the same cross-validation scheme, classification accuracy increased substantially to approximately 98%. Across all tested p-value thresholds, one region pair consistently appeared, as shown in Fig. 4.

Table 11. SVM AAL ATLAS-based classification accuracy, mean k-fold cross-validation loss obtained for corresponding *p* value of DFC (*p*DFC) ceiling (maximum).

Maximum pDFC	Mean k-fold Loss	Classification Accuracy	Number of Pairs
0.001	0.3543	0.6457	1
0.002	0.3934	0.6066	2
0.005	0.1792	0.8208	4
0.01	0.1932	0.8068	12
0.02	0.0628	0.9372	32
0.03	0.0600	0.9400	62
0.04	0.0358	0.9642	87
0.05	0.0191	0.9809	113

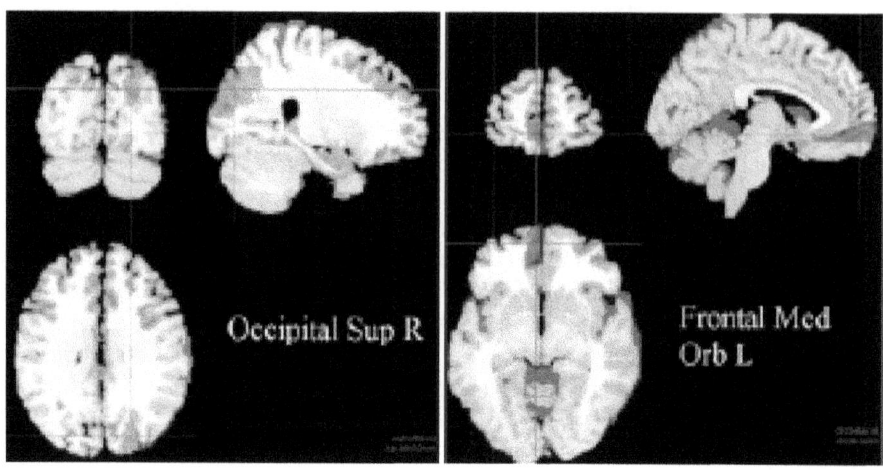

Fig. 4. The two AAL regions -right Superior Occipital Gyrus (AAL #50, shown in red) and left Medial Orbital Superior Frontal Gyrus (AAL #25, shown in blue)- formed the region pair with the strongest discriminative ability between the GWI-PTSD and NC groups based on stdDFC. The mean standard deviation of the dynamic functional connectivity (stdDFC) between these regions was significantly reduced in GWI-PTSD compared to NC (p < 0.001). Classification performance reached approximately 98% accuracy (52/53), with only one participant misclassified. [42].

4.5   Classification Results Using FC and SVM Classifier

Just as we did in the prior subsection, we implemented the same classifier fitcsvm, this time utilizing the FC features. Results obtained from some of them are explained below. In this process again we used unpaired t-test with unequal group variance (ttest2) to find out the suitable pairs to be used for classification.

SVM-based classification results of FC for different p values. For analysis, AAL atlas-based FC metrics were calculated separately for the two groups, with differences in mean and overall patterns observed. Histograms were used to examine the distribution of standard deviations across features. A t-test (similar to the prior subsection) produced a 116×116 matrix of p-values for each brain region pair. Classification using all features with 10-fold cross-validation repeated 100 times yielded a mean loss of 0.332, or ~ 66.8% accuracy. While SVM improved results slightly, using all features in either SVM or CNN showed no substantial gains. Subsequently, various t-test p-values were applied to identify brain region-pairs with significant group differences.

SVM-based classification for FC for p < 0.001. Utilizing t-tests, only those features were selected whose chance probability were less than 0.001. With this probability, only one region-pair existed. This pair was identified as having AAL brain region # 74 (Putamen_R) and # 116 (Vermis_10). Classification was calculated based on FC data with default 10-fold-cv and repeated by 100 × and mean cross-validation loss of 0.332 was obtained. This means classification accuracy of around 66.8% was found.

SVM-based classification for FC for $p < 0.05$. To include more features, the p-value threshold was relaxed, as detailed in Table 12. We began with the strictest p value ($p < 0.001$) and found only one pair, prompting us to increase the threshold. Cross-validation accuracy was then computed for several p values, summarized in Table 12. At $p < 0.05$, 234 pairs were identified. Classification was performed using FC with default 10-fold CV, repeated 100 times, yielding a mean cross-validation loss of 0.317, corresponding to ~ 68.3% accuracy. Ten p values were tested with FC (see Table 12), and the best result, ~ 72% accuracy, occurred at $p < 0.005$. Across all p-value thresholds, only one pair—number 74 (Putamen_R) and 116 (Vermis_10)—was consistently found.

Table 12. Mean k-fold cross-validation loss, classification accuracy and number of brain region pairs obtained for corresponding p value of FC (pFC) ceiling (maximum), with AAL ATLAS brain regions and SVM classifier.

Maximum pFC	Mean k-fold Loss	Classification Accuracy	Number of Pairs
0.001	0.3319	0.6681	1
0.002	0.3108	0.6892	5
0.005	0.2764	0.7236	13
0.01	0.3026	0.6974	31
0.02	0.2885	0.7115	84
0.03	0.2962	0.7038	131
0.04	0.3055	0.6945	180
0.05	0.3174	0.6826	234

4.6 Important Brain Regions Found Involved in Classification

A p-value for stdDFC < 0.005 resulted into four AAL brain region pairs. These were: 'Superior frontal gyrus, medial orbital' and 'Superior occipital gyrus'; 'Superior frontal gyrus, medial orbital' and 'Superior parietal gyrus'; 'Olfactory cortex' and 'Caudate nucleus'; and 'Olfactory cortex' and 'Lenticular nucleus, putamen'. There was one additional pair, 'Inferior frontal gyrus, triangular part' and 'Superior frontal gyrus, dorsolateral', which also appeared as the fifth pair when the maximum p-value threshold was raised to 0.01. AAL region for the top single pair was already shown in Fig. 4. The other four pairs are presented in Figures below.

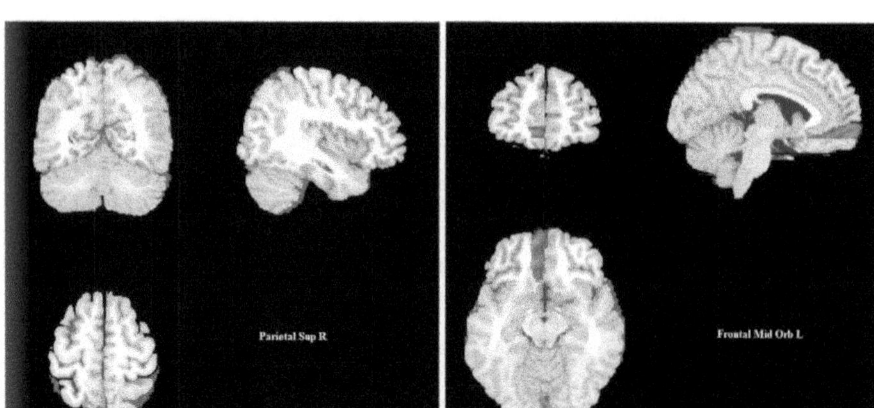

Fig. 5. Region Pair: Parietal Sup R and Frontal Mid Orb L (AAL Regions 60 and 25).

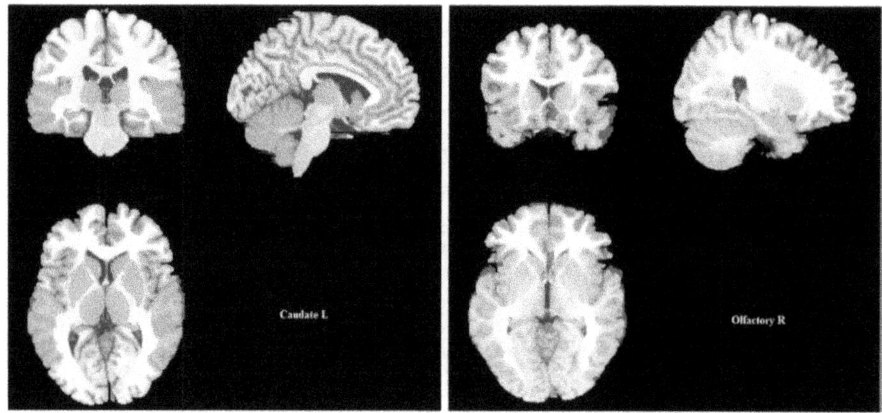

Fig. 6. Region Pair: Left Caudate and Right Olfactory (AAL Regions 71 and 22).

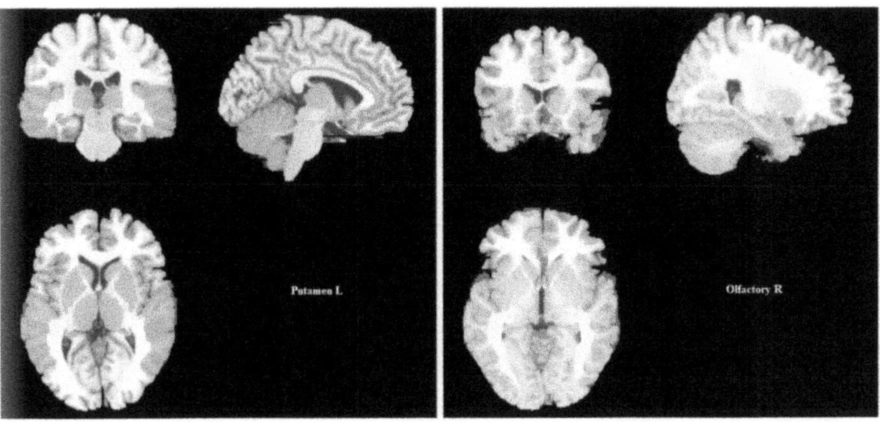

Fig. 7. Region Pair: Left Putamen and Right Olfactory (AAL Regions 73 and 22).

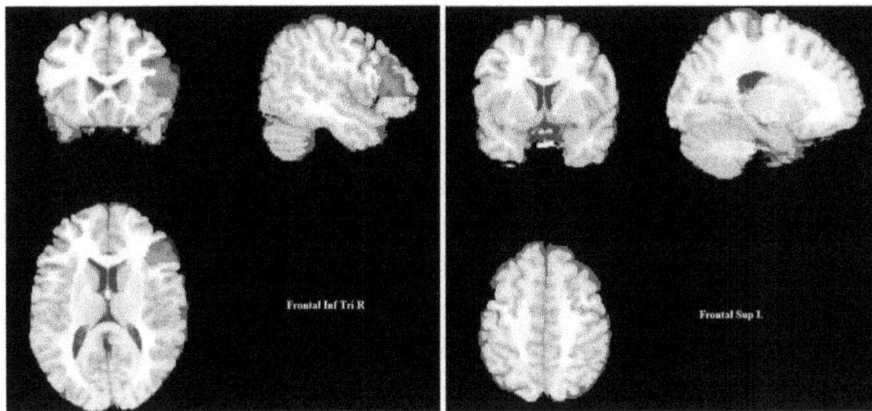

Fig. 8. Region Pair: R Frontal Inferior and L Frontal Superior (AAL Regions 14 and 3).

Table 13 below lists AAL [44] region-pairs obtained utilizing FCs for pFC < 0.001 and < 0.002 and AAL region-pairs obtained utilizing stdDFCs for pDFC < 0.001, < 0.002 and < 0.005. The lists of AAL region pairs obtained utilizing the stdDFCs for maximum pDFC values of 0.01, 0.02, 0.03, 0.04, and 0.05 will be presented in a related subsequent publication and omitted here for brevity.

Table 13. List of AAL [44] region pairs for pFC < 0.001 and < 0.002; and, for pDFC < 0.001, < 0.002, and < 0.005.

Maximum p	AAL Region Pair(s): (AAL Region$_i$, AAL Region$_j$)
pFC < 0.001	(74,116)
pFC < 0.002	(74,116); (36,74); (56,116); (92,116); (100,116)
pDFC < 0.001	(25,50)
pDFC < 0.002	(25,50); (22,73)
pDFC < 0.005	(25,50); (22,73); (22,71); (25,60)

Figure 9 presents classification accuracy performance of the FC-based and DFC-based methods using different maximum p-values of the group-discriminatory t-test, which results in different number of brain region pairs. It can be noticed that, for maximum p values of 0.005 and above, the DFC-based method yields greater accuracy with a smaller number of brain region pairs: e.g. for maximum p value of 0.005 it achieves an accuracy of 82% with only 5 region-pairs whereas the FC-based method achieves 72% accuracy with 13 region-pairs. The DFC-based method overwhelmingly outperforms the FC-based method as the p-value or the number of region-pairs is increased: it achieves a maximum accuracy of 98%, with 113 region-pairs, whereas the FC-based method achieves a maximum accuracy of 72%, with thirteen region-pairs. Increasing the number of features any further decreases the FC-based accuracy. A large difference of accuracy metric between the DFC-based and FC-based methods remains as different number of brain-regions are used, from a few to hundreds.

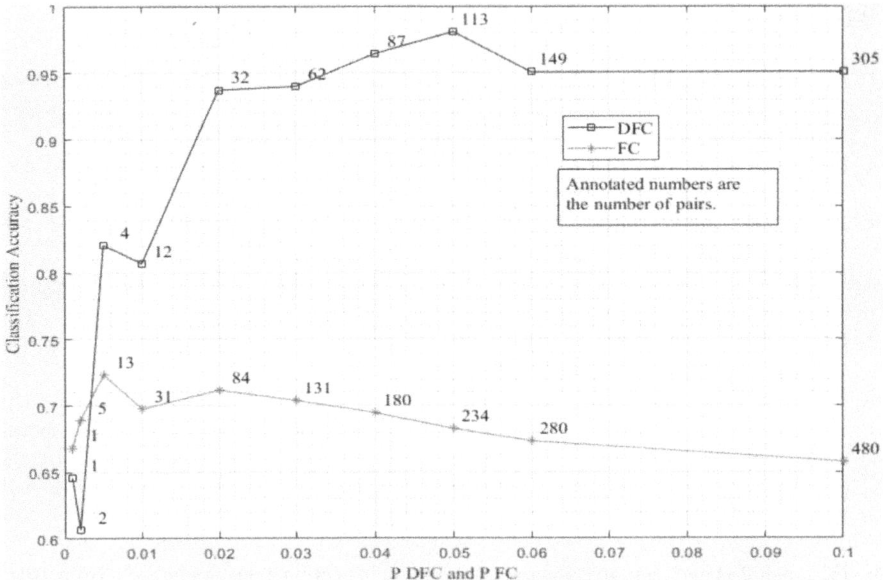

Fig. 9. Comparison of classification accuracy of DFC-based and FC-based methods versus different p value thresholds (maximum), where the p values were used to determine the brain region pairs whose FC or DFC features were used for classification.

5 Conclusions and Discussions

In this study, SVM outperformed deep learning methods such as CNN in classifying Gulf War Illness (GWI) type PTSD versus normal control veterans. Dynamic functional connectivity (DFC) features showed substantially greater discriminative power than static FC, achieving classification accuracies up to 98%, compared to a maximum of 72% with static FC. Notably, the standard deviation of DFC (stdDFC) between the Left Medial Orbital Superior Frontal Gyrus and the Right Superior Occipital Gyrus was significantly lower (17%, $p < 0.001$) in GWI-PTSD participants, suggesting impaired connectivity in regions associated with visual, sensorimotor, and semantic processing. These findings align with reported cognitive and motor deficits in GWI-PTSD veterans [36, 42, 43], though further interpretation of the involved networks is needed.

Overall, the results support prior evidence of widespread resting-state FC impairments in GWI-PTSD. DFC-based measures, including the proposed stdDFC, show promise as neuroimaging biomarkers for GWI-PTSD and potentially for other neurological conditions. While deep learning approaches such as U-Net may be explored, their effectiveness is limited by small sample sizes and high feature dimensionality in fMRI studies. Future work includes dimensionality reduction techniques such as principal component analysis (PCA) and/or linear discriminant analysis (LDA) to help mitigate these challenges.

References

1. TReNDS Neuroimaging Prediction Challenge. https://www.kaggle.com/c/trends-assessment-prediction. Accessed 21 Nov 2025
2. Neuroimaging. https://en.wikipedia.org/wiki/Neuroimaging. Accessed 29 Nov 2025
3. Huettel, S., Song, A., McCarthy, G.: Functional Magnetic Resonance Imaging. Sinauer Associates, Sunderland, MA (2004)
4. BOLD fMRI. https://en.wikipedia.org/wiki/Blood-oxygen-level-dependent_imaging, 29 Nov 2025
5. Biswal, B., VanKylen, J., Hyde, J.: Simultaneous assessment of flow and BOLD signals in resting-state functional connectivity maps. NMR Biomed. 165–170 (1997)
6. Friston, K.: Causal modelling and brain connectivity in functional magnetic resonance imaging. PLOS Biol., 220–225 (2009)
7. Fernandez-Seara, M.: Effects on resting cerebral blood flow and functional connectivity induced by metoclopramide: a perfusion MRI study in healthy volunteers. Br J. Pharmacol. (2011)
8. Smith, S.M.: The future of FMRI connectivity. Neuroimage **62**(2), 1257–1266 (2012)
9. Resting State fMRI. https://en.wikipedia.org/wiki/Resting_state_fMRI. Accessed 29 Nov 2025
10. Friston, K.J.: Functional and effective connectivity in neuroimaging: a synthesis. Hum. Brain Mapp. **2**, 56–78 (1994)
11. Sakoglu, U., Pearlson, G.D., Kiehl, K.A., Wang, Y.M., Michael, A.M., Calhoun, V.D.: A method for evaluating dynamic functional network connectivity and task-modulation: application to schizophrenia. Magn. Reson. Mater. Phys., Biol. Med. **23**(5), 351–366 (2010)
12. Sakoglu, U., Huisa-Garate, B., Rosenberg, G., Sood, R.: Application of FT-based MMSE deconvolution method for cerebral blood flow measurement in patients with Leukoaraiosis. Magnetic Resonance Imaging, Elsevier **27**, 625–630 (2009)
13. Pai, M., Sakoglu, U., Peterson, S., Lyons, C., Sood, R.: Characterization of BBB permeability in a preclinical model of cryptococcal meningoencephalitis using magnetic resonance imaging. J. Cereb. Blood Flow Metab. **29**, 545–553 (2009)
14. Sakoglu, U., Pearlson, G., Kiehl, K., et al.: A method for evaluating dynamic functional network connectivity and task-modulation: application to schizophrenia. Magnet. Reson. Mater. Phys. Med. (MAGMA) **23**(5), 351–366 (2010)
15. Sakoglu, U., Calhoun, V.: Temporal dynamics of functional network connectivity at rest: a comparison of schizophrenia patients and healthy controls. Neuroimage **47** (2009)
16. Jampana, S.: Novel Time-Series Classification Analysis of EEG Data. MS Thesis, ETAMU (2015)
17. Bhamidipati, S.: Machine Learning Applications to Classify Events from EEG Data. MS Thesis, UHCL (2016)
18. Sakoglu, U., Calhoun, V. D.: Dynamic windowing reveals task-modulation of functional connectivity in schizophrenia patients vs healthy controls. In: ISMRM Honolulu (2009)
19. Sakoglu, U., Michael, A., VD, C.: Classification of schizophrenia patients vs healthy controls with dynamic functional network connectivity. In: Human Brain Mapping, San Francisco (2009)
20. Sakoglu, U., Mete, M., Esquivel, J., Rubia, K., Briggs, R., Adinoff, B.: Classification of cocaine dependent subjects with dynamic functional connectivity from functional magnetic resonance imaging data. J. Neurosci. Res. **97**(7), 790–803 (2019)
21. Hutchisonet, R., et al.: Dynamic functional connectivity: promise, issues, and interpretations. Neuroimage, 360–378 (2013)

22. Dynamic Functional Connectivity. https://en.wikipedia.org/wiki/Dynamic_functional_con nectivity. Accessed 29 Nov 2025
23. Majeed, W., Magnuson, M., Keilholz, S. D.: Spatiotemporal dynamics of low frequency fluctuations in BOLD fMRI of the rat. J. Magnet. Reson. Imag. 384–393 (2009)
24. Majeed, W., et al.: Spatiotemporal dynamics of low frequency BOLD fluctuations in rats and humans. NeuroImage. 1140–1150 (2011)
25. Tagliazucchi, E., Balenzuela, P., Fraiman, D., Montoya, P., Chialvo, D. R.: Spontaneous BOLD event triggered averages for estimating functional connectivity at resting state. Neurosci. Lett. 158–163 (2011)
26. Tagliazucchi, E., Balenzuela, P., Fraiman, D., Chialvo, D. R.: Criticality in large-scale brain fMRI dynamics unveiled by a novel point process analysis. Front. Physiol. (2012)
27. Tagliazucchi, E., Carhart-Harris, R., Leech, R., Nutt, D., Chialvo, D. R.: Enhanced repertoire of brain dynamical states during the psychedelic experience. Human Brain Mapping (2014)
28. Tagliazucchi, E., Siniatchkin, M., Laufs, H., Chialvo, D. R.: The voxel-wise functional connec- tome can be efficiently derived from co-activations in a sparse spatio-temporal point-process. Front. Neurosci. (2016)
29. Petridou, N., Gaudes, C.C., Dryden, I.L., Francis, S.T., Gowland, P.A.: Periods of rest in fMRI contain individual spontaneous events which are related to slowly fluctuating spontaneous activity. Hum. Brain Mapp. **34**(6), 1319–1329 (2013)
30. Liu, X., Duyn, J. H.: Time-varying functional network information extracted from brief instances of spontaneous brain activity. In: Proceedings of the National Academy of Sciences, pp. 4392–4397 (2013)
31. Chen, J. E., Chang, C., Greicius, M. D., Glover, G. H.: Introducing co-activation pattern metrics to quantify spontaneous brain network dynamics. NeuroImage **111**, 476–488, 2015
32. Chang, C., Glover, G. H.: Time–frequency dynamics of resting-state brain connectivity measured with fMRI. NeuroImage, **50**(1), 81–98 (2010)
33. Dynamic Functional Connectivity. [Online]. Available: https://en.wikipedia.org/wiki/Dyn amic_functional_connectivity
34. Sakoglu, U., Pearlson, G. D., Kiehl, K. A., Wang, Y. M., Michael, A. M., Calhoun, V. D.: A method for evaluating dynamic functional network connectivity and task-modulation: applica- tion to schizophrenia. In: Magnetic Resonance Materials in Physics and Medicine (MAGMA) (2010)
35. Jones, D. T., et al.: Non-stationarity in the resting brain's. Modul. Archit. PLOS ONE (2012)
36. Gopinath, K., Sakoglu, U., Crosson, B., Haley, R.: Exploring brain mechanisms underlying gulf war illness with group ICA based analysis of fMRI resting state network. Neurosci. Lett. Elsevier **710**, 136–141 (2019)
37. CCK-Law. [Online]. Available: https://cck-law.com/types-of-va-disabilities/gulf-war-syn drome/
38. Akgun, D., Sakoglu, U., Esquivel, J., Adinoff, B., Mete, M.: GPU accelerated dynamic func- tional connectivity analysis for functional MRI data, Computerized Medical Imaging and Graphics. Elsevier **43**, 53–63 (2015)
39. Region Based Convolutional Neural Networks. [Online]. Available: https://en.wikipedia.org/ wiki/Region_Based_Convolutional_Neural_Networks
40. [Online]. Available: http://matlab.mathworks.com
41. [Online]. Available: https://www.ibm.com/docs/it/spss-modeler/SaaS?topic=models-how- svm-works
42. Sakoglu, U., Mishra, A., Gopinath, K., Crosson, B., Haley, R.: Classification of gulf war illness patients vs control veterans using fMRI dynamic functional connectivity. In: ISMRM, London (2022)

43. Research Advisory Committee on Gulf War Veterans' Illnesses, J. Binns et al. "Gulf War Illness and the Health of Gulf War Veterans: Scientific Findings and Recommendations", Washington, D.C.: U.S. Government Printing Office, November 2008
44. Tzourio-Mazoyer, N., et al.: Automated Anatomical Labeling of activations in SPM using a Macroscopic Anatomical Parcellation of the MNI MRI single-subject brain. NeuroImage,**15**, 273–289 (2002)
45. Ronneberger O, Fischer P, Brox T: U-Net: convolutional networks for biomedical image segmentation. arXiv:1505.04597 (2015)
46. Pruim, R.H.R., Mennes, M., van Rooij, D., Llera, A., Buitelaar, J.K., Beckmann, C.F.: ICA-AROMA: a robust ICA-based strategy for removing motion artifacts from fMRI data. Neuroimage **112**, 267–277 (2015)

PanACRpred: Predicting Accessible Chromatin Regions in Pangenomes Using Motif Chaining

Madelyn J. Warr[1], Trung Dinh[2], Bella Root[3], Elyse Onstott[3], Kevin Yu[4],
Joann Mudge[3], Thiruvarangan Ramaraj[5], Indika Kahanda[2],
and Brendan Mumey[6(✉)]

[1] University of Utah, Salt Lake City, UT 84112, USA
[2] University of North Florida, Jacksonville, FL 32224, USA
{n01382743,indika.kahanda}@unf.edu
[3] National Center for Genome Resources, Santa Fe, NM 87505, USA
{broot,eonstott,jm}@ncgr.org
[4] Tufts University, Boston, MA 02155, USA
[5] DePaul University, Chicago, IL 60614, USA
tramaraj@depaul.edu
[6] Montana State University, Bozeman, MT 59717, USA
brendan.mumey@montana.edu

Abstract. In this work, we investigate whether motif subsequence features can predict accessible chromatin regions (ACR) in a genome, i.e. regions that are accessible to regulatory proteins, thus enabling transcription of associated genes. We focus on plants, whose agricultural and ecological importance make them interesting and important organisms to study, and whose complex genomes provide important stress tests for our algorithm. We show that regulatory motif sequence similarity can be found efficiently using co-linear chaining. The similarity scores found are then used as features in machine learning models to explore the feasibility of effectively predicting ACRs in genome assemblies.

Keywords: accessible chromatin regions · co-linear chaining · machine learning

1 Introduction

As DNA sequencing technologies and assembly algorithms have improved, many high-quality genome assemblies are often being generated in a single species [9, 20]. This has allowed us to move away from analyzing the genetic variation in a single species through comparison of sparse sequence data from multiple individuals back to a single reference, which introduces bias. Instead, we can now create pangenome data structures that uniformly capture genetic variation from individuals to enable capture and analysis of all the genetic variation in

K. L. Kabir et al. (Eds.): BICOB 2026, CCIS 2977, pp. 221–232, 2026.
https://doi.org/10.1007/978-3-032-26028-4_16

a species. While the availability of multiple genomes per species is relatively common, other sequencing data types are lagging behind.

In addition to understanding the genetic variation present in a species, it is also important to understand how that variation is expressed. *ATAC-seq* is a technology that sequences regions of the genome that are accessible to proteins, known as accessible chromatin regions (ACRs). These regions bind regulatory factors, thus marking clusters of *cis-regulatory elements* (CREs) such as promoters and enhancers that are actively driving transcription of their corresponding genes [14,30,32]. CREs are noncoding DNA regions in, around, or distal to genes that affect expression timing, tissue-specificity, and environmental response [3,4,18,21]. CREs affect expression by binding transcription factors, small RNAs, enzymes affecting DNA modifications, and/or modified histones [28], bringing them into physical contact with the gene they control, often through chromosome looping [26,31].

ATAC-seq is a powerful tool for investigating the role of ACRs and CREs in regulating gene expression. Additionally, ATAC-seq can be used to identify the short, conserved DNA sequences that characterize CREs referred to as *motifs*. However, ATAC-seq data are still limited, often focused on multiple tissues of the reference genome, rather than being assessed across multiple individuals. Therefore, in a pangenomic setting, where genome assemblies of many individuals are available, ATAC-seq data and the accessible chromatin regions they identify may only be available for one or two lines.

In this work, we investigate whether the CRE content in a region can be used to predict chromatin accessibility in a given condition. We focus on plants, whose agricultural and ecological importance, combined with their complex genomes make them interesting and important organisms to study [27]. We show that limited ATAC-seq data combined with CRE information can be used to predict ACRs in genome assemblies. Accessible chromatin regions associated with the same genes have some short regions of CRE conservation [15,16,29]. As a first step toward chaining CREs across pangenomes (Fig. 1), we quantify CRE similarity between genome regions using *co-linear chaining* [8,17]. Co-linear chaining has primarily been used in more traditional bioinformatics tasks such as read

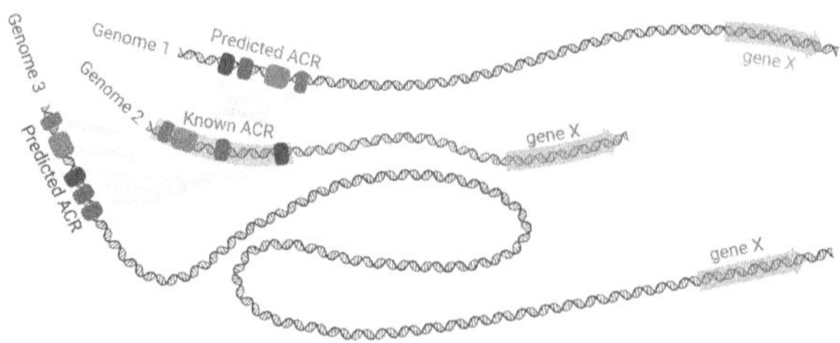

Fig. 1. Co-linear chaining can quickly find motif subsequence matching scores between known ACRs and candidate regions; these scores are used to train a machine learning classifier to predict accessibility of candidate regions in a pangenomic data set. Created by BioRender.

mapping and can efficiently find shared subsequences between two strings. In this work, we are interested in testing whether shared subsequences of CRE motifs can help characterize ACRs; to our knowledge this is the first work to investigate this question. Specifically we use chaining scores against known ACRs as a *feature vector* to predict accessibility of unknown regions. We show this strategy is effective at accurately predicting novel ACR regions.

2   Related Work

Early machine learning methods to predict regulatory elements such as gkm-SVM [5] relied on hand-crafted features, primarily k-mers. DeepSEA [33] was one of the first deep learning models, which used a Convolutional Neural Network (CNN) to predict chromatin effects directly from raw DNA sequences, outperforming SVMs by capturing hierarchical features. Then, researchers began using specialized architectures to capture the "grammar" of the genome, which introduced (a) Basset [11], which applies CNNs specifically to ATAC-seq data across multiple cell types simultaneously, (b) Deopen [13], which combined CNNs with k-mer features to improve accuracy in predicting continuous DNase-seq signals, and (c) DanQ [24], which combined CNNs with LSTMs (Long Short-Term Memory) to capture both local motifs and their long-range spatial relationships. Current State-of-the-Art models focus on high-resolution (base-pair level) and long-range context. For example, Basenji [10] uses dilated convolutions allowing the model to look at sequences up to 131kb long to predict accessibility tracks. Similarly, Enformer [1] can integrate information across 200kb+, significantly improving predictions for distal regulatory elements (enhancers).

While much progress is being made, many of these algorithms were originally designed and tuned for human or mouse [1,10,23]. Models that address regulatory elements in plants are beginning to emerge, including AgroNT, which was trained on data from 48 plant species (mostly crops) [19]. It is unclear how useful these models would be for the many unrelated, non-model plant species for which little information is known about their regulatory elements and where obtaining experimental validation of regulatory elements is prohibitive. To overcome these limitations, we propose the use of co-linear chaining, a computational technique that allows for the identification of chains of small but clustered regions of similarity, such as those found within accessible chromatin regions, within and across genomes. This method combines the use of ATAC-seq peaks that highlight active regulatory elements, any available information on known regulatory motifs, and sequence similarity to predict active regulatory elements within and across genomes. To the best of our knowledge, our work is the first attempt at developing a framework for ACR prediction within genomic data using co-linear chaining.

3   Methods

As depicted in Fig. 2, we developed a pipeline to test the utility of combining co-linear chaining with machine learning to predict chromatin accessibility.

First, we obtain the sequence and ATAC-seq data of an *Arabidopsis thaliana* genome, which we use to identify ACRs. Next, we label motifs (potential CREs) from the *A. thaliana* DAPv1 database [22], which contains binding sites for hundreds of transcription factors, in the sequences. Then, we assign positive regions as the HELD-OUT set, and we assign both positive and negative regions as the EXPERIMENT set. We further split the EXPERIMENT set into TRAIN, VALIDATE, and TEST sets. We then obtain a feature vector for every region in EXPERIMENT by colinear-chaining a given region in EXPERIMENT with every region in HELD-OUT. Finally, we use these feature vectors in machine learning, masking TEST until after the model training and tuning to evaluate the accessibility predictions. Each of the steps depicted in Fig. 2 are described with detail in Subsects. 3.1-3.4. Our software tool, PanACRpred, is freely available at: https://github.com/msu-alglab/pan_ACR_pred.

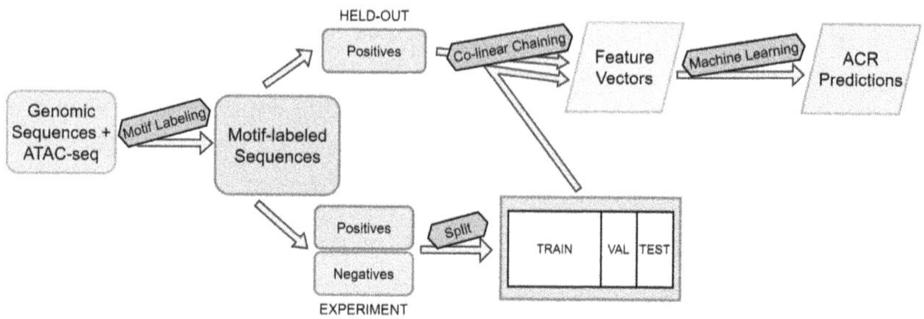

Fig. 2. Pipeline to process genomic data and predict ACRs via co-linear chaining and machine learning.

3.1   Genomic Sequence Data

We obtained genome coordinates for publicly-available ACRs from 78 different bioprojects that were identified across tissues in the *Arabidopsis thaliana* reference genome (Columbia-0) as well as other ecotypes and mutants. We extracted the underlying sequence data from the TAIR10 assembly version using bedtool's getfasta tool [12,25]. These are the positive examples.

To obtain negative regions (inaccessible regions of the genome) for machine learning, we employ two separate generation methods: *random* and *upstream*. In the random method, we select random regions of the genome not identified as ACRs, fixing the lengths to match the length distribution of the positive set used as samples for machine learning. In the upstream method, we select proximal regions not identified as ACRs upstream of known genes, once again matching the length distribution of the positive set. The distance upstream from the gene is randomized, but does not exceed 4000 base pairs.

To label motifs as potential CREs in these positive and negative sequences, we used the DAPv1 database [22] of known *Arabidopsis thaliana* CREs. To avoid overcounting CREs, we clustered similar motifs in the database using a motif comparison tool, Tomtom [7], and hierarchical clustering. We then define a consensus motif for each cluster and use FIMO [6] to find these consensus motifs in the genomic sequences. Both Tomtom and FIMO are part of the MEME suite [2].

After labeling candidate CRE motifs in the genome as described above, we randomly choose 12,806 positive regions as our HELD-OUT set. Next, we combine 11,758 positive regions with 11,758 negative regions to form our EXPERI-MENT set. We then perform a 60% / 20% / 20% split on EXPERIMENT into TRAIN / VALIDATE / TEST sets, respectively, ensuring a 50% / 50% split between positive and negative samples in each dataset.

3.2 Co-linear Chaining

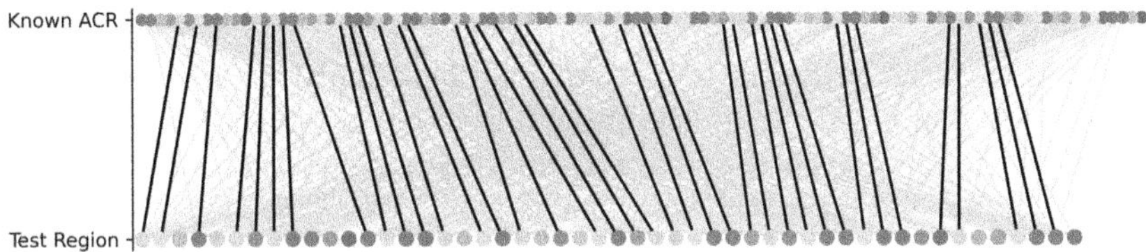

Fig. 3. Illustration of co-linear chaining between a HELD-OUT ACR and an EXPERI-MENT region. Dots represent motifs; grey lines connect pairs of matching motifs; black lines connect pairs of matching motifs in some set which achieves the global chain score. The highest scoring pair is shown.

We define a string to be a sequence of characters in the alphabet $\Sigma = \{A, C, G, T\}$. A motif in a string S is a subsequence of the characters in S. For some motif M in S which begins at position i in S, let INDEX(M)$= i$.

Suppose we have a string A with a sequence of motifs $M_A = a_1 \ldots a_n$, with INDEX(a_i) $\leq$ INDEX(a_{i+1}) for all $1 \leq i \leq n$. Likewise, suppose we have a string B with a sequence of motifs $M_B = b_1 \ldots b_m$, with INDEX(b_k) $\leq$ INDEX(b_{k+1}) for all $1 \leq i \leq m$. Let $X = \{(a_i, b_k) \mid a_i \in M_A, b_k \in M_B, a_i = b_k\}$. We call some element $x \in X$ a *motif match*. Now we define a precedence relation $\prec$ on X, where $(a_i, b_k) \prec (a_j, b_l)$ if and only if $i < j$ and $k < l$.

Definition 1. *We define the* global chain length *to be the cardinality of the largest set $S \subseteq X$ with $\prec$ as a total relation on S. That is, for all $s, t \in S$ with $s \neq t$, either $s \prec t$ or $t \prec s$.*

Additionally, we can assign some weight w_x to the motif associated with every $x \in X$. For some $S \subseteq X$, let $W_S = \sum_{s \in S} w_s$.

Definition 2. *We define the* global chain score *to be the maximum W_S over all $S \subseteq X$ with $\prec$ as a total relation on S.*

The global chain length and global chain score between two sequences with identified motif matches can be computed in $\mathcal{O}(|X| \log |X|)$ time with existing co-linear chaining methods methods, e.g. [17]. Figure 3 shows the top-scoring example from the tested data set.

3.3   Generating Feature Vectors

Given some set H (the HELD-OUT set from Fig. 2) of known ACRs, with $|H| = n$, we can create a feature vector for some DNA region r of unknown accessibility. First, we arbitrarily assign an order to the ACRs in H. Then, we use co-linear chaining to find the global chain score c_i between r and h_i for every $h_i \in H$. This allows us to define a feature vector $v_r = \begin{bmatrix} c_1 & c_2 & \dots & c_n \end{bmatrix}$. This feature vector gives a quantification of the relationship between r and the set of known ACRs.

To calculate the global chain score, we assign a weight w_m to every motif m identified as a potential CRE. The weight is based on the frequency at which m appears in our set of known ACRs as opposed to the whole genome. Let $\mathrm{prob}(m_{\mathrm{ACR}})$ be the probability that an identified motif in a known ACR is m. Likewise, let $\mathrm{prob}(m_{\mathrm{BG}})$ be the probability that an identified motif in the background (in this case, the entire genome) is m. We calculate w_m as follows, setting $\lambda = 0.8$ based on initial experiments.

$$w_m = \log_2 \left(\frac{\mathrm{prob}(m_{\mathrm{ACR}})}{\mathrm{prob}(m_{\mathrm{BG}})} \right) - \lambda \log_2(\mathrm{prob}(m_{\mathrm{BG}}))$$

The first term in the equation scales with the frequency of a motif's occurrence in known ACRs as opposed to the background. The second term adds an additional boost to the weight which scales with the rarity of the motif in the entire genome. The logarithmic scale ensures the weights are confined to a more narrow range. The intention behind this weighting scheme is to incorporate additional predictive power into the feature vectors by enhancing motifs which are more unique to ACRs. Thus, unknown regions which contain motifs more often found in ACRs will likely have a higher global chain score with some of the known ACRs.

3.4   Machine Learning Models and Experimental Setup

The task of predicting ACRs was formulated as a supervised binary classification problem, in which instances are sequences, and features are the chaining scores (i.e., scores between instances and sequences in the HELD-OUT set). A label of 1 indicates that the region is chromatin accessible, and a label of 0 indicates that the region is not chromatin accessible. We use co-linear chaining to obtain a feature vector for each sample, as described in Sect. 3.3. The results are TRAIN, VALIDATE, and TEST sets containing 14108, 4704, and 4704 samples, respectively, with each sample represented by 12,806 features.

To evaluate the performance of our models, we utilized hold-out validation, where TRAIN is used for developing the model, VALIDATE is used for tuning model hyperparameters, and TEST is used for calculating performance metrics. We use Area Under the Receiver Operating Characteristic (AUROC) as our primary performance evaluation metric. AUROC is a key machine learning metric evaluating a model's ability to rank a given positive example above negative example, essentially showing its classification performance across all possible thresholds.

We evaluated the following classifiers: Multi-layer Perceptron (MLP), eXtreme Gradient Boosting (XGBoost), Gradient Boosting Classifier (GBC), and Histogram Gradient Boosting Classifier (HistGBC). MLP was implemented using Keras libarary (https://keras.io/), and XGBoost was implemented with XGBoost library (https://xgboost.readthedocs.io/en/latest/index.html). Both GBC and HistGBC were implemented via scikit-learn library (https://scikit-learn.org/stable/). In addition to these classifiers, we also developed an in-house Threshold-Based Classifier (TBC) that uses only a threshold on the chaining scores to make predictions (i.e. without using machine learning) to provide a baseline comparator.

The tuned hyperparameters are summarized in Table 1. For GBC and Hist-GBC, we performed tuning on their *learning rates, max depth*, and *number of estimators* used via *GridSearchCV* in scikit-learn. For XGBoost, we performed tuning on *learning rates, number of estimators, max depth, min_child_weight, gamma, subsample, colsample_bytree*, and *reg_alpha* via *RandomizedSearchCV* in scikit-learn. For MLP, the hyperparameter tuning explored the *dropout rate* between layers and the *number of neurons* in each layer, as well as the ordering of the layers. For TBC, we tuned the *threshold*. For MLP, we used 3 hidden layers with 128, 64, and 32 neurons. Dropout layers were placed between each of these hidden layers (dropout rate of 0.2 was used) and *relu* activation was used for all layers, with the exception of the output layer, which used *sigmoid* activation. This architecture was used for experiments with both the random and upstream negative sets.

For GBC, we used a learning rate of 0.1 and a max depth of 3 for both the random and upstream negative set experiments. For HistGBC, used a learning rate of 0.1 and no max depth for the random negative set experiment, while the upstream negative set experiment used a learning rate of 0.01 and a max depth of 8. For XGBoost, the random negative set experiment utilized a learning rate of 0.05, alpha of 0.05, colsample_bytree of 0.7, 300 estimators, and all other parameters were kept to their default values. The upstream negative set experiment utilized a learning rate of 0.05, max depth of 5, minimum child weight of 5, gamma of 0.2, subsample of 0.9, colsample_bytree of 0.9, 300 estimators, and default values for all other parameters.

For TBC, we set the threshold to be the 25th percentile over the maximum chaining scores of the positive samples in the TRAIN set. Every test sequence with a maximum chaining score greater than or equal to the threshold is predicted to be positive (i.e. ACR).

4   Experimental Results

Table 1. Hyperparameters tuned for each model.

Classifier	Hyperparameters Tuned
TBC	threshold
MLP	model layers / architecture, dropout rate
XGBoost	learning rate, number of estimators, max depth, min child weight gamma, subsample, colsample_bytree, reg_alpha
GBC	learning rate, max depth
HistGBC	learning rate, max depth

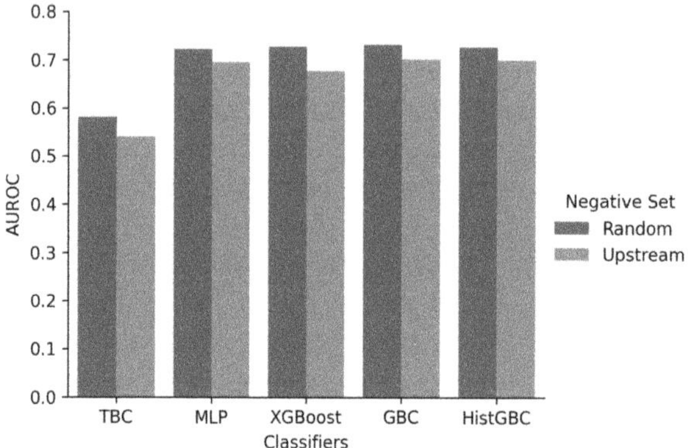

Fig. 4. Bar graph showing the classifier performances based on AUROC values. AUROC (Area Under the Receiver Operating Characteristic Curve) evaluates a model's ability to distinguish between positive and negative classes. It's a single number (0 to 1) summarizing the ROC curve, where 1 is perfect, and 0.5 is random guessing. Negative set explains the methodology for generating the negative samples for the data e.g. Random negative set describes the generation of negative samples via random selection.

All classifiers (excluding the TBC classifier) produced very similar AUROC values within their respective test data set, differing by at most .015 and .009 for the random negative set and upstream negative set (See Fig. 4 and Tables 2 and 3). Their AUROC values range between 0.70 – 0.74, showing a noticeable improvement over a random classifier (which would provide an AUROC of 0.5 for perfectly balanced data). Interestingly, TBC classifier performs only slightly better than a random classifier (i.e. AUROC values of 0.581 and 0.540 for random and upstream negative data, respectively). The experiment with randomly generated negative regions gave higher AUROC values across all models, including the TBC classifier (Fig. 4).

The runtimes to train TBC, MLP, XGBoost, and HistGBC were at most 4 min. GBC runtimes, on the other hand, were significantly longer at 57 min for the random negative set experiment and 59 min for the upstream negative set experiment (Tables 2 and 3). All timing data was collected on Google Colab Pro with an AMD EPYC 7B12 processer (2.25 GHz), equipped with 4 cores and 50 GB of available RAM.

Table 2. Summary of various models' performance across experiments using the random negative set. Runtime depicts the time taken to train the classifiers.

	Random Negative Set			
Classifier	AUROC	Accuracy	F1 Score	Runtime
TBC	0.581	0.572	0.634	<1 min
MLP	0.723	0.656	0.657	2 min
XGBoost	0.738	0.669	0.678	2 min
GBC	0.732	0.674	0.683	57 min
HistGBC	0.727	0.653	0.658	1 min

Table 3. Summary of various models' performance across experiments using the upstream negative set. Runtime depicts the time taken to train the classifiers.

	Upstream Negative Set			
Classifier	AUROC	Accuracy	F1 Score	Runtime
TBC	0.540	0.537	0.615	<1 min
MLP	0.696	0.642	0.660	2 min
XGBoost	0.705	0.643	0.642	4 min
GBC	0.702	0.642	0.639	59 min
HistGBC	0.700	0.633	0.628	4 min

5   Discussion

Here, we have shown that the conservation of clusters of short regulatory motifs in conjunction with ML models can be informative for predicting ACRs, identifying genomic regions that are likely to be important in controlling gene expression and the resulting phenotypes. One of the next steps is find the most important/informative features used by the trained classifiers for inference (e.g. using "feature importance scores"), and look at which CREs are most frequent in the top-K features. This will provide insight into the underlying biological mechanisms governing gene regulation. Our ultimate goal is to predict ACRs across multiple genomes within a pangenome, identify both known and novel motifs within the ACRs, and use these as features to predict phenotypes and better understand which genomic variants are driving important phenotypic traits.

In the future, this project will also include the implementation of Positive Unlabeled (PU) learning. In our current work, the gold-standard data is composed of positive examples (i.e., known ACR regions), and a set of either random or upstream sequences. However, due to the incompleteness of the ACRs, negatively labeled sequences may include examples that are positive in other tissues or conditions, giving rise to concerns about whether all our negative data are truly negative. However, with the use of PU learning in future work, we can explore its ability to circumvent this underlying concern by requiring the PU-enabled classifiers to implicitly discover truly negative examples, potentially leading to improved performance.

Being able to predict ACRs in a species allows researchers to focus in on the genomic regions whose sequence variation is most likely driving differences in important phenotypes. Prediction is especially important in species where little ATAC-seq data are available or species where the data are available only for the reference genome. Being able to predict ACRs in genomes of related individuals or in a pangenome context, despite low sequence similarity between individuals, allows us to correlate variability in these important genotypic regions with phenotypic variability.

Acknowledgments. Support for this project came from NSF awards 2414134 and 2243010.

References

1. Avsec, Ž, et al.: Effective gene expression prediction from sequence by integrating long-range interactions. Nat. Methods **18**(10), 1196–1203 (2021)
2. Bailey, T.L., Johnson, J., Grant, C.E., Noble, W.S.: The meme suite. Nucleic Acids Res. **43**(W1) (2015)
3. Cherry, T.J., et al.: Mapping the cis-regulatory architecture of the human retina reveals noncoding genetic variation in disease. Proc. Natl. Acad. Sci. **117**(16), 9001–9012 (2020)
4. Cruz, M.A.D., Lund, D., Szekeres, F., Karlsson, S., Faresjö, M., Larsson, D.: Cis-regulatory elements in conserved non-coding sequences of nuclear receptor genes indicate for crosstalk between endocrine systems. Open Med. **16**(1), 640–650 (2021)
5. Ghandi, M., Lee, D., Mohammad-Noori, M., Beer, M.A.: Enhanced regulatory sequence prediction using gapped K-MER features. PLoS Comput. Biol. **10**(7), e1003711 (2014)

6. Grant, C.E., Bailey, T.L., Noble, W.S.: Fimo: scanning for occurrences of a given motif. Bioinformatics **27**(7), 1017–1018 (2011)
7. Gupta, S., Stamatoyannopoulos, J.A., Bailey, T.L., Noble, W.S.: Quantifying similarity between motifs. Genome Biol. **8**(2) (2007)
8. Jain, C., Gibney, D., Thankachan, S.V.: Algorithms for colinear chaining with overlaps and gap costs. J. Comput. Biol. **29**(11), 1237–1251 (2022)
9. Jiao, W.B., Schneeberger, K.: The impact of third generation genomic technologies on plant genome assembly. Curr. Opin. Plant Biol. **36**, 64–70 (2017)
10. Kelley, D.R.: Cross-species regulatory sequence activity prediction. PLoS Comput. Biol. **16**(7), e1008050 (2020)
11. Kelley, D.R., Snoek, J., Rinn, J.L.: Basset: learning the regulatory code of the accessible genome with deep convolutional neural networks. Genome Res. **26**(7), 990–999 (2016)
12. Lamesch, P., et al.: The Arabidopsis information resource (Tair): improved gene annotation and new tools. Nucleic Acids Res. **40**(D1), D1202–D1210 (2012)
13. Liu, Q., Xia, F., Yin, Q., Jiang, R.: Chromatin accessibility prediction via a hybrid deep convolutional neural network. Bioinformatics **34**(5), 732–738 (2017)
14. Lu, R.J.H., Liu, Y.T., Huang, C.W., Yen, M.R., Lin, C.Y., Chen, P.Y.: Atacgraph: profiling genome-wide chromatin accessibility from ATAC-seq. Front. Genetics **11** (2021)
15. Lu, Z., Marand, A.P., Ricci, W.A., Ethridge, C.L., Zhang, X., Schmitz, R.J.: The prevalence, evolution and chromatin signatures of plant regulatory elements. Nature Plants **5**(12), 1250–1259 (2019)
16. Lu, Z., Ricci, W.A., Schmitz, R.J., Zhang, X.: Identification of cis-regulatory elements by chromatin structure. Curr. Opin. Plant Biol. **42**, 90–94 (2018)
17. Mäkinen, V., Sahlin, K.: Chaining with overlaps revisited. In: 31st Annual Symposium on Combinatorial Pattern Matching (CPM 2020), pp. 25:1–25:12. Schloss Dagstuhl–Leibniz-Zentrum für Informatik (2020)
18. Marand, A.P., Zhang, T., Zhu, B., Jiang, J.: Towards genome-wide prediction and characterization of enhancers in plants. Biochim. Biophys. Acta **1860**(1), 131–139 (2017)
19. Mendoza-Revilla, J., et al.: A foundational large language model for edible plant genomes. Commun. Biol. **7**(1), 835 (2024)
20. Michael, T.P., VanBuren, R.: Building near-complete plant genomes. Curr. Opin. Plant Biol. **54**, 26–33 (2020)
21. Minnoye, L., et al.: F Chromatin accessibility profiling methods. Nat. Rev. Methods Primers **1**(1), 1–24 (2021)
22. O'Malley, R.C., et al.: Cistrome and Epicistrome features shape the regulatory DNA landscape. Cell **165**(5), 1280–1292 (2016)
23. Pampari, A., et al.: Chrombpnet: bias factorized, base-resolution deep learning models of chromatin accessibility reveal cis-regulatory sequence syntax. Nature Genetics (2022). also see updated bioRxiv 2024: https://doi.org/10.1101/2024.12.25.630221
24. Quang, D., Xie, X.: Danq: a hybrid convolutional and recurrent deep neural network for quantifying the function of DNA sequences. Nucleic Acids Res. **44**(11), e107–e107 (2016)
25. Quinlan, A.R.: Bedtools: the Swiss-army tool for genome feature analysis. Curr. Protoc. Bioinform. **47**(1), 11–12 (2014)
26. Ricci, W.A., et al.: Widespread long-range cis-regulatory elements in the maize genome. Nat. Plants **5**(12), 1237–1249 (2019)

27. Schatz, M.C., Witkowski, J., McCombie, W.R.: Current challenges in de novo plant genome sequencing and assembly. Genome Biol. **13**(4), 243 (2012)
28. Schmitz, R.J., Grotewold, E., Stam, M.: Cis-regulatory sequences in plants: their importance, discovery, and future challenges. Plant Cell **34**(2), 718–741 (2022)
29. Tu, X., et al.: Reconstructing the maize leaf regulatory network using chip-seq data of 104 transcription factors. Nat. Commun. **11**(1), 1–13 (2020)
30. Wang, S., Qian, Y.Q., Zhao, R.P., Chen, L.L., Song, J.M.: Graph-based pangenomes: increased opportunities in plant genomics. J. Exp. Botany **74**(1), 24–39 (2022)
31. Weber, B., Zicola, J., Oka, R., Stam, M.: Plant enhancers: a call for discovery. Trends Plant Sci. **21**(11), 974–987 (2016)
32. Zanini, S.F., et al.: Pangenomics in crop improvement–from coding structural variations to finding regulatory variants with pangenome graphs. Plant Genome **15**(1), e20177 (2022)
33. Zhou, J., Troyanskaya, O.G.: Predicting effects of noncoding variants with deep learning-based sequence model. Nat. Methods **12**(10), 931–934 (2015)

scFiLM: Dynamic Fusion of Gene Identity and Expression for Single-Cell Analysis

Xingze Li, Xinghai Zeng, Tao Tang, and Yingbo Cui[✉]

College of Computer Science and Technology, National University of Defense Technology, Changsha, China
`yingbocui@nudt.edu.cn`

Abstract. Single-cell sequencing has transformed the study of cellular heterogeneity and its role in health and disease. Current models predominantly adopt expression-driven learning frameworks that encode cellular states through the joint modeling of genes and their associated expressions. Current methods treat gene expression in various ways, such as using a fixed gene order, ranking genes based on expression levels, or combining gene IDs with expression values through simple addition. However, these approaches overlook the biological context of genes and rely on static representations that fail to capture gene activity's functional dynamics. Here, we present scFiLM, a dynamic fusion framework built upon the Feature-wise Linear Modulation (FiLM) paradigm, designed to integrate prior biological knowledge encoded in pretrained semantic gene embeddings derived from NCBI gene descriptions with expression-driven feature modulation. Specifically, for each geneexpression pair, scFiLM adaptively modulates the corresponding semantic embedding through FiLM layers, utilizing expression-derived scaling and shifting parameters. The resulting modulated embeddings are subsequently fed into a bidirectional Mamba encoder to capture long-range dependencies, thereby generating informative cell-level representations for downstream analyses. Comprehensive cell-level evaluations demonstrate that scFiLM not only effectively mitigates batch effects, but also achieves state-of-the-art performance in B cell type classification. At gene level, scFiLM retains the intrinsic functionality of genes and enables effective categorization of gene functions, thereby enhancing the interpretability of learned representations. The results highlight the potential of scFiLM as a robust and generalizable framework for single-cell transcriptomic analysis.

Keywords: Single-cell foundation model · Feature-wise Linear Modulation · Prior knowledge · Cell type annotation · Single-cell atlas

1 Introduction

Single-cell RNA sequencing [1,2] has profoundly transformed modern biology by enabling high-resolution characterization of cellular heterogeneity and revealing its essential roles in development, homeostasis, and disease progression [3].

© The Author(s), under exclusive license to Springer Nature Switzerland AG 2026
K. L. Kabir et al. (Eds.): BICOB 2026, CCIS 2977, pp. 233–243, 2026.
https://doi.org/10.1007/978-3-032-26028-4_17

With the rapid increase in the scale and complexity of single-cell transcriptome data [4], the need for building fundamental models that can generalize to multiple tasks and datasets is becoming increasingly urgent in the field of single-cell biology [5]. Modeling genes and their expression patterns using deep neural networks has become a core approach to extracting high-quality cell representations from massive single-cell datasets. Accurate and biologically interpretable cell embeddings are crucial for advancing key downstream tasks such as cell type annotation, perturbation response prediction, inference of drug or gene editing effects, and developmental trajectory reconstruction [6,7]. However, modeling single cells is inherently challenging due to the unique characteristics of single-cell data.

First, the quadratic computational complexity of many deep learning architectures, especially those Transformer-based models, hinders their scalability when applied to single-cell datasets, which are typically high-dimensional and sparse and may contain millions of cells or tens of thousands of genes [8,9].

Second, single-cell data inherently lack a natural sequence structure due to the unordered nature of gene expression profiles, posing a fundamental challenge for modeling, particularly for Transformer-based architectures that typically rely on sequential input. To address this, existing methods impose artificial sequence representations: scBERT [10] and scTab [11] employs a fixed gene order, Geneformer [12] ranks genes by expression levels as a positional prior, and scGPT [13] combines gene identifiers with binned expression values via additive embeddings. However, these approaches, while enabling model processing, fundamentally treat genes as mere position-like tokens [14], which fail to fully exploit the rich gene-expression relationships and the biologically grounded prior knowledge about gene function or semantics.

To address these challenges, we introduce scFiLM, a gene semantic and gene expression fusion method based on feature linear modulation, designed to simultaneously integrate gene functional semantics with expression dynamics. Unlike traditional modal fusion methods [15] that treat gene identity and expression as separate, static components, scFiLM leverages the feature linear modulation mechanism [16] to bridge gene semantics and expression dynamics in a context-dependent manner. scFiLM derives semantic gene embeddings directly from NCBI gene function description [17,18], infusing each gene token with explicit biological information. For each geneexpression pair, scFiLM applies FiLM-based modulation: the expression value generates channel-wise scaling and shifting parameters that dynamically reshape the gene embedding, allowing expression to selectively emphasize relevant semantic dimensions of the gene rather than merely contributing additive magnitude. The resulting dynamically modulated sequence is then processed by the Mamba2 encoder [19,20], which captures long-range dependencies with linear-time complexity, making efficient modeling of large-scale gene sets feasible. By unifying gene semantics with expression dynamics, scFiLM provides biologically grounded cell embeddings that enhance performance across multiple downstream tasks.

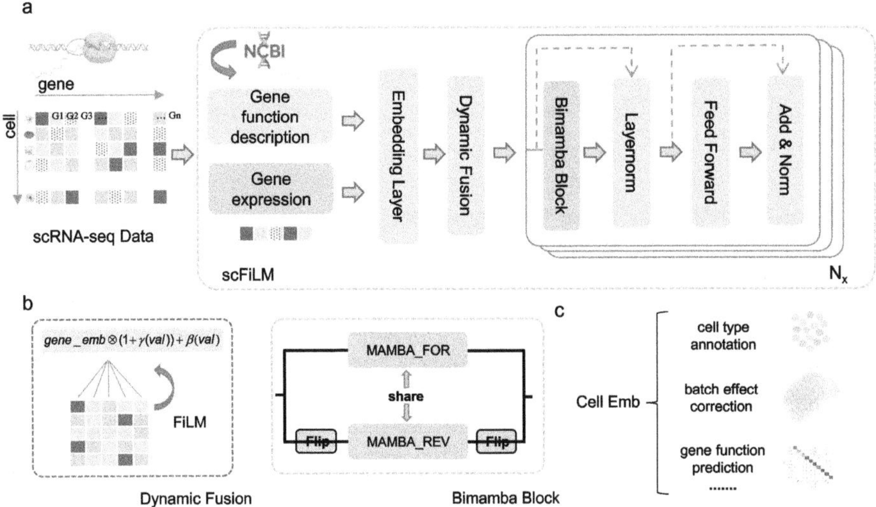

Fig. 1. Overview of the scFiLM framework. (a) For each single-cell data containing a pair of scRNA-seq data and genes: 1) During preprocessing, for genes identified in the scRNA-seq data, corresponding abstracts were obtained from NCBI. The functional descriptions in these abstracts were encoded into semantic embeddings via a text encoder. 2) Semantic embeddings of genes and embeddings of expression are dynamically fused through the film mechanism. 3) A bidirectional Mamba encoder was employed to model long-range dependencies in the dynamic transitions of single-cell gene expression profiles. (b) the architecture of Dynamic Fusion and Bimamba Block. (c) Downstream applications of scFiLM, including cell type annotation, batch effect correction, and gene function prediction, showcasing its versatility in various single-cell tasks.

2   Methods

2.1   Overall of scFiLM

scFiLM is a single-cell foundation model that integrates gene semantics with single-cell transcriptomic modeling. As illustrated in Fig. 1a, the overall framework consists of three interconnected parts: gene embeddings enriched with functional semantics derived from textual descriptions of gene functions, a dynamic fusion module based on the FiLM mechanism that adaptively combines gene identity embeddings with functional gene embeddings, and a bidirectional state space model built upon Mamba2 designed to capture long-range dependencies and generalize co-expression patterns across cells.

The functional gene embeddings are initialized using GenePT [21], a simple and effective embedding model for single-cell biology which aggregates gene lists commonly utilized in leading single-cell foundational models. Utilizing the GPT-3.5 text embedding model [22], GenePT generate dense vector representations that encapsulate the functional semantics of each gene. Subsequently, we employ the FiLM mechanism to modulate the functional embeddings based on gene expression levels using learned scale and shift parameters. The detailed architecture of this dynamic fusion process and the subsequent BiMamba block for sequence modeling are depicted in Fig. 1b. These combined representations are biologically meaningful, as demonstrated by various downstream applications shown in Fig. 1c.

2.2   Feature-Wise Linear Modulation

To effectively bridge static gene semantics with dynamic cellular contexts, scFiLM introduces a dynamic fusion mechanism based on Feature-wise Linear Modulation. FiLM is a lightweight yet expressive conditioning approach that adaptively adjusts pretrained gene embeddings using observed gene expression values. Specifically, FiLM takes as input a set of static gene embeddings encoding biological prior knowledge and a corresponding set of gene expression values for each cell. It then generates modulation parameters including a scaling factor γ and a shifting factor β by passing the expression values through a small neural network. These parameters are used to modulate the gene embeddings according to the following transformation:

$$\mathbf{Z} = \mathbf{G} \cdot (1 + \gamma) + \beta$$

where $\mathbf{G}$ denotes the original gene embeddings and $\mathbf{Z}$ represents the modulated token that integrates both semantic information and cellular expression dynamics. This mechanism enables the model to emphasize or suppress the contribution of individual genes based on their expression levels in each cell, improving the biological relevance of the learned representations.

2.3   Dynamic Gene Embedding Encoding with Bidirectional Mamba2

The bidirectional Mamba encoder takes as input the gene tokens modulated by FiLM, and applies state-space-based sequence modeling to capture long-range dependencies and global expression patterns among genes within each cell. By leveraging bidirectional processing, the model aggregates both past and future contextual information, producing a comprehensive and discriminative cell embedding that reflects the overall transcriptional state.

The Mamba architecture utilizes a structured state space model (SSM). The continuous-time SSM maps an input signal $x(t) \in \mathbb{R}$ to an output $y(t) \in \mathbb{R}$ through a latent state $h(t) \in \mathbb{R}^N$:

$$\dot{h}(t) = \mathbf{A}h(t) + \mathbf{B}x(t), \quad y(t) = \mathbf{C}h(t), \tag{1}$$

where $\mathbf{A} \in \mathbb{R}^{N \times N}$ governs the state dynamics, $\mathbf{B} \in \mathbb{R}^{N \times 1}$ projects the input, and $\mathbf{C} \in \mathbb{R}^{1 \times N}$ produces the output.

For practical implementation, the system is discretized using a zero-order hold (ZOH) scheme with step size Δ:

$$h_t = \overline{\mathbf{A}}h_{t-1} + \overline{\mathbf{B}}x_t, \quad y_t = \mathbf{C}h_t. \tag{2}$$

With the discretized parameters given by:

$$\overline{\mathbf{A}} = e^{\Delta \mathbf{A}}, \tag{3}$$

$$\overline{\mathbf{B}} = \left(\int_0^{\Delta} e^{\mathbf{A}\tau} d\tau \right) \mathbf{B}. \tag{4}$$

When $\Delta \mathbf{A}$ is invertible, $\overline{\mathbf{B}}$ can be approximated as $(\Delta \mathbf{A})^{-1}(e^{\Delta \mathbf{A}} - \mathbf{I})\mathbf{B}$.

Mamba2 enhances this basic framework by introducing data-dependent parameterization, dynamically generating the parameters $\overline{\mathbf{A}}_t, \overline{\mathbf{B}}_t, \mathbf{C}_t$ (and Δ_t) as functions of the input x_t. This enables the model to selectively propagate or forget information based on the current input. Mamba2 further improves this approach through the State Space Duality (SSD) framework, simplifying the state transition matrix $\mathbf{A}$ and enabling optimized parallel algorithms.

The core computation in a Mamba2 block involves:

$$h_t = \overline{\mathbf{A}}_t h_{t-1} + \overline{\mathbf{B}}_t x_t, \quad y_t = \mathbf{C}_t h_t. \tag{5}$$

To capture bidirectional context for genomic interactions, an explicit bidirectional encoding strategy is introduced. The input sequence $\mathbf{X} = \{\mathbf{x}_{\text{cls}}, \mathbf{gene}_1, \dots, \mathbf{gene}_n, \mathbf{pad}_1, \dots, \mathbf{pad}_m\}$ is processed in the forward direction, and a reversed sequence $\mathbf{X}_{\text{rev}} = \{\mathbf{gene}_n, \dots, \mathbf{gene}_1, \mathbf{x}_{\text{cls}}, \mathbf{pad}_1, \dots, \mathbf{pad}_m\}$ is created by flipping the gene embeddings. Both sequences are processed independently through identical stacks of Mamba2 blocks. The final cell embedding $\mathbf{c}$ is obtained by summing the output [CLS] token embeddings from both directions:

$$\mathbf{c} = \text{Mamba2}(\mathbf{X})_{[\text{CLS}]} + \text{Mamba2}(\mathbf{X}_{\text{reverse}})_{[\text{CLS}]}. \tag{6}$$

3   Results

3.1   scFiLM Removes Batch Effect and Enables Accurate Annotation of B Cells

Cell annotation is the cornerstone of single-cell omics research. Particularly, precision B cell annotation is challenging and crucial [23]. B cells exhibit significant functional diversity in the tumor microenvironment, and their precise annotation is reshaping our understanding of tumor immunity, providing key scientific evidence for precision medicine approaches [24, 25]. To rigorously benchmark the performance of our proposed foundation model, scFiLM, we conducted a large independent evaluation using a carefully curated dataset [26], which contains 126,101 rigorously quality-controlled B cells and plasma cells, including 10 subtypes in the B cell differentiation process. scFiLM was benchmarked against a suite of representative single-cell foundation models, including Geneformer, scGPT, scFoundation, and scTab. These baselines cover both recent transformer-based architectures and semantically irrelevant foundation models.

We first assess whether cell embeddings are robust to batch-dependent technical artefacts such as patient variability. Specifically, we performed the following analysis to quantify batch effects at the patient level: (i) First, the data (either raw RNA-seq data or one of the pre-trained embeddings) were projected onto the top 50 principal components; (ii) Then, k-means clustering was applied, where k represents the number of distinct patients; (iii) The ARI values between cell clusters and patient clusters were calculated [27]. Higher ARI values indicate stronger batch effects at the patient level. Due to its refined data processing, the raw single-cell RNA sequencing data in the B-cell dataset has an ARI of 0.12, indicating moderate batch effect. Nevertheless, we were able to reduce the ARI to 0.05, compared to 0.06 for Geneformer and scGPT, 0.07 for scFoundation, and 0.10 for scTab. We visualized the UMAP plots of the 20 patients with the highest cell count in the B cell dataset. As shown in Fig. 2a, the left plot shows the raw RNA, exhibiting significant batch effect, while the embeddings generated by scFiLM demonstrate robustness to batch effect. To further validate this, we used 10% random samples from the cardiomyopathy dataset provided by genept. With all settings consistent, scFiLM reduced the ARI from 0.33 to 0.01, and the right plot in Fig. 2a displays the UMAP projection of cardiomyocytes, colored by individual patient.

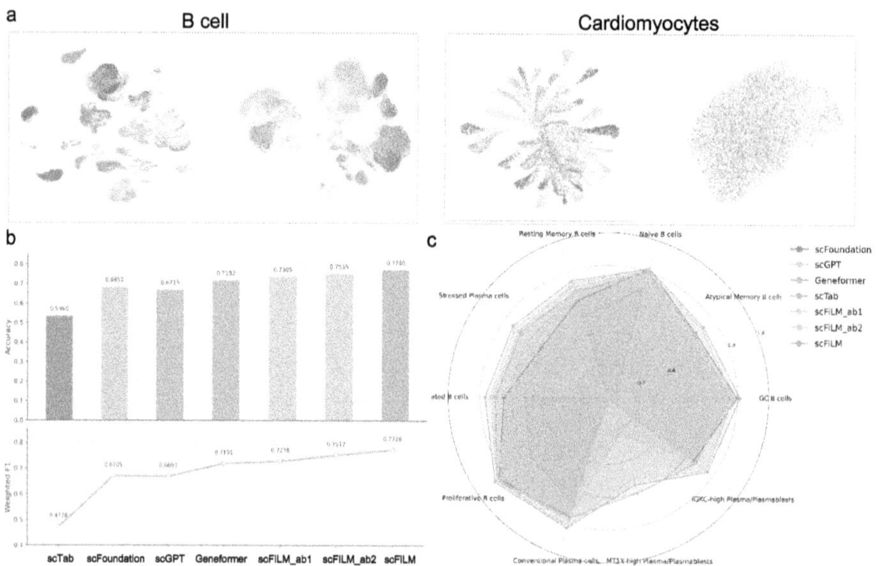

Fig. 2. scFiLM lead to better batch integration and cell type annotation. (a) UMAP visualizations of B cell dataset(left) and cardiomyopathy dataset(right): generated using both raw RNA-seq and scFiLM embeddings, demonstrate scFiLM's effectiveness in mitigating batch effects. (b) Cell type classification performance on the B cell test set, reported as accuracy and weighted F1-score. The results include scFiLM, other single-cell foundation models, and ablation variants designed to isolate the contribution of gene semantics (scFiLM_ab1) and the FiLM mechanism (scFiLM_ab2). (c) Radar plots of F1-scores for B cell subtypes. scFiLM demonstrates superior discriminative power across nearly all cell types compared to other models.

We then divided the B cell dataset into training, testing, and validation sets according to the donor level. Performance was primarily assessed using cell type classification accuracy (ACC) and weighted F1-score (WF1). As illustrated in Fig. 2b, scFiLM achieved state-of-the-art performance on the test set, attaining an accuracy of 0.7740 and a weighted F1-score of 0.7728, outperforming all baseline methods. Specifically, it outperformed Geneformer (ACC = 0.7191, WF1 = 0.7191), the second-best method, as well as scFoundation (ACC = 0.6852, WF1 = 0.6705), scGPT (ACC = 0.6715, WF1 = 0.6691), and scTab (ACC = 0.5360, WF1 = 0.4728) by substantial margins. scFiLM utilizes shallow neural networks for continuous value projection, thereby preserving transcriptional continuity while avoiding information loss. Furthermore, by directly integrating semantic gene embeddings into the learning process, the feature representations generated by scFiLM inherently possess cell type annotation discriminativeness, as shown in Fig. 2c, scFiLM achieved the highest F1 score in almost all B cell subtypes.

To further analyze the contributions of our model, we conducted ablation experiments to validate the effectiveness of the FiLM mechanism and gene semantics. Specifically, a scFiLM variant using a simple add mechanism without the FiLM mechanism showed a performance drop, with an accuracy of 0.7535. Meanwhile, another variant without semantic gene emb achieved an accuracy of 0.7385. Together, these findings demonstrate that integrating gene semantic information with the FiLM mechanism can significantly enhance the performance of foundation models in single-cell analysis.

3.2   scFiLM Keeps Underlying Gene Functionality

In this section, we conduct a comprehensive evaluation of the quality of gene embeddings generated by scFiLM. Unlike conventional approaches that freeze pre-trained semantic gene embeddings, scFiLM leverages the Masked Language Modeling (MLM) objective to dynamically integrate gene co-expression patterns directly into the semantic space. This allows the model to refine the initial gene representations, creating a more biologically informed embedding.

To quantify this improvement, we selected ten distinct gene functional classes from the GenePT database, encompassing a diverse set of 33,537 genes. We then evaluated the organization of genes from different functional classes within the embedding space by calculating the average silhouette coefficient (Fig. 3a). The scFiLM embeddings achieved a higher score of 0.3312, compared to just 0.1681 for the original semantic embeddings. The resulting plot also reveals a better separated clusters with scFiLM embeddings. This demonstrates that scFiLM successfully retains the foundational semantic knowledge while layering on crucial biological context derived from co-expression, resulting in a more robust representation of gene function.

To move beyond qualitative assessment, we performed a quantitative validation using a supervised learning framework. We partitioned the genes into a 70%/30% training/test split and trained a logistic regression classifier to predict gene functional categories based on their embeddings. The results show that

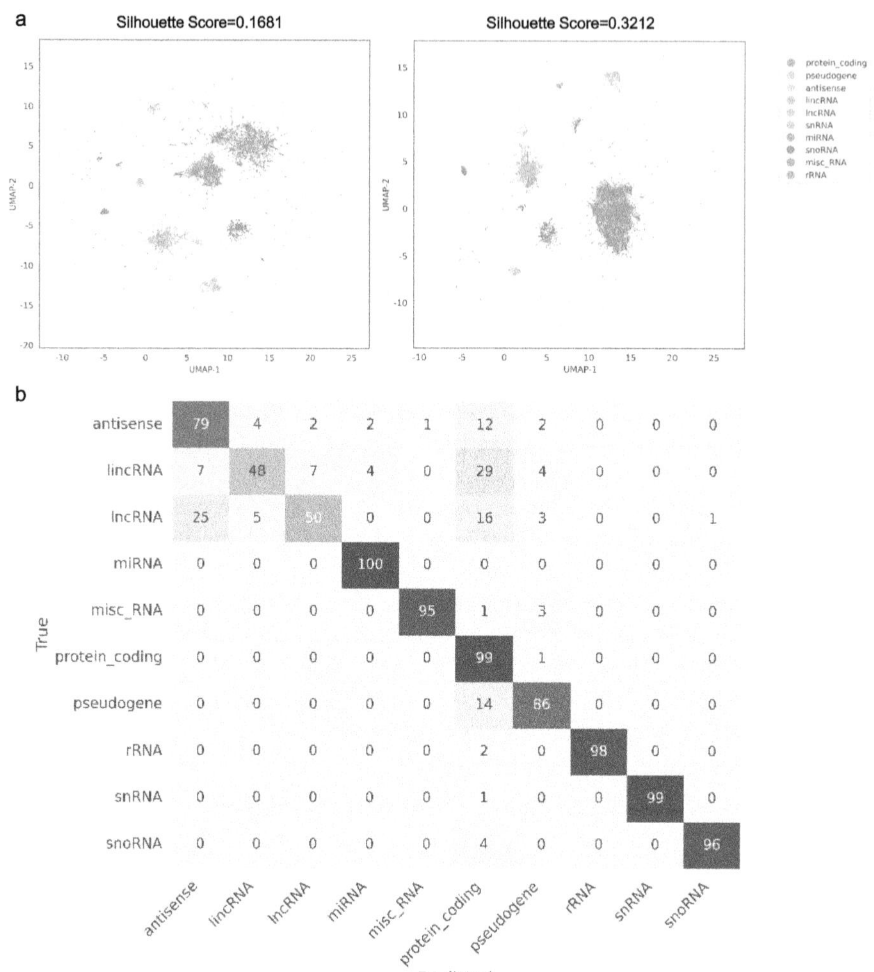

Fig. 3. scFiLM enhances gene embedding semantics and biological relevance. (a) UMAP visualization of gene embeddings colored by functional class, comparing original semantic embeddings (left) and scFiLM-pretrained embeddings (right). Embeddings were derived from 33,537 genes across 10 functional classes. (b) Confusion matrix of functional category predictions on the test set, using logistic regression trained on scFiLM-pretrained embeddings.

the classifier performs exceptionally well with both embedding types, achieving overall accuracies of 93% (original) and 95% (scFiLM). Crucially, the confusion matrix (Fig. 3b) reveals that the few misclassifications for the scFiLM embeddings occur predominantly between highly related functional groups, such as different classes of non-coding RNAs (lincRNA, lncRNA).

4   Conclusion

In this study, we present scFiLM, a dynamic fusion framework that integrates gene functional semantics with single-cell expression profiles to enhance downstream analysis. Our model first encodes gene identity as semantic embeddings and cell states as expression-based representations. To align these heterogeneous modalities, we employ a FiLM mechanism that modulates expression features conditioned on gene semantics, enabling context-aware feature refinement. For efficient modeling of the fused representations, we further incorporate a bidirectional Mamba encoder to capture long-range dependencies in cellular state transitions. In our benchmarks, scFiLM achieves state-of-the-art performance.

Despite these advances, scFiLM still presents several limitations that call for further improvement. First, its gene functional semantics, which are currently derived from NCBI summaries and embedded using a GPT-3.5 text encoder, could benefit from a biologically specialized language model capable of generating more context-relevant gene embeddings. Second, although the FiLM mechanism provides an efficient and lightweight fusion module for aligning gene identity with gene expression, the use of more sophisticated strategies may support deeper multimodal interactions. Third, the current training objective relies solely on a masked language modeling (MLM) loss, which primarily encourages local contextual reconstruction but may not fully capture the hierarchical dependencies and functional relationships among genes. A more comprehensive loss design could be introduced, combining multiple complementary objectives—such as contrastive alignment between gene and expression spaces, cell-type classification or clustering consistency regularization, and reconstruction-based objectives for denoising and manifold preservation. Such a composite optimization framework would encourage the model to learn richer and more biologically faithful representations that generalize better across tissues and perturbation conditions.

Implementation Details

scFiLM consists of a dynamic fusion layer and 5 stacked Bi_mamba blocks. During pre-training, we used 15 million cells for training scFiLM. The model was optimized using the Adam optimizer with a starting learning rate of 1e-6 and trained for a total of 2 epochs. The total batch size was 96, distributed equally across 2 A6000 servers, with each GPU utilizing 48 GB memory.

Data Availability

scFiLM was pre-trained on a comprehensive single-cell RNA sequencing dataset with a total of 15,240,192 high-quality single-cell samples. The complete dataset, including pre-processed training, validation, and test subsets, is publicly available for download at: https://pklab.med.harvard.edu/felix/data/merlin_cxg_2023_05_15_sf-log1p.tar.gz.

For the task of B cell type annotation, the corresponding dataset is accessible via https://zenodo.org/records/13385122. This dataset encompasses 10 B cell subsets, including GC B cells, Atypical Memory B cells, Naive B cellsResting Memory B cells, Stressed Plasma cells, Activated B cells, Proliferative B cells, Conventional Plasma cells, MT1X-high Plasma/Plasmablasts and IGKC-high Plasma/Plasmablasts.

Acknowledgments. This work was supported by the National Natural Science Foundation of China No. 62102427, the Science and Technology Innovation Program of Hunan Province No. 2024RC3115, the Innovative Talent Program of the National University of Defense Technology.

Author contributions. Y.C. conceived and supervised the project. X.L. and X.Z. contributed to the algorithm implementation. X.L. wrote the manuscript. Y.C., T.T. and X.L. were involved in the discussion and proofreading.

Conflict of Interest. The authors declare that no competing interests exist.

References

1. Xia, Z., et al.: A review of parallel implementations for the smith–waterman algorithm. Interdisc. Sci. Comput. Life Sci. 1–14 (2021)
2. Wang, Z., Cui, Y., Peng, S., Liao, X., Yu, Y.: Minimapr: a parallel alignment tool for the analysis of large-scale third-generation sequencing data. Comput. Biol. Chem. **99**, 107735 (2022)
3. Han, X., et al.: Construction of a human cell landscape at single-cell level. Nature **581**(7808), 303–309 (2020)
4. Yanai, I., Haas, S., Lippert, C., Kretzmer, H.: Cellular atlases are unlocking the mysteries of the human body. Nature Publishing Group UK London (2024)
5. Yiu, T., Chen, B., Wang, H., Feng, G., Fu, Q., Hu, H.: Transformative advances in single-cell omics: a comprehensive review of foundation models, multimodal integration and computational ecosystems. J. Transl. Med. **23**(1), 1176 (2025). https://doi.org/10.1186/s12967-025-07091-0
6. Morris, S.A.: The evolving concept of cell identity in the single cell era. Development **146**(12), 169748 (2019). https://doi.org/10.1242/dev.169748
7. Erfanian, N., et al.: Deep learning applications in single-cell genomics and transcriptomics data analysis. Biomed. Pharmacother. **165**, 115077 (2023). https://doi.org/10.1016/j.biopha.2023.115077
8. Szałata, A., et al.: Transformers in single-cell omics: a review and new perspectives. Nat. Methods **21**(8), 1430–1443 (2024). https://doi.org/10.1038/s41592-024-02353-z
9. Xu, C., et al.: Automatic cell-type harmonization and integration across human cell atlas datasets. Cell **186**(26), 5876–5891 (2023)
10. Yang, F., et al.: scBERT as a large-scale pretrained deep language model for cell type annotation of single-cell RNA-seq data. Nat. Mach. Intell. **4**(10), 852–866 (2022)
11. Fischer, F., et al.: scTab: scaling cross-tissue single-cell annotation models. Nat. Commun. **15**(1), 6611 (2024)
12. Theodoris, C.V., et al.: Transfer learning enables predictions in network biology. Nature **618**(7965), 616–624 (2023). https://doi.org/10.1038/s41586-023-06139-9
13. Cui, H., et al.: scGPT: toward building a foundation model for single-cell multiomics using generative AI. Nat. Methods 1–11 (2024)

14. Lan, W., He, G., Liu, M., Chen, Q., Cao, J., Peng, W.: Transformer-based single-cell language model: a survey. Big Data Mining Anal. **7**(4), 1169–1186 (2024). https://doi.org/10.26599/BDMA.2024.9020034

15. Cui, Y., et al.: Spabalance: balanced learning for efficient spatial multi-omics decoding. Adv. Sci., 12973 (2025). https://doi.org/10.1002/advs.202512973

16. Perez, E., Strub, F., Vries, H., Dumoulin, V., Courville, A.: Film: visual reasoning with a general conditioning layer. In: Proceedings of the AAAI Conference on Artificial Intelligence, vol. 32. AAAI Press, (2018). https://arxiv.org/abs/1709.07871

17. Wheeler, D.L., et al.: Database resources of the national center for biotechnology information. Nucleic Acids Res. **28**(1), 10–14 (2000). https://doi.org/10.1093/nar/28.1.10. https://academic.oup.com/nar/article-pdf/28/1/10/9895099/280010.pdf

18. Brown, G.R., et al.: Gene: a gene-centered information resource at NCBI. Nucleic Acids Res. **43**(D1), 36–42 (2014). https://doi.org/10.1093/nar/gku1055. https://academic.oup.com/nar/article-pdf/43/D1/D36/7314925/gku1055.pdf

19. Dao, T., Gu, A.: Transformers are SSMs: generalized models and efficient algorithms through structured state space duality. In: Proceedings of the International Conference on Machine Learning (ICML) (2024). arXiv preprint arXiv:2405.21060. https://arxiv.org/abs/2405.21060

20. Zhao, Y., Zhao, B., Zhang, F., He, C., Wu, W., Lai, L.: Sc-mamba2: leveraging state-space models for efficient single-cell ultra-long transcriptome modeling. bioRxiv (2024). https://doi.org/10.1101/2024.09.30.615775. https://www.biorxiv.org/content/early/2024/10/26/2024.09.30.615775.full.pdf

21. Chen, Y.T., Zou, J.: Genept: a simple but hard-to-beat foundation model for genes and cells built from chatgpt. bioRxiv (2023)

22. Chen, Q., et al.: A comprehensive benchmark study on biomedical text generation and mining with chatgpt. bioRxiv (2023). https://doi.org/10.1101/2023.04.19.537463. https://www.biorxiv.org/content/early/2023/04/20/2023.04.19.537463.full.pdf

23. Chen, J., et al.: Transformer for one stop interpretable cell type annotation. Nat. Commun. **14**(1), 223 (2023)

24. Garaud, S., et al.: Tumor-infiltrating B cells signal functional humoral immune responses in breast cancer. JCI Insight **4**(18), 129641 (2019)

25. Glass, D.R., et al.: An integrated multi-omic single-cell atlas of human B cell identity. Immunity **53**(1), 217–232 (2020)

26. Fitzsimons, E., et al.: A pan-cancer single-cell RNA-seq atlas of intratumoral B cells. Cancer Cell **42**(10), 1784–1797 (2024). https://doi.org/10.1016/j.ccell.2024.09.011

27. Hubert, L., Arabie, P.: Comparing partitions. J. Classif. **2**(1), 193–218 (1985)

Computational Discovery of CRISPR-Cas13b Guide RNAs for Broad-Spectrum Dengue Virus Targeting

Syed Muhammad Ali Naqvi[✉][iD], Farhan Khan, and Syed Muhammad Muslim

Habib University, Karachi, Pakistan
{sn07590,sh07730}@st.habib.edu.pk, farhan.khan@sse.habib.edu.pk
http://www.habib.edu.pk/

Abstract. Dengue (DENV), an RNA virus, remains a significant global health threat, particularly in developing regions, with no widely effective antiviral therapy available. The CRISPR-Cas13b system, specifically the *PspCas13b* subtype, has emerged as a promising programmable antiviral tool capable of targeting viral RNA with high specificity. However, the efficacy of Cas13b-based interventions relies heavily on the design of potent and conserved CRISPR RNA (crRNA) spacer sequences, a task complicated by high viral genetic diversity. Unlike CRISPR-Cas9, which targets double-stranded DNA in eukaryotic genomes, Cas13b directly targets single-stranded RNA, making it ideally suited for RNA virus therapeutics; however, existing computational tools predominantly focus on Cas9 DNA targeting or Cas13d for mammalian transcript knockdown, leaving a significant gap for Cas13b-specific viral antiviral design.

In this paper, we develop a computational pipeline and machine learning framework for the rational design of high-efficacy Cas13b guide RNAs targeting all four Dengue serotypes. Our approach integrates large-scale genomic data extraction, conservation analysis, and a novel *in silico* optimization module for guide RNA (gRNA) sequences, based on recently reported Cas13b design rules (e.g., 5' GG motif preference, Cytosine penalties). To predict targeting efficiency, we benchmark classical machine learning models (Random Forest, XGBoost) against foundation model-based predictors (Nucleotide Transformer, RNA-FM) using a dataset of experimentally validated spacers and find that classical feature-engineered models outperform deep learning approaches when trained on experimentally validated gRNA datasets in low-data regimes.

We identify highly conserved, optimized crRNA candidates, including several pan-serotype guides with predicted high potency. This work establishes a reproducible, open pipeline and prioritized candidate resource to accelerate experimental testing of Cas13b-based antivirals against Dengue and other RNA viruses.

Code: https://github.com/muhammadali74/CAS13b_pipeline.

Supplementary Information The online version contains supplementary material available at https://doi.org/10.1007/978-3-032-26028-4_18.

K. L. Kabir et al. (Eds.): BICOB 2026, CCIS 2977, pp. 244–259, 2026.
https://doi.org/10.1007/978-3-032-26028-4_18

Keywords: CRISPR-Cas13b · Dengue Virus · Machine Learning · Guide RNA Design · Computational Biology

1   Introduction

Dengue Virus (DENV) remains one of the most critical mosquito-borne viral pathogens worldwide, posing a severe public health burden, particularly in developing nations such as Pakistan, India, and Brazil [2, 24]. With four distinct serotypes (DENV-1 through DENV-4), the virus causes millions of infections annually, ranging from mild fever to fatal hemorrhagic fever. Despite the prevalence of the disease, therapeutic options remain limited. Developing effective antivirals or vaccines is complicated by the high genetic diversity of the virus and the phenomenon of antibody-dependent enhancement (ADE), where sub-optimal antibodies can worsen symptoms upon secondary infection [14]. Consequently, there is an urgent need for precise, molecular-level interventions that can target conserved viral regions across all serotypes.

The CRISPR-Cas13 system has emerged as a revolutionary tool for RNA targeting. Unlike the DNA-targeting Cas9, Cas13 enzymes (such as Cas13a, Cas13b, and Cas13d) target single-stranded RNA, making them ideal candidates for combating RNA viruses like Dengue. Among these, the Cas13b subtype (specifically *PspCas13b*) has demonstrated high specificity and minimal off-target effects, rendering it a promising candidate for therapeutic and diagnostic applications [11]. However, the efficacy of Cas13b is highly dependent on the design of the CRISPR RNA (crRNA) spacer sequence. Selecting a 30-nucleotide (nt) spacer that is both conserved across viral mutations and highly efficient in triggering Cas13b activity is a multi-objective optimization problem that is computationally non-trivial.

Traditional selection of guide RNAs often relies on simple sequence alignment or tedious manual screening in the wet lab, which is cost-prohibitive and time-consuming. While Artificial Intelligence (AI) and Machine Learning (ML) have been successfully applied to Cas9 and Cas13d guide design [5, 6, 22], there is a distinct lack of computational frameworks specifically tailored for Cas13b, particularly models trained to navigate the constraints of viral conservation.

In this work, we develop a systematic computational pipeline to identify and optimize Cas13b guide RNA candidates targeting Dengue Virus. By applying known design principles, conservation-based filtering, and machine learning-based ranking, we compile a prioritized catalog of candidates suitable for experimental validation. This resource is intended to accelerate wet-lab screening and therapeutic development. We address the challenge of limited experimental data by benchmarking classical ML models against large language model (LLM) embeddings (such as Nucleotide Transformer and RNA-FM), establishing a robust baseline for Cas13b efficiency prediction.

The main contributions of this work are as follows:

1. A comprehensive data processing pipeline that identifies highly conserved 30-nt spacers across all four Dengue serotypes from public genomic databases.

2. An optimization module that enhances crRNA efficacy by introducing strategic mismatches based on specific Cas13b design rules.
3. A comparative analysis of machine learning approaches for efficiency prediction on limited experimental data.
4. A catalog of top-ranked, optimized crRNA candidates provided as a resource for experimental validation.

2 Literature Review

2.1 CRISPR-Cas13 as an Antiviral Strategy

The discovery of CRISPR-Cas systems adapted for RNA targeting has opened new avenues for antiviral therapy. While Cas9 targets DNA, the Class 2 Type VI CRISPR systems, characterized by the effector protein Cas13, possess RNase activity capable of cleaving single-stranded RNA [1]. Early applications utilized Cas13a (formerly C2c2) for pathogen detection, such as the SHERLOCK platform. More recently, focus has shifted toward Cas13b and Cas13d for therapeutic silencing due to their lack of protospacer flanking site (PFS) requirements in eukaryotic cells [7]. In the context of RNA viruses, Cas13 has been shown to effectively inhibit Influenza A, SARS-CoV-2, and Dengue in cell cultures, positioning it as a programmable antiviral agent [10].

2.2 Design Rules for PspCas13b

For a CRISPR system to be effective, the guide RNA must hybridize efficiently with the target. While Cas13 systems generally exhibit tolerance to mismatches, recent studies have elucidated specific design constraints for *PspCas13b*. Hu et al. [13] conducted a systematic screening of Cas13b crRNAs, revealing that secondary structure has a moderate effect on efficacy, while specific nucleotide motifs drive potency. Specifically, a "GG" motif at the 5' end of the spacer (positions 1–2) typically enhances activity, whereas Cytosine (C) residues at positions 1–4 and the central region (11–17) act as negative predictors. These findings imply that simply selecting guide RNAs based on conservation of the target viral sequence is insufficient; a pipeline must optimize sequences to satisfy these biochemical preferences, even if it requires introducing artificial mismatches, which Cas13b is known to tolerate [13].

2.3 Machine Learning in CRISPR Guide Design

The application of Machine Learning to predicting CRISPR guide efficiency is well-established for Cas9. Tools such as DeepCRISPR [6] and CRISPR-scan [21] utilize deep neural networks and extensive feature engineering to predict on-target efficiency and off-target risks. However, these models are not transferable to Cas13 due to fundamental differences in the cleavage mechanism and target substrate (RNA vs. DNA).

For Cas13 systems, recent tools like CASowary [15], and prediction models for Cas13d [5,25] have begun to emerge. These typically rely on Convolutional Neural Networks (CNNs) trained on high-throughput screening data. However, for Cas13b specifically, public datasets remain scarce. The challenge is further compounded when applying large genomic foundation models. While Transformer-based models like Nucleotide Transformer [8] and RNA-FM [4] excel at capturing long-range genomic dependencies, their application to regression tasks on small, specialized datasets (such as Cas13b efficiency data) often results in overfitting or poor generalization compared to classical ensemble methods using domain-specific engineered features. This paper addresses this gap by developing a specialized pipeline that combines bio-informatic filtering with an ML model optimized for the specific constraints of Cas13b.

3   Methodology

Our proposed pipeline integrates bioinformatics filtering, conservation analysis, sequence optimization, and machine learning-based efficacy prediction to identify high-potency Cas13b crRNA candidates for Dengue Virus. The workflow consists of five main stages described below.

3.1   Data Acquisition and Preprocessing

Dengue Virus genome sequences were retrieved from the Bacterial and Viral Bioinformatics Resource Center (BV-BRC) database. The raw dataset comprised genomic sequences spanning all four Dengue serotypes (DENV-1 through DENV-4) as well as unclassified strains. To ensure data quality, sequences were filtered based on two criteria:

1. **Ambiguity threshold:** Sequences containing more than 5% ambiguous nucleotides (denoted as 'N') were excluded to prevent spurious spacer extraction.
2. **Minimum length:** Only genomes exceeding 1,000 base pairs were retained, as shorter fragments are unlikely to represent complete viral coding regions.

Each genome was annotated with its corresponding serotype by parsing the genome name metadata. The final filtered dataset was converted into FASTA format for downstream processing. Post-filtering statistics are presented in Table 1.

3.2   Spacer Extraction and Conservation Analysis

To identify candidate crRNA spacers, we employed a sliding window approach to extract all possible 30-nucleotide (nt) subsequences from each viral genome. A memory-efficient SQLite-based storage system was implemented to handle the large combinatorial space. For each unique 30-nt spacer:

– We tracked the set of genome identifiers in which it appeared to avoid redundant counting.

Table 1. Genome Statistics After Quality Filtering

Serotype	Count
DENV-1	14,906
DENV-2	11,975
DENV-3	4,897
DENV-4	2,868
Unclassified	3,411
Total	**38,057**

– Conservation percentage was computed as:

$$\text{Conservation} = \frac{\text{No. of genomes containing spacer}}{\text{Total genomes}} \times 100 \qquad (1)$$

– Subtype-specific occurrence counts were calculated by mapping genome IDs to their serotype annotations.

To focus on the most promising candidates, we selected the top-300 most frequently occurring spacers per serotype, ensuring broad coverage across all viral variants.

3.3   Sequence Filtering and Quality Control

Each candidate spacer was annotated with biochemical properties known to influence Cas13b activity. Key metrics included: (1) **GC Content**, where extreme values can affect stability; and (2) **Homopolymer Runs** (> 4 consecutive identical nucleotides), which can cause transcriptional artifacts. Strict filtering was not applied at this stage, as subsequent optimization would further refine selection.

3.4   Cas13b-Specific Sequence Optimization

Recent functional studies [13] have revealed that PspCas13b activity is strongly influenced by position-specific nucleotide preferences. We adapted the scoring metric from Hu et al. for our optimization algorithm: (1) Guanine (G) at positions 1 or 2 confers +20 points; (2) Cytosine (C) at positions 1–4 incurs -20 points; and (3) Cytosine at positions 11, 12, 15–17 incurs -5 points.

Based on this scheme, we developed an optimization algorithm that operates as follows:

1. Compute the baseline score for the original spacer.
2. Identify all positions where beneficial mutations exist (e.g., replacing C with G).
3. Generate all combinations of 1, 2, or 3 mutations.

4. Score each variant and retain those achieving the maximum score.
5. If multiple variants share the top score, all are preserved.

This strategy leverages the known mismatch tolerance of Cas13b [13], which accommodates up to 3–4 non-consecutive mismatches without significant loss of activity.

3.5   Machine Learning Model for Efficiency Prediction

Training Dataset. A curated dataset of 245 Cas13b spacer sequences with experimentally validated mean targeting efficiencies was used for model training. The data were extracted from the supplementary data of [13], and the specific source table is provided in Supplementary Table S1

Feature Engineering. Given the limited sample size ($n = 245$), we employed extensive feature engineering to encode domain-specific knowledge:

- **Position-Specific One-Hot Encoding:** 120 binary features representing A, T, G, C at each position.
- **Motif Features:** Boolean features for critical motifs (e.g., 'GG' at positions 1–2).
- **Sequence Composition:** GC content, nucleotide fractions, and dinucleotide frequencies.

A complete list of the 157 engineered features is provided in Supplementary Table S2

Model Selection and Training. We compared multiple regression models suitable for small, high-dimensional datasets: Random Forest, Gradient Boosting, XGBoost, SVR, and Ridge Regression. Hyperparameter tuning was performed using 5-fold cross-validation with grid search, optimizing for R^2. The final model was trained on all 245 samples.

Deep Learning Experiments. To explore whether large pre-trained genomic foundation models could improve upon classical ML approaches, we conducted two additional experiments. We experimented with two foundation models:

1. **Nucleotide Transformer (NT):** A BERT-style transformer pre-trained on genomic DNA used to extract 512-dimensional embeddings.
2. **RNA-FM:** A foundation model pre-trained on RNA sequences used for embedding extraction.

Both approaches used a lightweight regression head trained with heavy regularization (dropout $= 0.3$, weight decay $= 0.01$) to mitigate overfitting.

3.6   Candidate Ranking and Selection

The trained machine learning model was applied to all optimized spacer sequences. Candidates were ranked by: (1) Predicted targeting efficiency; (2) Conservation across serotypes; and (3) Cas13b position score. For each serotype, we identified the top most conserved spacers and all spacers with predicted efficiency ≥ 0.8.

4   Experiments and Results

4.1   Spacer Extraction and Conservation

The sliding window extraction across all filtered genomes yielded 3.78 million unique 30-nt spacers. After subtype-specific conservation ranking and selection of the top-300 per serotype, we retained 1,740 candidates for optimization.

4.2   Sequence Optimization Results

The Cas13b-specific optimization algorithm significantly improved spacer quality. Of the 1,740 input spacers, 1,714 sequences (98.5%) received at least one beneficial mutation, while 26 sequences (1.5%) were already optimal.

The optimization yielded substantial improvements in predicted potency:

- **Average Score Improvement:** +47.17 points (SD $= 20.37$)
- **Maximum Score Improvement:** +100 points

The majority of spacers required multiple mutations to achieve maximum potency: 9.9% required one mutation, 23.7% required two mutations, and 64.9% required all three allowed mutations. Score improvement correlated monotonically with mutation count, yielding average improvements of 17.8, 40.2, and 55.6 points for one, two, and three mutations, respectively.

Position-specific analysis revealed that 73.3% of all mutations occurred at positions 1–2 (introducing the favorable 'GG' motif), while the remaining mutations targeted penalized Cytosine residues in positions 3–4 and the central region (11–12, 15–17). Notably, conservation percentage did not predict optimization potential: highly conserved spacers ($\geq$80%) benefited equally from optimization as less conserved candidates, indicating that sequence conservation and biochemical potency represent orthogonal selection criteria. Figure 1 illustrates these optimization dynamics.

4.3   Machine Learning Model Performance

Classical ML Models. Table 2 summarizes the 5-fold cross-validation performance. XGBoost achieved the highest cross-validated R^2 of 0.235, the lowest MAE of 0.161, and a high Spearman correlation of 0.529, indicating moderate predictive power suitable for ranking candidate spacers.

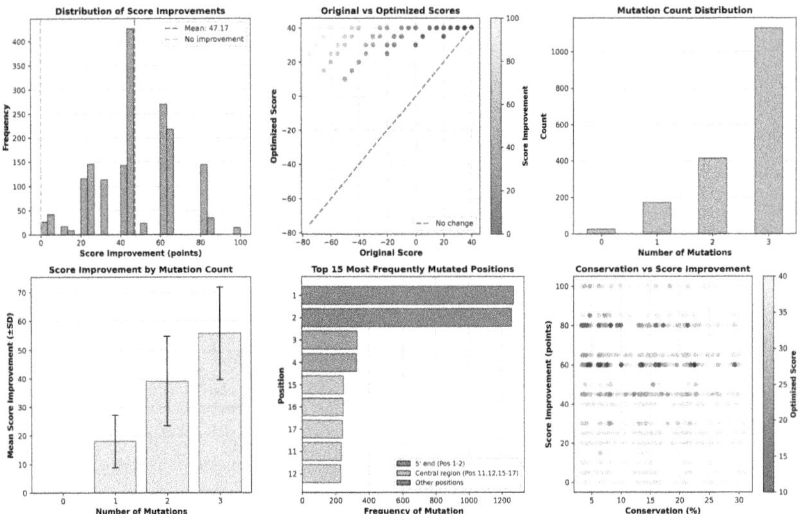

Fig. 1. Comprehensive analysis of Cas13b-specific sequence optimization. (a) Distribution of score improvements showing mean of +47.17 points. (b) Scatter plot of original vs optimized scores colored by improvement magnitude. (c) Distribution of mutation counts per spacer. (d) Mean score improvement stratified by mutation count. (e) Top 15 most frequently mutated positions. (f) Conservation vs score improvement.

Table 2. Machine Learning Model Comparison (5-Fold CV)

Model	Mean R^2	Mean MAE	Mean Spearman
XGBoost	0.235	0.161	0.529
Gradient Boosting	0.199	0.220	0.497
Random Forest	0.192	0.221	0.480
SVR	0.227	0.217	0.533
Ridge Regression	0.139	0.231	0.494

Deep Learning Model Performance. Despite leveraging large pre-trained models, the deep learning approaches underperformed relative to classical ML, likely due to insufficient training data. For both Nucleotide Transformer and RNA-FM embeddings, we trained a MLP with hidden layers of sizes 256, 128 and 64, learning rates 1e-3, 5e-4, and L2 penalties 1e-4, 1e-5, selecting the best configuration via 5-fold cross-validation. Deeper heads or larger hidden sizes consistently led to overfitting and worse validation performance. Given the small sample size, we restricted the complexity and hyperparameter search space of the regression heads to avoid overfitting since extensive hyperparameter searches are unlikely to improve generalization under these data constraints.

- **Nucleotide Transformer (NT-DL):** Mean R^2: 0.0181, Mean Spearman: 0.2607

– **RNA-FM:** Mean R^2: 0.1131, Mean Spearman: 0.3429

RNA-FM showed modest improvement over NT, likely due to its RNA-specific pre-training, but both failed to match the engineered-feature baseline. This outcome is consistent with the literature: foundation models require substantial task-specific data to outperform domain-engineered features in low-data regimes [3, 8, 12].

4.4 Feature Importance Analysis

SHAP (SHapley Additive exPlanations) analysis revealed that Cas13b efficiency is driven by a hierarchy of sequence composition features rather than single-position identities alone. The most influential features (see Fig. 2) were:

1. **CC dinucleotide frequency** (`dinuc_CC`): The single most dominant predictor. High values had a massive negative impact on predicted efficiency (SHAP value ≤ -0.04).
2. **Trinucleotide context:** Specific trimers such as `AAT`, `GTT`, and `GAG` were highly predictive.
3. **Overall Cytosine count** (`count_C`): Consistently showed a negative correlation with efficacy.
4. **G at Position 1** (`G_pos1`): Confirmed as a top positive predictor.
5. **Specific positional nucleotides:** Features such as `G_pos7`, `T_pos13`, and `A_pos2` appeared as key modulators.

These findings align with experimental results from Hu et al. [13].

4.5 Final Candidate CrRNAs

Post-optimization, we filtered the 1,740 unique spacers to identify the top-performing candidates specific to each Dengue serotype. Figure 3 illustrates the relationship between conservation coverage and predicted targeting efficiency. We observed distinct landscape differences across serotypes:

– **DENV-3 and DENV-4 (High Convergence, Limited Representation):** Although underrepresented (22.8% and 21.5% of candidates), these serotypes exhibited promising dual-purpose candidates. For DENV-4, the most conserved spacer (`GGAACA...`) covers 80.51% of genomes with efficiency 0.836. Similarly, the top DENV-3 candidate covers 69.88% with efficiency 0.859.
– **DENV-1 (Balanced Representation):** DENV-1 (20.1% of pool) showed a dense cluster of high-efficiency spacers. The most conserved candidate covers 68.42% with efficiency 0.804. The highest-efficiency candidate achieves 0.881.
– **DENV-2 (Trade-off with Compensatory Options):** DENV-2 (20.2% of pool) presented a divergence between conservation and potency. The most conserved spacer (68.75% coverage) had lower efficiency (0.613). However, the highest-efficiency candidate (`GGAGAC...`, 0.867) covers 41.90%. Importantly, the fourth-ranked candidate offers a middle ground (efficiency 0.832, reasonable coverage).

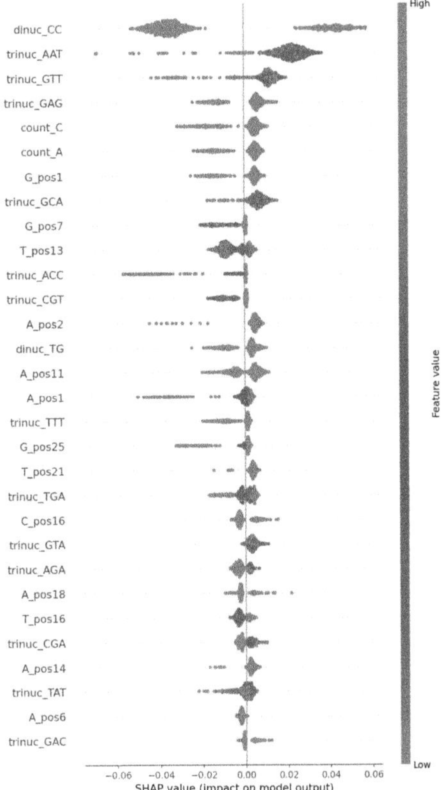

Fig. 2. SHAP summary plot ranking the top 30 features by mean absolute SHAP value. Red dots indicate high feature values, while blue dots indicate low values. (Color figure online)

Table 3 lists the specific sequences for the optimal candidates. More candidate rankings are provided in Supplementary Table S3.

5   Discussion and Analysis

5.1   Biological Significance of Identified Candidates

The computational pipeline successfully identified serotype-specific crRNA spacers with distinct biological properties tailored to Cas13b targeting:

– **Highly conserved targets:** Most candidates maintain $\geq 45\%$ representation within their respective serotype genomes, with optimal candidates reaching 68% coverage. This high conservation reduces the likelihood of viral escape through mutation and suggests that targeted regions encode functionally constrained viral elements.

Table 3. Best Candidate crRNAs per Serotype

Serotype	Selection Criteria	Sequence (Optimized)	Pred. Eff.	Coverage
DENV-1	Most Conserved	GGAAGAACAAGACGGGAACTTTGTGTGTCG	0.804	68.42%
	Best Efficiency	GGAGTAGTCGTACTAGGATCACAAGAAGGA	0.881	45.94%
DENV-2	Most Conserved	GGGAGTCAACATAGAAGCAGAACCTCCATT	0.613	68.75%
	Best Efficiency	GGAGAAACACAACATGGAACAATAGTTATC	0.867	41.90%
DENV-3	Most Conserved	GGTGCAAGGCGGATGGGCATCTTGGGAGAC	0.859	69.88%
	Best Efficiency	GGGGTTGGATCGCAAGAGGGAGCAATGCAC	0.882	41.41%
DENV-4	Most Conserved	GGAACAAGATGTCCAAGGCAAGGAGAGCCT	0.836	80.51%
	Best Efficiency	GGGGTAGAACATGGAGGATGCGTCACAACC	0.888	47.77%

– **Optimized for Cas13b biochemistry:** The extensive optimization phase
  (yielding 98.5% improvement rate with mean +47.17 points) demonstrates
  that even moderately suboptimal natural sequences can be substantially
  enhanced through strategic nucleotide substitutions. The algorithm-generated
  guides exploit known Cas13b preferences while maintaining specificity
  through the model's learned features.
– **Predicted to be highly effective:** Machine learning predictions identified top candidates informed by 158 engineered features validated against a
  curated dataset of 245 experimentally characterized Cas13b spacers.

5.2   Genomic Mapping

To evaluate the biological relevance of identified spacer candidates and determine
which viral genes were targeted, we performed comprehensive genomic mapping
of the top-ranked spacers across all four DENV serotypes. Spacer sequences
were aligned to complete reference genomes using a matching algorithm that
permitted up to 4 mismatches, enabling detection of near-perfect matches while
accounting for minor sequence variations. Mapping was performed independently
for each serotype using serotype-specific reference genomes obtained from the
NCBI Nucleotide database:

– DENV-1: NC_001477.1 (10,735 bp)
– DENV-2: NC_001474.2 (10,723 bp)
– DENV-3: DQ863638.1 (10,707 bp)
– DENV-4: NC_002640.1 (10,649 bp)

Genomic coordinates were annotated with serotype-specific gene boundaries
to assign each spacer to its cognate viral protein/region. Both forward and
reverse complement strands were searched to identify matches on either strand.

Genomic mapping revealed a consistent pattern: the highest-ranked spacer
candidates (≥ 0.8) consistently targeted the envelope (E) protein across all four
DENV serotypes. The envelope (E) protein mediates viral entry via receptor

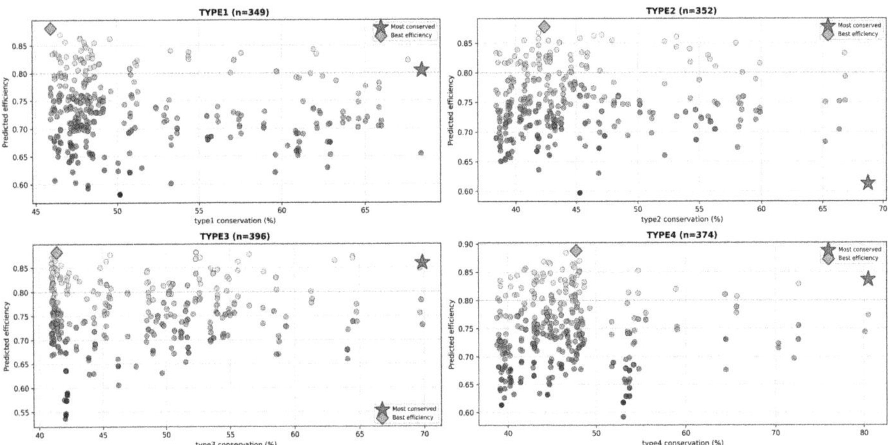

Fig. 3. Predicted efficiency versus serotype-specific conservation for the top candidate crRNAs. Each point represents a unique spacer optimized for DENV-1 ($n = 349$), DENV-2 ($n = 352$), DENV-3 ($n = 396$), and DENV-4 ($n = 374$). The **Green Diamond** marks the candidate with the highest predicted efficiency, while the **Red Star** marks the candidate with the highest conservation coverage. The red dashed line indicates the high-efficiency threshold (≥ 0.8). (Color figure online)

binding by domain III and low-pHtriggered membrane fusion by the fusion loop in domain II [16,17]. The fusion loop is one of the most conserved regions of E; key hydrophobic residues are strictly conserved and required for efficient membrane fusion and infectivity, and many substitutions in this motif severely impair or abolish fusion activity [19,20]. This structurally constrained region is a major target of cross-reactive and broadly neutralizing antibodies and has been exploited in E-based vaccine designs [9,23]. However, the absence of spacer candidates from other important conserved regions like NS3, NS5 is due to the limitation of the genome sequences retrived from the BV-BRC database. Specifically, only 532 of the genome sequences used in the study were reported to contain these proteins, limiting the ability to identify spacers across these regions.

5.3   Off-Target Assessment Against Human Transcriptome

To assess potential safety concerns arising from unintended targeting of human transcripts, optimized spacer sequences with predicted efficency (≥ 0.8) were aligned against the human reference transcriptome (GENCODE v49) using BLASTN with parameters optimized for short oligonucleotide queries. The analysis identified 12195 alignments across 224 unique spacer sequences, with a minimum of 4 mismatches and a median of 22 mismatches.

Risk assessment was performed based on established PspCas13b characteristics, specifically its tolerance of up to 4 mismatches and requirement for approximately 26-nucleotide base pairing for catalytic activity [13]. Alignments were categorized as High risk (perfect or near-perfect matches with $\geq$26nt base pairing), Moderate risk (3–4 mismatches with substantial base pairing), or Low/Negligible risk (>4 mismatches or insufficient base pairing).

Risk classification revealed no High/Moderate-High risk alignments, 0.01% Moderate risk alignments, and 99.99% Low/Negligible risk alignments. The single moderate-risk alignment (4 mismatches) and two low-risk alignments (5 mismatches) were distributed across different transcript types, with no clustering to essential protein-coding genes. Among all identified alignments, 63.5% mapped to protein-coding transcripts, 20.1% to long non-coding RNAs (lncRNAs), 6.3% to retained introns, and 5.6% to nonsense-mediated decay transcripts, reflecting the composition of the human transcriptome. Importantly, the vast majority of alignments (99.97%) contained $\geq$7 mismatches, substantially exceeding the activity threshold and indicating minimal predicted collateral targeting. The distribution of mismatches across all alignments reveals that the vast majority cluster well above the activity threshold (Fig. 4), with a median of 22 mismatches and a primary mode spanning 15–27 mismatches.

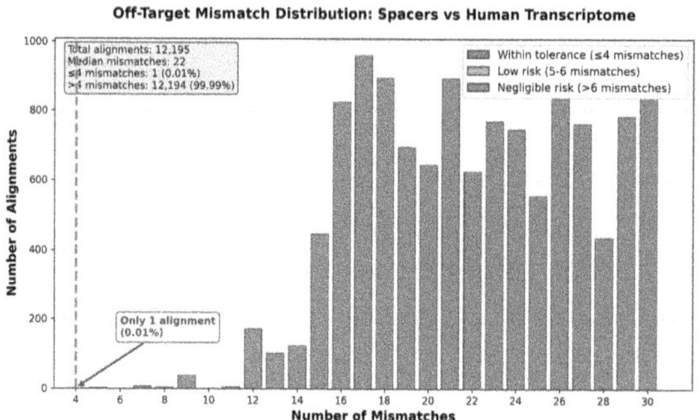

Fig. 4. Off-target mismatch distribution for optimized CRISPR-Cas13b spacer sequences.

These findings demonstrate that the computationally optimized spacer sequences exhibit high specificity for Dengue viral targets with negligible predicted off-target activity against human transcripts. The near-absence of alignments within the PspCas13b tolerance threshold provides preliminary computational evidence for the safety profile of the designed crRNA candidates. While experimental validation through transcriptome-wide RNA-seq remains necessary to confirm the absence of collateral activity in cellular contexts, the computational off-target analysis supports the further development of these crRNA candidates for antiviral applications.

5.4   Performance of Classical ML vs. Deep Learning

Our results demonstrate that in data-scarce regimes ($n = 245$), feature-engineered classical ML models significantly outperform deep learning approaches. The underperformance of foundation-model embeddings relative to feature-engineered tree-based models is consistent with prior observations on tabular and low-sample-size data, where models that explicitly encode domain priors often outperform generic embeddings. In our setting, the limited availability of training data to fine-tune the models for Cas13b explains the inferior performance of NT and RNA-FM-based predictors. This finding has three important implications: (1) **Sample efficiency:** Random Forest and XGBoost effectively leverage domain knowledge, whereas foundation models (NT, RNA-FM) rely on large-scale patterns that may not align with task-specific objectives in small datasets; (2) **Interpretability:** SHAP analysis of tree-based models directly reveals biochemical drivers; and (3) **Future potential:** Foundation models may become competitive only as larger datasets become available.

5.5   Comparison with Existing Guide Design Tools

To our knowledge, this is the first computational framework specifically tailored for Cas13b crRNA design against viral genomes. Existing tools like CASowary [15] target Cas13a/d for mammalian transcript knockdown, while platforms such as CHOPCHOP [18] focus on Cas9. Our pipeline specifically addresses: (1) conservation analysis across viral serotypes; (2) mismatch tolerance and optimization for viral diversity; and (3) integration of virus-specific design rules.

5.6   Limitations and Future Work

(1) **Experimental validation** is required to confirm efficacy in infected cells; (2) **Off-target analysis** against the human transcriptome is recommended before therapeutic application; (3) **Expanded training data** is needed to unlock the potential of foundation models; (4) **Expanded genome data** is required to enable unbiased identification of conserved spacer targets across all viral genomic regions; and (5) **Delivery mechanisms** for Cas13b need development for clinical translation which is beyond the scope of this work.

6   Conclusion

We have presented a computational pipeline for systematic identification and optimization of Cas13b guide RNA targeting Dengue Virus, addressing a critical gap in antiviral therapeutic development. By integrating genomic data extraction, sequence optimization, and machine learning-based efficacy prediction, we identified a catalog of high-confidence candidate guide RNAs suitable for immediate experimental validation. Our work establishes the first baseline for Cas13b efficiency prediction using classical machine learning, demonstrating superior performance over large language model embeddings in small-data settings. Future efforts will focus on experimental validation of top candidates and extension of the pipeline to other RNA viruses such as Chikungunya and SARS-CoV-2.

Acknowledgment. We extend our sincere gratitude to Dr. Syed Faraz Ahmed, Department of Electrical and Electronic Engineering, University of Melbourne, for his careful review of this manuscript, constructive suggestions for improvement, and invaluable support in data curation and analysis.

Supplementary Material. The following supplementary tables and figures are provided as a separate PDF document:
- Supplementary Table S1: 245 crRNA spacers used for model training
- Supplementary Table S2: 157 engineered features for the ML model
- Supplementary Table S3: Top candidate crRNAs per Dengue serotype.

References

1. Abudayyeh, O.O., et al.: C2C2 is a single-component programmable RNA-guided RNA-targeting CRISPR effector. Science **353**(6299), aaf5573 (2016)
2. Ali, S., et al.: Dengue outbreak and fragile healthcare system: doctors at the verge of mental and physical stress. Front. Public Health **10**, 966042 (2022)
3. Brigato, L., Iocchi, E.: A close look at deep learning with small data. In: 25th International Conference on Pattern Recognition (ICPR), pp. 2490–2497. IEEE (2021)
4. Chen, J., et al.: RNA-FM: a foundation model for RNA structure and function prediction. arXiv preprint arXiv:2204.00300 (2022)
5. Cheng, X., et al.: Prediction of on-target and off-target activity of CRISPR-Cas13d guide RNAs using deep learning. Nat. Commun. **14**(1), 1–14 (2023)
6. Chuai, G., et al.: DeepCRISPR: optimized CRISPR guide RNA design by deep learning. Genome Biol. **19**(1), 1–18 (2018)
7. Cox, D.B., et al.: RNA editing with CRISPR-Cas13. Science **358**(6366), 1019–1027 (2017)
8. Dalla-Torre, H., et al.: The Nucleotide Transformer: Building and evaluating robust foundation models for human genomics. bioRxiv (2023). https://doi.org/10.1101/2023.01.11.523679
9. Dejnirattisai, W., et al.: A new class of highly potent, broadly neutralizing antibodies isolated from viremic patients infected with dengue virus. Nat. Immunol. **16**(2), 170–177 (2015). https://doi.org/10.1038/ni.3058
10. Freije, C.A., et al.: Programmable inhibition and detection of RNA viruses using Cas13. Mol. Cell **76**(5), 826–837 (2019)
11. Gootenberg, J.S., Abudayyeh, O.O., Kellner, M.J., Joung, J., Collins, J.J., Zhang, F.: Multiplexed and portable nucleic acid detection platform with Cas13, Cas12a, and Csm6. Science **360**(6387), 439–444 (2018)
12. Grinsztajn, L., Oyallon, E., Varoquaux, G.: Why do tree-based models still outperform deep learning on typical tabular data? Adv. Neural. Inf. Process. Syst. **35**, 507–520 (2022)

13. Hu, W., et al.: Single-base tiled screen unveils design principles of PspCas13b for potent and off-target-free RNA silencing. Nat. Struct. Mol. Biol. **31**, 1702–1716 (2024). https://doi.org/10.1038/s41594-024-01336-0

14. Katzelnick, L.C., et al.: Antibody-dependent enhancement of severe dengue disease in humans. Science **358**(6365), 929–932 (2017)

15. Kim, H.K., et al.: CASowary: CRISPR-Cas13 guide RNA predictor for transcript depletion. Nucleic Acids Res. **49**(3), e16 (2021)

16. Klein, D.E., Choi, J.L., Harrison, S.C.: Structure of a dengue virus envelope protein late-stage fusion intermediate. J. Virol. **87**(4), 2287–2293 (2013). https://doi.org/10.1128/JVI.02957-12

17. Kuhn, R.J., et al.: Structure of dengue virus: implications for flavivirus organization, maturation, and fusion. Cell **108**(5), 717–725 (2002). https://doi.org/10.1016/S0092-8674(02)00660-8

18. Labun, K., Montague, T.G., Krause, M., Torres Cleuren, Y., Beskemeier, H., Valen, E.: CHOPCHOP v3: expanding the CRISPR web toolbox beyond genome editing. Nucleic Acids Res. **47**(W1), W171–W174 (2019)

19. Lee, H., Ren, J., Pesavento, R.P., Sun, L., Smith, T.J.: On the role of surrounding regions in the fusion peptide in dengue virus infection. Virology **554**, 68–78 (2021). https://doi.org/10.1016/j.virol.2020.11.010

20. Modis, Y., Ogata, S., Clements, D., Harrison, S.C.: Variable surface epitopes in the crystal structure of dengue virus type 3 envelope glycoprotein. J. Virol. **79**(2), 1223–1231 (2005). https://doi.org/10.1128/JVI.79.2.1223-1231.2005

21. Moreno-Mateos, M.A., et al.: CRISPRscan: designing highly efficient sgRNAs for CRISPR-Cas9 targeting in vivo. Nat. Methods **12**(10), 982–988 (2015)

22. Pan, Y., et al.: Transitioning from wet lab to artificial intelligence: a systematic review of AI predictors in CRISPR. Briefings Bioinform. **26**(1) (2025), pMCID: PMC11796103

23. Rouvinski, A., et al.: Recognition determinants of broadly neutralizing human antibodies against dengue viruses. Nature **520**(7545), 109–113 (2015). https://doi.org/10.1038/nature14130

24. Stanaway, J.D., et al.: The global burden of dengue: an analysis from the global burden of disease study 2013. Lancet. Infect. Dis. **16**(6), 712–723 (2016)

25. Wessels, H.H., Méndez-Mancilla, A., Guo, X., Legut, M., Daniloski, Z., Sanjana, N.E.: Massively parallel Cas13 screens reveal principles for guide RNA design. Nat. Biotechnol. **38**(6), 722–727 (2020)

Graph-Regularized Embedding Refinement for Spatial Domain Identification

Bohao Mo, Zeyao Chen, and Hongdong Li[✉]

Hunan Provincial Key Lab on Bioinformatics, School of Computer Science and Engineering, Central South University, Changsha, Hunan, China
hongdong@csu.edu.cn

Abstract. Spatial domain identification is a fundamental task in spatial transcriptomics, aiming to partition tissues into coherent regions that reflect both transcriptional similarity and spatial organization. Existing methods typically generate low-dimensional embeddings through predefined transformations or neural network encoders, and treat these representations as fixed inputs for downstream clustering. This limits the model's ability to adapt to multiple spatial and expression-derived constraints. To address this limitation, we propose Graph-Regularized Embedding Refinement for Spatial Transcriptomics (GRER-ST), a refinement framework that enhances spatial domain identification by explicitly optimizing the latent embedding. GRER-ST first integrates spatial coordinates and gene expression profiles to construct an initial low-dimensional representation that captures local spatial context. A trainable latent embedding, initialized from this representation, is then refined under a unified objective that incorporates spatial smoothness, expression similarity, and neighborhood-based contrastive regularization. By refining the embedding rather than relying on a fixed representation, GRER-ST produces spatial domains that are both locally coherent and discriminative. Extensive experiments on multiple spatial transcriptomics datasets demonstrate that GRER-ST consistently outperforms baseline methods in spatial domain identification, yielding more spatially coherent domain structures. These findings suggest that embedding refinement is an effective and flexible strategy for integrating heterogeneous spatial and transcriptional information in spatial transcriptomics.

Keywords: Spatial transcriptomics · Spatial domain identification · Representation learning · Graph regularization · Contrastive learning

1 Introduction

Spatial transcriptomics (ST) has emerged as a transformative technology that retains the spatial coordinates of gene expression measurements [1], thereby enabling systematic investigation of how molecular profiles vary across tissue

© The Author(s), under exclusive license to Springer Nature Switzerland AG 2026
K. L. Kabir et al. (Eds.): BICOB 2026, CCIS 2977, pp. 260–273, 2026.
https://doi.org/10.1007/978-3-032-26028-4_19

architecture and how such variation underlies biological function and disease mechanisms. Within the ST analysis pipeline, spatial domain identification—partitioning tissues into contiguous regions with coherent gene expression and structural properties—constitutes a critical step toward interpreting spatially resolved biological patterns. Accurate delineation of spatial domains facilitates characterization of tissue microenvironments, discovery of spatially localized gene-expression programs, and elucidation of how cellular organization underpins functional specialization across anatomically distinct regions. For instance, spatial domain identification can reveal localized activation of cell–cell signaling pathways [2–5]. Moreover, spatial domain information supports downstream analyses, including differential expression analysis to identify genes with spatially distinct patterns [6–8], trajectory analysis to reconstruct spatially guided cellular differentiation paths [16–19], integration with other omics modalities to gain multi-layered insights into tissue organization [9–11], and the investigation of spatial patterns associated with cancer [12–15].

Many computational methods have been developed for spatial domain identification. Early approaches adapted single-cell clustering methods, correcting gene expression by averaging over local neighbors before applying standard clustering algorithms. While these methods improved performance over non-spatial clustering, their coarse use of spatial information limited the accuracy and continuity of identified domains. More recent approaches explicitly integrate spatial information into representation learning and domain identification. SEDR [20] combines a deep autoencoder with masked self-supervised learning and a variational graph autoencoder to jointly embed gene expression and spatial context, enabling scalable and robust spatial clustering. PROST [21] introduces a quantitative framework that links spatially variable gene characterization with domain segmentation through a self-attention mechanism, allowing unified analysis of spatial patterns across resolutions. GraphST [22] further integrates graph neural networks with self-supervised contrastive learning to learn discriminative spot representations and supports multisample integration while correcting batch effects. Together, these methods demonstrate the effectiveness of incorporating spatial structure and advanced representation learning strategies for spatial domain analysis in spatial transcriptomics.

Despite these advances, most existing methods rely on fixed low-dimensional embeddings generated by encoders or predefined transformations, which may not fully integrate multiple sources of spatial and transcriptional information. Developing methods that explicitly refine latent embeddings under complementary constraints can provide more accurate and discriminative representations for spatial domain identification. In this study, we propose Graph-Regularized Embedding Refinement for Spatial Transcriptomics (GRER-ST), which iteratively refines embeddings by jointly considering spatial proximity, expression similarity, and neighborhood-based contrastive regularization. The refined embedding captures local coherence while preserving discriminative structure, providing a flexible foundation for downstream spatial analysis.

2   Methods

Spatial domain identification aims to partition spatially resolved transcriptomic profiles into coherent regions that reflect both transcriptional similarity and spatial organization. While existing spatial domain identification methods commonly obtain low-dimensional representations through predefined feature transformations or neural network encoders, the learned representations are typically produced as the output of a model and subsequently treated as fixed inputs for downstream clustering or community detection algorithms. As a result, the representation itself is not explicitly optimized with respect to the spatial domain identification objective once it is obtained.

To address this limitation, we propose Graph-Regularized Embedding Refinement for Spatial Transcriptomics (GRER-ST), a framework that introduces a trainable latent representation directly optimized under multiple spatial and expression-derived constraints. By explicitly refining the embedding rather than relying on a fixed representation, GRER-ST enables progressive adaptation of the representation for spatial domain identification. The overall framework is illustrated in Fig. 1.

Conceptually, GRER-ST is built upon three tightly coupled components that together enable adaptive embedding refinement for spatial domain identification. First, a spatially informed low-dimensional representation is initialized by integrating intrinsic gene expression with neighborhood-aggregated expression features. Second, spatial and expression-aware neighborhood graphs are constructed to capture local spatial proximity and transcriptional similarity among spots. Third, the latent embedding is iteratively refined by optimizing a unified objective that combines embedding preservation, graph-based smoothness regularization, and neighborhood-level contrastive constraints, followed by downstream clustering for spatial domain identification. For clarity, the detailed methodological components of GRER-ST are presented in the following subsections according to their functional roles, rather than strictly following a stagewise pipeline.

2.1   Data Preprocessing

Given a spatial transcriptomics dataset with n spots, we denote the spatial coordinates as $\mathbf{P} \in \mathbb{R}^{n \times 2}$ and the preprocessed gene expression matrix as $\mathbf{E} \in \mathbb{R}^{n \times g}$, where g represents the number of highly variable genes.

Following the preprocessing strategy adopted in MNMST [23], we construct a spatial neighborhood graph based on Euclidean distances in the coordinate space, yielding a distance-weighted adjacency matrix $\mathbf{W}_{\text{spatial}}$, and propagate gene expression features over this graph to capture local spatial context. The neighborhood-aggregated expression features are then concatenated with the original expression profiles.

Principal component analysis (PCA) is subsequently applied to the augmented features to obtain a low-dimensional representation $\mathbf{X}_0 \in \mathbb{R}^{n \times d}$ ($d = 100$), which serves as the initial embedding for subsequent refinement.

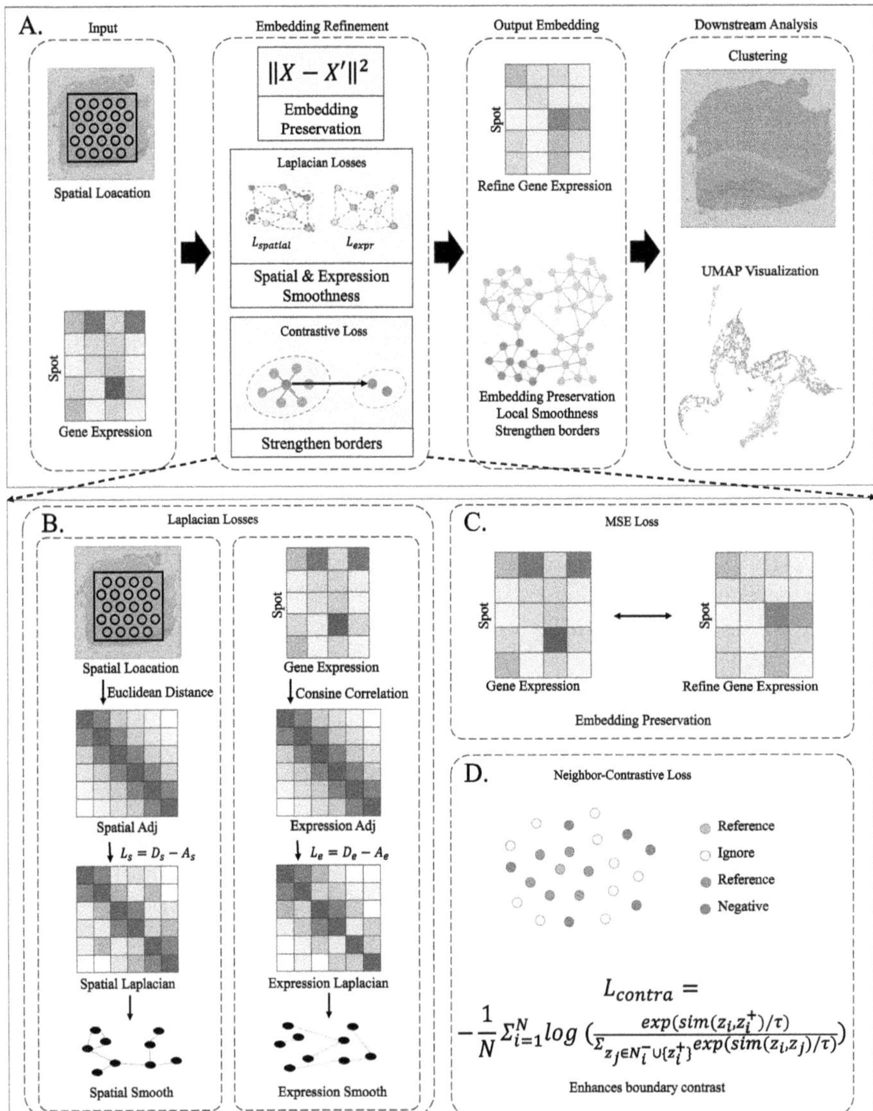

Fig. 1. Overview of the GRER-ST framework. (A) Overall workflow of GRER-ST. Spatial coordinates and gene expression profiles are integrated to obtain an initial embedding, which is further refined by jointly optimizing multiple regularization terms before downstream clustering. (B) Laplacian losses derived from spatial and expression graphs enforce local smoothness in spatial and transcriptional neighborhoods. (C) An embedding preservation loss constrains the refined embedding to remain close to the initial representation. (D) A neighborhood-based contrastive loss enhances boundary discrimination by pulling embeddings toward spatial neighbors while pushing them away from non-neighbor spots, where z_i^+ is the mean embedding of node i's neighbors (positive samples), $\mathcal{N}_i^-$ is the set of negative samples for node i, and τ is the temperature parameter.

2.2 Expression Similarity Graph Construction

In addition to spatial proximity, transcriptional similarity provides complementary information for identifying coherent spatial domains. Based on the initial embedding $\mathbf{X}_0$, we construct an expression-similarity graph by computing cosine similarity between all pairs of spots. For each spot, the top m most similar neighbors ($m = 15$) are selected to form a k-nearest-neighbor graph in the expression space.

To better capture higher-order co-occurrence patterns among transcriptionally similar spots, we further transform the expression-similarity graph into a shifted positive pointwise mutual information (SPPMI) matrix [30], denoted as $\mathbf{A}_{\mathrm{expr}}$. This matrix emphasizes meaningful co-expression relationships while suppressing spurious similarities, and serves as an additional source of neighborhood structure in GRER-ST.

2.3 Trainable Embedding Refinement

Rather than directly clustering on the initial embedding $\mathbf{X}_0$, GRER-ST introduces a trainable latent representation $\mathbf{X} \in \mathbb{R}^{n \times d}$, initialized as $\mathbf{X}_0$. The embedding is optimized by minimizing a unified objective function that integrates multiple complementary regularization terms. We formulate the following objective function:

$$\mathcal{L} = \|\mathbf{X} - \mathbf{X}_0\|_{\mathrm{F}}^2 + \mathrm{Tr}(\mathbf{X}^\top \mathbf{L}_{\mathrm{spatial}} \mathbf{X}) + \mathrm{Tr}(\mathbf{X}^\top \mathbf{L}_{\mathrm{expr}} \mathbf{X}) + \mathcal{L}_{\mathrm{contra}}, \qquad (1)$$

where $\mathbf{L}_{\mathrm{spatial}}$ and $\mathbf{L}_{\mathrm{expr}}$ are the graph Laplacian matrices derived from $\mathbf{W}_{\mathrm{spatial}}$ and $\mathbf{A}_{\mathrm{expr}}$, respectively. The first term preserves fidelity to the initial embedding, preventing excessive deviation during optimization. The second and third terms enforce smoothness of the embedding over spatial and expression-derived neighborhoods, encouraging nearby or transcriptionally similar spots to have similar representations.

2.4 Neighborhood-Based Contrastive Regularization

While Laplacian regularization promotes local smoothness, it may also lead to over-smoothing and reduced discriminability between spatial domains. To mitigate this issue, GRER-ST incorporates a neighborhood-based contrastive loss $\mathcal{L}_{\mathrm{contra}}$.

For each spot, we treat the average embedding of its spatial nearest neighbors as a positive sample, while randomly selected non-neighbor spots are treated as negative samples. A contrastive-learning objective is applied to encourage the embedding to remain close to its true spatial context while being distinguishable from unrelated regions.

This contrastive regularization complements the graph-based smoothness constraints and enhances the robustness of the learned embedding for downstream clustering.

2.5 Spatial Domain Identification

After optimization, the refined embedding $\mathbf{X}$ is used as input for standard clustering algorithms, such as leiden [27], to identify spatial domains. The resulting clusters correspond to spatially coherent regions that integrate both transcriptional and spatial information.

2.6 Evaluation Metrics

To quantitatively evaluate the performance of spatial domain identification, we adopt the Adjusted Rand Index (ARI) as the primary evaluation metric. ARI measures the similarity between predicted cluster assignments and ground-truth spatial domain annotations, while correcting for agreement by chance. It takes values in the range $[-1, 1]$, with higher values indicating better agreement, and an ARI of 0 corresponding to random clustering.

Formally, given two partitions U and V of the same set of samples, ARI is defined as

$$\mathrm{ARI} = \frac{\sum_{ij} \binom{n_{ij}}{2} - \left[\sum_i \binom{a_i}{2} \sum_j \binom{b_j}{2}\right] / \binom{n}{2}}{\frac{1}{2}\left[\sum_i \binom{a_i}{2} + \sum_j \binom{b_j}{2}\right] - \left[\sum_i \binom{a_i}{2} \sum_j \binom{b_j}{2}\right] / \binom{n}{2}},$$

where n_{ij} denotes the number of samples shared by cluster i in U and cluster j in V, $a_i = \sum_j n_{ij}$, $b_j = \sum_i n_{ij}$, and n is the total number of samples.

In this work, ARI is computed independently for each tissue section using the standard implementation provided by the `scikit-learn` [31] library. For datasets consisting of multiple sections, the mean ARI is reported by averaging ARI scores across all sections.

3 Experiments and Results

3.1 Dataset Selection and Rationale

We chose the DLPFC [28] and Stereo-seq mouse embryo (MOE) [29] datasets for benchmarking and ablation studies for the following reasons.

First, the DLPFC dataset is a well-established and widely adopted benchmark for spatial domain identification, and has been extensively used to evaluate a broad range of spatial clustering and representation learning methods, including DeepST [24], GraphST [22], SEDR [20], PROST [21], MNMST [23], and SLGCA [25]. The dataset provides manually curated cortical layer annotations across multiple tissue sections, enabling reliable quantitative comparison across methods. Its moderate spatial resolution and relatively stable tissue architecture further facilitate systematic benchmarking.

Second, the Stereo-seq MOE dataset represents a complementary and more challenging evaluation scenario. This dataset is characterized by ultra-high spatial resolution, large-scale measurements, and complex spatial patterns arising from embryonic development, resulting in pronounced biological heterogeneity. Such properties make MOE a stringent testbed for assessing the robustness and generalization ability of spatial domain identification methods under complex spatial and transcriptional settings. Notably, GraphST has previously adopted the Stereo-seq mouse embryo dataset for spatial domain analysis, demonstrating its applicability in evaluating advanced graph-based spatial models [22].

Together, these two datasets cover both well-structured adult tissue with established benchmarks and highly heterogeneous developmental tissue with increased spatial complexity, enabling a comprehensive evaluation of GRER-ST across different spatial transcriptomics platforms, resolutions, and biological contexts.

3.2   Benchmarking GRER-ST on the DLPFC Dataset

Human dorsolateral prefrontal cortex (DLPFC) [28] dataset consists of manually annotated cortical layers (L1–L6) and white matter (WM) of 12 slices with gene markers and cytoarchitecture. These data are generated using $10\times$ Visium platform, serving as a publicly available benchmarking dataset for spatial domain identification (Fig. 2A).

We evaluated GRER-ST on the DLPFC dataset and compared it with several representative baseline methods, including Scanpy [26], GraphST [22], SEDR [20], DeepST [24], MNMST [23], SLGCA [25], and PROST [21]. Among these methods, Scanpy serves as a non-spatial baseline based solely on gene expression features, while all other methods explicitly incorporate spatial information during representation learning or clustering.

Overall, GRER-ST demonstrates strong and robust performance on the DLPFC dataset. As shown in Table 1 and Fig. 2B, GRER-ST achieves competitive or superior ARI scores across most slices compared with baseline methods, outperforming several state-of-the-art spatial domain identification approaches. This result indicates that GRER-ST provides more consistent and effective spatial domain recognition across different tissue sections. In contrast, the non-spatial Scanpy baseline yields lower ARI scores on most slices (Table 1), highlighting the importance of incorporating spatial information for accurate spatial domain identification.

To further examine the qualitative behavior of GRER-ST, we visualize the spatial domain identification results on slice 151670 in Fig. 2C. This slice contains five cortical layers, each represented by a distinct color. GRER-ST produces spatially coherent domains that closely align with the ground-truth layer structure. Notably, for Layer 3, most baseline methods fragment this layer into multiple smaller regions, indicating over-segmentation. In contrast, GRER-ST successfully identifies Layer 3 as a single coherent domain, demonstrating its ability to preserve both spatial continuity and transcriptional consistency.

These results suggest that refining the latent embedding under joint spatial and expression-derived constraints enables GRER-ST to better capture fine-grained yet coherent spatial structures in complex tissue architectures.

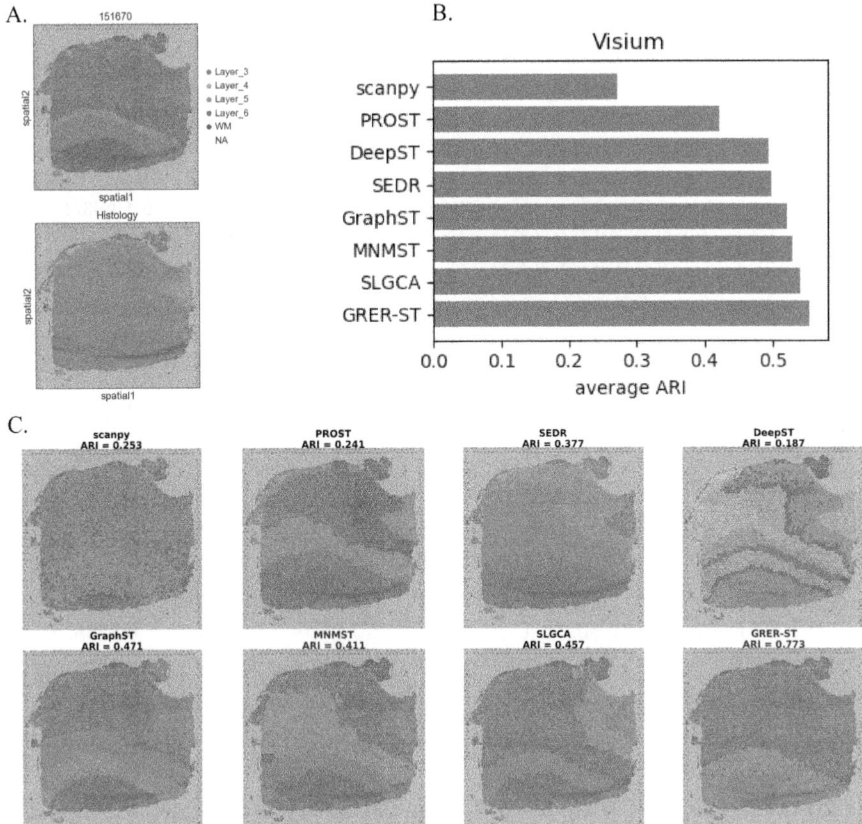

Fig. 2. Performance of GRER-ST on the DLPFC dataset. (A) Histology image and ground-truth spatial domain annotations. (B) Bar plots comparing the mean performance of GRER-ST with baseline methods. (C) Spatial domain identification results on slice 151670.

3.3   Benchmarking **GRER-ST** on the Stereo-Seq Mouse Embryo Dataset

The Stereo-seq mouse embryo (MOE) dataset [29] consists of spatial transcriptomics profiles collected from mouse embryos at embryonic day E9.5, covering multiple tissue sections with complex spatial organization. This dataset provides a challenging benchmark for spatial domain identification due to its high spatial resolution and pronounced biological heterogeneity across sections (Fig. 3A).

We evaluated the performance of GRER-ST on the MOE dataset and compared it with several representative baseline methods, including Scanpy, GraphST, SEDR, DeepST, MNMST, SLGCA, and PROST. Consistent with the DLPFC experiments, Scanpy serves as a non-spatial baseline, while the remaining methods explicitly leverage spatial information during modeling.

Table 1. ARI scores on DLPFC datasets for different methods

Method	151507	151508	151509	151510	151669	151670	151671	151672	151673	151674	151675	151676	Average
GraphST	0.4146	0.4975	0.4999	0.5196	0.4244	0.4709	0.6043	0.5532	0.6277	0.5387	0.5294	0.5741	0.5212
SEDR	0.5460	0.5033	0.5379	0.4040	0.3492	0.3774	0.5425	0.5655	0.5473	0.5628	0.4884	0.5539	0.4982
DeepST	0.5140	0.5154	0.4725	0.5480	0.4719	0.1873	0.5323	0.4984	0.5600	0.5463	0.5208	0.5482	0.4929
MNMST	0.5122	0.5105	0.5266	0.5350	0.4958	0.4114	0.5551	0.5241	0.5559	0.5959	0.6164	0.5024	0.5284
SLGCA	0.4443	0.5181	0.5023	0.5024	0.5045	0.4566	0.6453	0.6462	0.6490	0.6180	0.5641	0.4374	0.5407
Scanpy	0.3204	0.2705	0.3173	0.2821	0.2175	0.2531	0.2541	0.2514	0.2868	0.2712	0.2953	0.2450	0.2721
PROST	0.5114	0.4729	0.4717	0.4207	0.2589	0.2408	0.4880	0.5548	0.3983	0.3972	0.4287	0.4158	0.4216
GRER-ST	0.4977	0.4672	0.4021	0.5334	0.7493	0.7729	0.3934	0.6935	0.4832	0.5805	0.5293	0.5397	0.5535

Table 2. ARI scores on MOSTA E9.5 datasets

Method	E9.5_E1S1	E9.5_E2S1	E9.5_E2S2	E9.5_E2S3	E9.5_E2S4	Average
MNMST	0.2577	0.3434	0.5042	0.4764	0.4276	0.4019
Scanpy	0.3269	0.3823	0.4539	0.4318	0.4358	0.4055
GraphST	0.3034	0.2452	0.3784	0.3518	0.3213	0.3200
SEDR	0.3403	0.3418	0.3791	0.4123	0.3655	0.3660
SLGCA	0.2915	0.3454	0.4008	0.3498	0.4221	0.3619
PROST	0.3696	0.2264	0.3218	0.4526	0.2536	0.3248
DeepST	0.2735	0.3613	0.4476	0.5304	0.4667	0.4096
GRER-ST	0.3193	0.4058	0.4952	0.4990	0.4729	0.4385

Overall, GRER-ST demonstrates competitive performance on the MOE dataset. As shown in Table 2 and Fig. 3B, GRER-ST achieves competitive or superior ARI scores across most MOE sections, resulting in one of the highest average performances among the compared methods. These results suggest that GRER-ST is able to capture spatial domain structures in heterogeneous embryonic tissues at an aggregate level. Notably, the non-spatial Scanpy baseline also attains competitive ARI scores on several slices (Table 2), reflecting the intrinsic difficulty of the MOE dataset and suggesting that improvements from incorporating spatial information are not always uniformly reflected across all slices.

To qualitatively assess the spatial domain identification results, we visualize the clustering outcomes on the *E9.5_E2S1.MOSTA* section in Fig. 3C. GRER-ST produces spatially coherent and biologically plausible domains that align well with the annotated structures. Compared with baseline methods, GRER-ST yields smoother and more contiguous spatial regions, reducing fragmentation while preserving meaningful spatial boundaries.

Together, these quantitative and qualitative results demonstrate that GRER-ST can effectively identify coherent spatial domains in complex embryonic tissues, highlighting its robustness and generalization capability across diverse spatial transcriptomics platforms.

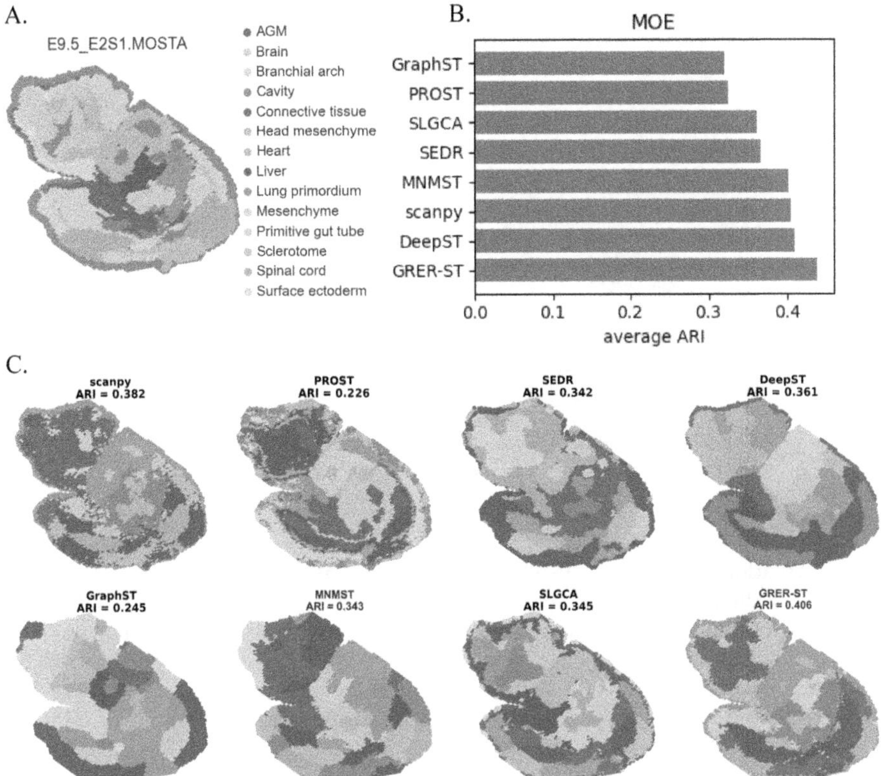

Fig. 3. Performance of GRER-ST on the Stereo-seq mouse embryo dataset. (A) Ground-truth spatial domain annotations for representative embryonic sections. (B) Bar plots comparing the mean clustering performance of GRER-ST with baseline methods across multiple MOE samples. (C) Spatial domain identification results on the *E9.5_E2S1.MOSTA* section.

3.4 Ablation Study

To assess the effect of different components in GRER-ST, we perform ablation studies on both the DLPFC and Stereo-seq mouse embryo (MOE) datasets. Specifically, we construct three ablated variants by removing the neighbor contrastive loss (NO_Neighbor_Contra), the spatial smoothing constraint (NO_Spatial_Smooth), and the expression smoothing constraint (NO_Expression_Smooth), respectively. The quantitative results are summarized in Fig. 4 and Tables 3–4.

As shown in Fig. 4A and Table 3, removing any component of GRER-ST leads to reduction in mean clustering performance on the DLPFC dataset, indicating that all three components contribute to the overall effectiveness of the model. Among the ablated variants, removing the expression smoothing constraint results in the largest decrease in average performance, highlighting its important role in preserving transcriptional consistency during embedding refinement. In comparison, removing either the neighbor contrastive loss or the spatial smoothing constraint leads to smaller but noticeable reductions in mean performance, suggesting that both components provide complementary benefits for modeling local neighborhood relationships and spatial coherence.

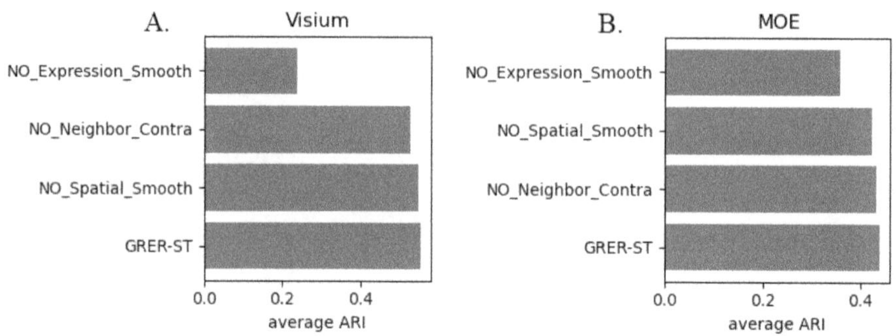

Fig. 4. Ablation study of GRER-ST. (A) Bar plots comparing the mean clustering performance of the full GRER-ST model and its ablated variants on the DLPFC dataset. (B) Bar plots comparing the mean clustering performance of GRER-ST and its ablated variants on the Stereo-seq mouse embryo dataset. NO_Neighbor_Contra removes the neighbor contrastive loss, NO_Spatial_Smooth removes the spatial smoothing constraint, and NO_Expression_Smooth removes the expression smoothing constraint.

Table 3. Ablation study of GRER-ST on DLPFC datasets

Method	151507	151508	151509	151510	151669	151670	151671	151672	151673	151674	151675	151676	Average
GRER-ST	0.4977	0.4672	0.4021	0.5334	0.7493	0.7729	0.3934	0.6935	0.4832	0.5805	0.5293	0.5397	0.5535
No Neighbor Contrastive	0.4834	0.4673	0.4009	0.5308	0.5028	0.7729	0.4824	0.6938	0.4561	0.5755	0.4832	0.5023	0.5293
No Spatial Smooth	0.5099	0.4718	0.4664	0.4839	0.6846	0.7195	0.5216	0.6189	0.5363	0.5240	0.6139	0.4445	0.5496
No Expression Smooth	0.3555	0.2247	0.2589	0.1183	0.2142	0.1762	0.2775	0.4194	0.1528	0.2281	0.2163	0.2040	0.2372

Similar observations are made on the MOE dataset (Fig. 4B and Table 4), where performance differences among ablation variants are generally smaller due to the increased complexity and heterogeneity of embryonic tissues. Nevertheless, NO_Expression_Smooth again exhibits the lowest average performance among the ablated variants, underscoring the importance of expression-based smoothing for robust spatial domain identification. Removing the spatial smoothing constraint or the neighbor contrastive loss also results in lower mean performance compared with the full GRER-ST model, indicating that these components still contribute positively at an aggregate level.

Overall, these ablation results demonstrate that the neighbor contrastive loss, spatial smoothing, and expression smoothing jointly contribute to the performance of GRER-ST, and that removing any individual component undermines its ability to identify coherent and biologically meaningful spatial domains.

Table 4. Ablation study of GRER-ST on MOSTA E9.5 datasets

Method	E9.5_E1S1	E9.5_E2S1	E9.5_E2S2	E9.5_E2S3	E9.5_E2S4	Average
GRER-ST	0.3193	0.4058	0.4952	0.4990	0.4729	0.4385
No Expression Smooth	0.2824	0.3125	0.3395	0.4238	0.4328	0.3582
No Spatial Smooth	0.3146	0.4031	0.4772	0.4550	0.4662	0.4232
No Neighbor Contrastive	0.3308	0.3971	0.4563	0.5017	0.4722	0.4316

4 Conclusion

In this work, we proposed GRER-ST, a graph-regularized embedding refinement framework for spatial domain identification in spatial transcriptomics. Unlike methods that treat low-dimensional representations as fixed outputs of pre-defined transformations or encoders, GRER-ST explicitly considers the latent embedding as an optimization target and progressively refines it under complementary spatial and expression-derived constraints.

By jointly incorporating spatial smoothness, transcriptional similarity, and neighborhood-based contrastive regularization, GRER-ST learns embeddings that are both locally coherent and structurally discriminative. This formulation enables the refined representation to capture fine-grained spatial organization while remaining robust to local heterogeneity and noise. Comparative experiments on spatial transcriptomics data demonstrate that GRER-ST produces spatially coherent domain structures and achieves competitive performance against representative baseline methods.

Acknowledgments. This work was supported in part by STI2030-Major Projects (No. 2022ZD0213700), Hunan Provincial Natural Science Foundation of China (2025JJ20068), the National Natural Science Foundation of China under Grants (No. 62473382).

References

1. Ståhl, P.L., Salmén, F., Vickovic, S., et al.: Visualization and analysis of gene expression in tissue sections by spatial transcriptomics. Science **353**(6294), 78–82 (2016)
2. Cang, Z., Zhao, Y., Almet, A.A., et al.: Screening cell–cell communication in spatial transcriptomics via collective optimal transport. Nat. Methods **20**, 218–228 (2023)
3. Shao, X., Li, C., Yang, H., et al.: Knowledge-graph-based cell-cell communication inference for spatially resolved transcriptomic data with SpaTalk. Nat. Commun. **13**, 4429 (2022)
4. Li, Z., Wang, T., Liu, P., et al.: SpatialDM for rapid identification of spatially co-expressed ligand–receptor and revealing cell–cell communication patterns. Nat. Commun. **14**, 3995 (2023)
5. Zhu, J., Wang, Y., Chang, W.Y., et al.: Mapping cellular interactions from spatially resolved transcriptomics data. Nat. Methods **21**, 1830–1842 (2024)

6. Ospina, O.E., Soupir, A.C., Manjarres-Betancur, R., et al.: Differential gene expression analysis of spatial transcriptomic experiments using spatial mixed models. Sci. Rep. **14**, 10967 (2024)

7. Qiu, Z., Li, S., Luo, M., et al.: Detection of differentially expressed genes in spatial transcriptomics data by spatial analysis of spatial transcriptomics: a novel method based on spatial statistics. Front. Neurosci. **16**, 1086168 (2022)

8. Cable, D.M., Murray, E., Shanmugam, V., et al.: Cell type-specific inference of differential expression in spatial transcriptomics. Nat. Methods **19**, 1076–1087 (2022)

9. Zhou, Y., Xiao, X., Dong, L., et al.: Cooperative integration of spatially resolved multi-omics data with COSMOS. Nat. Commun. **16**, 27 (2025)

10. Miao, J., Li, J., Xin, J., et al.: MultiGATE: integrative analysis and regulatory inference in spatial multi-omics data via graph representation learning. Nat. Commun. **16**, 9403 (2025)

11. Long, Y., Ang, K.S., Sethi, R., et al.: Deciphering spatial domains from spatial multi-omics with SpatialGlue. Nat. Methods **21**, 1658–1667 (2024)

12. Xiao, J., Yu, X., Meng, F., et al.: Integrating spatial and single-cell transcriptomics reveals tumor heterogeneity and intercellular networks in colorectal cancer. Cell Death Dis. **15**, 326 (2024)

13. Zhou, L., Liu, J., Yao, P., et al.: Spatial transcriptomics reveals unique metabolic profile and key oncogenic regulators of cervical squamous cell carcinoma. J. Transl. Med. **22**, 1163 (2024)

14. Dong, Z.R., Zhang, M.Y., Qu, L.X., et al.: Spatial resolved transcriptomics reveals distinct cross-talk between cancer cells and tumor-associated macrophages in intrahepatic cholangiocarcinoma. Biomark. Res. **12**, 100 (2024)

15. Arora, R., Cao, C., Kumar, M., et al.: Spatial transcriptomics reveals distinct and conserved tumor core and edge architectures that predict survival and targeted therapy response. Nat. Commun. **14**, 5029 (2023)

16. Song, Z., Zheng, C., Chen, J.: stTrace: detecting spatial-temporal domains from spatial transcriptome to trace developmental path. Brief. Bioinform. **26**(6), bbaf606 (2025)

17. Wang, W., Zheng, S., Shin, S.C., et al.: ONTraC characterizes spatially continuous variations of tissue microenvironment through niche trajectory analysis. Genome Biol. **26**, 117 (2025)

18. Pham, D., Tan, X., Balderson, B., et al.: Robust mapping of spatiotemporal trajectories and cell–cell interactions in healthy and diseased tissues. Nat. Commun. **14**, 7739 (2023)

19. Shen, X., Zuo, L., Ye, Z., et al.: Inferring cell trajectories of spatial transcriptomics via optimal transport analysis. Cell Syst. **16**(2), 101194 (2025)

20. Xu, H., Fu, H., Long, Y., et al.: Unsupervised spatially embedded deep representation of spatial transcriptomics. Genome Med. **16**, 12 (2024)

21. Liang, Y., Shi, G., Cai, R., et al.: PROST: quantitative identification of spatially variable genes and domain detection in spatial transcriptomics. Nat. Commun. **15**, 600 (2024)

22. Long, Y., Ang, K.S., Li, M., et al.: Spatially informed clustering, integration, and deconvolution of spatial transcriptomics with GraphST. Nat. Commun. **14**, 1155 (2023)

23. Wang, Y., Liu, Z., Ma, X.: MNMST: topology of cell networks leverages identification of spatial domains from spatial transcriptomics data. Genome Biol. **25**, 133 (2024)

24. Xu, C., Jin, X., Wei, S., et al.: DeepST: identifying spatial domains in spatial transcriptomics by deep learning. Nucleic Acids Res. **50**(22), e131 (2022)

25. Lu, X., Zhou, M., Wang, G. et al.: SLGCA: spatial cross-level graph contrastive autoencoder for multislice spatial domain identification and microenvironment exploration. Brief. Bioinform. **26**(6), bbaf574 (2025)
26. Wolf, F.A., Angerer, P., Theis, F.J.: SCANPY: large-scale single-cell gene expression data analysis. Genome Biol. **19**, 15 (2018)
27. Traag, V.A., Waltman, L., van Eck, N.J.: From Louvain to Leiden: guaranteeing well-connected communities. Sci. Rep. **9**, 5233 (2019)
28. Maynard, K.R., Collado-Torres, L., Weber, L.M., et al.: Transcriptome-scale spatial gene expression in the human dorsolateral prefrontal cortex. Nat. Neurosci. **24**, 425–436 (2021)
29. Chen, A., Liao, S., Cheng, M., et al.: Spatiotemporal transcriptomic atlas of mouse organogenesis using DNA nanoball-patterned arrays. Cell **185**(10), 1777–1792 (2022)
30. Li, Y., Sha, C., Huang, X., Zhang, Y.: Community detection in attributed graphs: an embedding approach. In: Proceedings of the AAAI Conference on Artificial Intelligence, vol. 32, pp. 1–8. AAAI Press, Palo Alto (2018)
31. Pedregosa, F., et al.: Scikit-learn: machine learning in Python. J. Mach. Learn. Res. **12**, 2825–2830 (2011)

SCOPE-PD: Explainable AI on Subjective and Clinical Objective Measurements of Parkinson's Disease for Precision Decision-Making

Md Mezbahul Islam[1], John Michael Templeton[2], Masrur Sobhan[1], Christian Poellabauer[1], and Ananda Mohan Mondal[1(✉)]

[1] Computing and Information Sciences, Florida International University, 11200 SW 8th St., Miami, FL 33199, USA
{misla093,msobh002,cpoellab,amondal}@fiu.edu
[2] Computer Science and Engineering, University of South Florida, 3820 USF Alumni Drive, Tampa, FL 33620, USA
jtemplet@usf.edu

Abstract. Parkinson's disease (PD) is a chronic and complex neurodegenerative disorder influenced by genetic, clinical, and lifestyle factors. Predicting this disease early is challenging because it depends on traditional diagnostic methods that face issues of subjectivity, which commonly delay diagnosis. Several objective analyses are currently in practice to help overcome the challenges of subjectivity; however, a proper explanation of these analyses is still lacking. While machine learning (ML) has demonstrated potential in supporting PD diagnosis, existing approaches often rely on subjective reports only and lack interpretability for individualized risk estimation. This study proposed SCOPE-PD, an explainable AI-based prediction framework by integrating subjective and objective assessments to provide personalized health decisions. Subjective and objective clinical assessment data are collected from the Parkinson's Progression Markers Initiative (PPMI) study to construct a multimodal prediction framework. Several ML techniques were applied to these data, and the best ML model was selected to interpret the results. Model interpretability was examined using SHAP-based analysis. The Random Forest algorithm achieved the highest accuracy of 98.66% while working with the combined features from both subjective and objective test data. Tremor, bradykinesia, and facial expression are identified as the top three contributing features from the MDS-UPDRS test in the prediction of PD.

Keywords: Parkinson's Disease · Machine Learning · Explainable Artificial Intelligence · Precision treatment · Clinical decision-making · Multi-modal Analysis

1   Introduction

Parkinson's disease (PD) has emerged as one of the most pressing neurodegenerative challenges worldwide, affecting an estimated 11.8 million people as of 2021—a number projected to rise dramatically to over 25 million by 2050 as the global population ages [1]. This rise not only puts a lot of stress on people and society, but it also makes the need for a quick and accurate diagnosis even more urgent, especially since there is still no cure that can improve the course of the disease. PD usually shows up as a mix of motor and non-motor symptoms caused by neurodegenerative changes, including the death of dopaminergic neurons and the buildup of alpha-synuclein aggregates [2]. The effects of PD on people and their families go beyond the emotional, cognitive, and physical effects. Additionally, the cost involved in diagnosing, monitoring, and treating PD is also rising, and it is now thought to be over USD 50 billion a year in the United States alone [3], which points to the need for earlier and more accurate diagnosis.

Even though neurological assessment has reached an advanced level, diagnostic accuracy is still low, at 70%-80%, especially in early stages. Clinical assessments are a big part of PD care and research, as they are the basis for diagnosis and the main way to record how the disease is getting worse. Traditionally, these assessments comprise a combination of subjective tests (e.g., patient-reported outcomes, questionnaires) and objective measures (e.g., clinician-rated motor and cognitive examinations). Although subjective and objective tools work well together to help doctors make decisions, their differences lead to ongoing debates about the best way to do things [4]. Finding a balance is important for both the accuracy of research and the care of patients. Recent evidence also suggests that patient-reported outcomes can meaningfully reflect underlying biological mechanisms, as demonstrated in REDONE-PD, where dopamine-related genetic mutations showed measurable associations with neurocognitive and functional outcomes in both PD patients and healthy controls [5].

Numerous studies have leveraged ML for the early detection of PD, harnessing everything from clinical diagnosis, sensor data, and voice characteristics to sophisticated neuroimaging. Grover et al. utilized voice recordings to accurately predict Parkinson's disease; however, medical history and cognitive data were excluded [6]. Templeton et al. found that perceived functionality by patients often differs from sensor-based metrics by applying ML while doing precise, stage-specific assessment and feature identification in PD [7]. Adebimp et al. developed an explainable ML pipeline for PD prediction using a multimodal dataset that combines demographics, medical history, lifestyle, symptoms, and standard clinical assessments (e.g., UPDRS, MoCA) with 93% prediction accuracy [8]. Pereira et al. effectively employed movement analysis, yet they did not consider lifestyle and demographic factors [9].

A central barrier to broad clinical implementation of ML models lies in their "black box" nature. Without clear, patient-specific rationales for predictions, both clinicians and patients may be reluctant to trust or act on ML-driven insights. EXplainable AI (XAI) techniques, such as SHapley Additive exPlanations (SHAP) [10], address this issue by clarifying how predictions are made in complex health sectors (e.g., cancer [11], depression [12]). Adebimp et al., who applied SHAP for PD prediction, did not consider explaining from patient-reported and expert-assessed aspects.

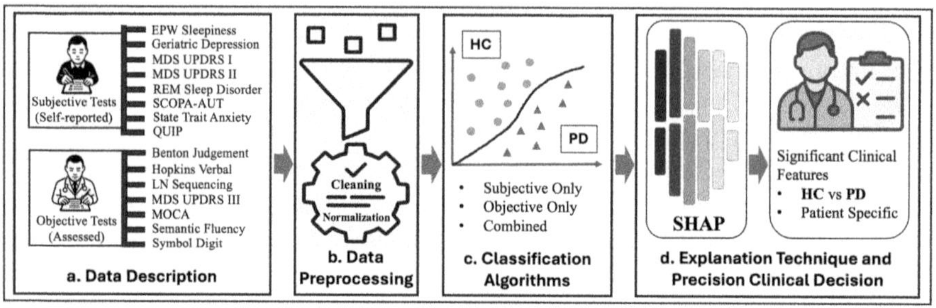

Fig. 1. Overall Study Framework including Data Description, Data Preprocessing, Classification Algorithms, and Explanation Technique and Precision Clinical Decision. EPW: Epworth Sleepiness Scale; MDS-UPDRS: Movement Disorder Society Unified Parkinson's Disease Rating Scale; REM: Rapid Eye Movement; SCOPA-AUT: Scales for Outcomes in Parkinson's disease - Autonomic Dysfunction; QUIP: Questionnaire for Impulsive-Compulsive Disorders; LN: Letter-Number; MOCA: Montreal Cognitive Assessment.

In response to these challenges, this study seeks to reconcile the disparities between conventional clinical practice and emerging computational techniques by developing and validating an interpretable machine learning framework for Parkinson's disease diagnosis. This study aims to achieve two primary objectives: (1) to enhance the predictive accuracy of PD by optimally integrating diverse measures from the well-known PPMI repository, including both subjective self-reports and objective expert assessments; and (2) to offer clear, clinically relevant explanations for each model's decision-making process, utilizing advanced SHAP tools.

The research illustrates, utilizing actual clinical data, that the integration of both data types enhances diagnostic accuracy. It identifies the most important features, like tremor (from subjective tests) and bradykinesia and facial expression (from objective tests), and also shows how they affect predictions. This approach not only makes ML more useful for diagnosing PD, but it also builds trust and makes it easier to understand, setting a standard for future explainable AI frameworks in neurodegenerative disease research.

The key contributions of this study are:

a. Assembled subjective and objective features from PPMI to train and compare state-of-the-art classifiers under nested cross-validation.

b. Embedded SHAP to quantify both local (patient-specific) and global (cohort-level) feature contributions in probability space, enabling intuitive statements like "this feature raised the PD probability by +0.05."

c. Separated global feature contributions for Healthy Control (HC) and PD cohorts and by data type, making it clear whether the subjective report or the objective assessment influences the differences.
d. Identified well-known motor determinants (e.g., tremor, bradykinesia) and quantified their relative influence alongside daily and autonomic functions

2   Dataset and Methodology

The data used in this study were obtained from the Parkinson's Progression Markers Initiative (PPMI) [13], a large-scale, longitudinal, and publicly available dataset designed to identify biomarkers of PD progression. The study methodology can be found at ppmi-info.org. The data were obtained from the Laboratory of NeuroImaging repository (LONI). The dataset includes a comprehensive collection of subjective (patient-reported) and objective (expert-assessed) neuropsychological and clinical assessments.

The overall framework of our study is divided into four principal sections as shown in Fig. 1: data source description, data preprocessing, classification algorithms, and explanation technique and precision clinical decision.

2.1   Data Description

In the PPMI dataset, each participant undergoes repeated evaluations over multiple visits. To ensure consistency and to avoid within-subject temporal dependencies, only the baseline (i.e., first-visit) data were used in this analysis. It also ensures the data collection alignment for the early prediction approach. The age range of participants is 26 to 86. The dataset encompasses a diverse range of subjective and objective assessments that collectively capture autonomic, behavioral, psychological, executive, memory, motor, sensory, sleep, and speech characteristics of individuals with PD and healthy controls. A detailed summary of the included tests, sample counts, number of survey questions, and feature distributions is provided in Table 1.

Subjective measurements include eight standardized self-reported questionnaires: the Epworth Sleepiness Scale (EPW) [14], Geriatric Depression Scale (GDS) [15], Movement Disorder SocietyUnified Parkinson's Disease Rating Scale Parts I and II (MDS-UPDRS III) [16], REM Sleep Behavior Disorder Questionnaire [17], Scales for Outcomes in Parkinson's DiseaseAutonomic Dysfunction (SCOPA-AUT) [18], StateTrait Anxiety Inventory (STAI) [19], and Questionnaire for Impulsive-Compulsive Disorders (QUIP) [20]. Together, these subjective tests yielded 79 features from 2,128 participants across eight tests.

Objective measurements consist of seven expert-assessed neuropsychological and motor tests: the Benton Judgment of Line Orientation (BJLO) [21], Hopkins Verbal Learning Test (HVLT) [22], LetterNumber Sequencing (LNS) [23], MDS-UPDRS Part III [16], Montreal Cognitive Assessment (MoCA) [24], Modified Semantic Fluency (MSF) [25], and the Symbol Digit Modalities Test (SDMT)

[26]. These objective evaluations provided 69 features across 1,862 participants present in all seven tests. After combining the subjective and objective tests, 1,786 participants were found in all tests, yielding a total of 148 features (i.e., 79 subjective + 69 objective), as shown in Table 1. The conversion of survey questions and responses into features and feature values is described in Sect. 2.2.2. Descriptions of these features are provided in the supplementary material.

Table 1. Summary of Subjective and Objective Tests. EPW: Epworth Sleepiness Scale; MDS-UPDRS: Movement Disorder Society Unified Parkinson's Disease Rating Scale; REM: Rapid Eye Movement; SCOPA-AUT: Scales for Outcomes in Parkinson's disease - Autonomic Dysfunction; QUIP: Questionnaire for Impulsive-Compulsive Disorders; LN: Letter-Number; MOCA: Montreal Cognitive Assessment; CSS: Common Subjective Samples; COS: Common Objective Samples; OCS: Overall Common Samples.

Test	Subjective Tests									Objective Tests								Overall
Test Name	EPW Sleepiness [14]	Geriatric Depression [15]	MDS UPDRS I [16]	MDS UPDRS II [16]	REM Sleep Disorder [17]	SCOPA-AUT [18]	State Trait Anxiety [19]	QUIP [20]	# of CSS / Total	Benton Judgement [21]	Hopkins Verbal [22]	LN Sequencing [23]	MDS UPDRS III [16]	MOCA [24]	Semantic Fluency [25]	Symbol Digit [26]	# of COS / Total	# of OCS / Total
# of Samples	3,832	3,761	2,212	2,212	3,839	3,830	3,760	3,829	**2,128**	3,813	3,819	3,813	3,817	3,813	2,562	2,370	**1,862**	**1,786**
# of Survey Questions	8	15	7	13	21	21	40	13	**138**	15	7	7	32	27	3	1	**92**	**230**
# of Features	1	1	7	13	21	21	2	13	**79**	1	4	1	32	27	3	1	**69**	**148**

2.2 Data Preprocessing

2.2.1 Handling Missing Data

The preprocessing starts with removing samples and features with missing values. The "VLTVEG" (total correct vegetables an individual can list in 60 s) and "VLTFRUIT" (total correct fruits an individual can list in 60 s) features of the MSF test were found to be missing in 422 samples out of 1,862. For that reason, these two features were removed from the objective test features. After this step, the number of samples with subjective tests reduced from 2,128 to 2,104, and objective tests reduced from 1,862 to 1,847. The number of samples after combining subjective and objective tests is 1,741. Figure 2 summarizes the distribution of four cohorts (PD, Prodromal, HC, and SWEED) for three test types (Subjective, Objective, and Combined) in the cleaned data. For example, in subjective tests, 1,050 are PD patients (i.e., participants with a clinical diagnosis of PD), 736 are Prodromal (i.e., participants without a PD diagnosis yet but who show risk factors or early non-motor symptoms), 253 are age-matched healthy controls (HC) (i.e., participants with no known neurological symptoms or risk factors), and 65 are Scans Without Evidence of Dopaminergic Deficit (SWEDD). This study examined the PD and HC groups from three assessment conditions (subjective, objective, and combined) to identify the most significant features that contribute to precise clinical decisions.

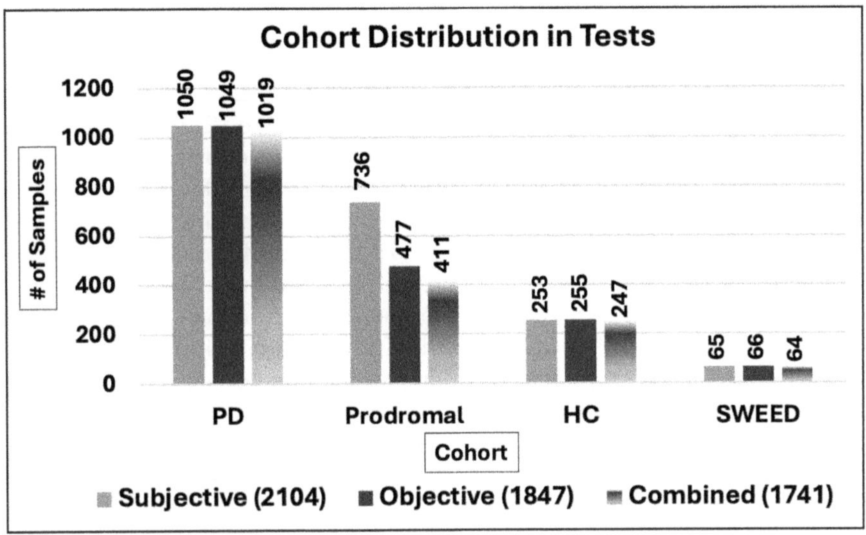

Fig. 2. Cohort description after data preprocessing.

2.2.2 Converting Survey Responses Into Feature Values

The features considered in this study represent a certain question from a test or a combination of responses from multiple questions. In the case of subjective tests, the EPW test includes eight different questions related to sleep, which are converted to a single value by adding the response values. The GDS test has 15 questions for patients related to depression, which are converted to a single value by adding them. The 20 raw responses from the MDS-UPDRS I and II are included as features. Similarly, the raw responses to 13 behavior/psychology-related questions from the QUIP test are considered as features. The REM test asks 21 questions related to sleep quality and the presence of certain diseases in a patient. One of them asks about the presence of "PARKISM" in a patient, which may bias the study for ML-based analysis. We dropped this feature and moved on with 20 features from the REM test. The SCOPA-AUT test asks 21 questions related to autonomic function, which are considered as features directly. The STAI test includes 40 questions, where 21 questions follow the natural flow of response value, where a higher score relates to a higher anxiety level by asking questions related to anxiety. The remaining 19 questions are related to reverse situations where a higher score relates to a lower anxiety level. To align the response values, we reversed the value of 19 questions. Among the 40 questions, the first 20 questions are related to state (how a person feels "right now" or "at this moment," reflecting temporary, situational anxiety), and the last 20 questions are related to trait (how a person generally feels, reflecting more stable, long-term anxiety tendencies). State and trait questions are converted to two single values by summing up the corresponding response values. So, in total, the subjective tests have 79 features to study.

Among the objective tests, the summation of 15 assessment scores related to executive function from the BJLO test is considered as a single feature. The HVLT test had seven assessment scores, ranging from 0 to 12, related to memory in the original assessment. To summarize the HVLT performance and reduce noise, they are converted to four different scores to consider as features. The first score established overall acquisition ability by summing up 3 individual recalls with a range of 0 to 36. The second one ranges from 0 to 12, reflecting the number of words recalled after a delay. The third one measures how much information in percentage is retained over delay. The final one scores the recognition discrimination index by subtracting false positives from true positives. The LNS test measures a patient's working memory and executive function with seven scores, which are converted to a single feature value by adding them. The MDS-UPDRS III test includes 32 questions related to motor function, which are directly considered as features. MOCA has 27 questions designed as a rapid screening instrument for mild cognitive dysfunction, which are directly used as features. MSF has three features, from which 2 of them are removed for missing values and only one is kept. The SDMT test assesses cognitive function by measuring how quickly a person can match symbols to numbers, and the total number of correct matches is used as a feature value. So, in total, the objective tests have 67 features to study. Here all the test response values except the MDS-UPDRS III follow the same pattern of higher scores representing better conditions of individual participants. To align all the objective and subjective tests, the scores are reversed so a higher score represents worse conditions.

2.2.3 Normalizing Feature Values

The clinical features used in this study originate from heterogeneous assessments with differing value ranges and statistical properties. For example, items from the MDS-UPDRS are scored on ordinal scales (04), whereas several cognitive and questionnaire-based measures follow binary or bounded scoring schemes. To ensure comparability across features while preserving relative variability, all numeric features were standardized using z-score normalization. Specifically, each feature was transformed using the StandardScaler, which rescales values to have zero mean and unit variance based on the training data distribution. This standardization reduces scale-induced bias, improves numerical stability, and facilitates effective model training, particularly in the presence of features with diverse ranges and variances. The same scaling parameters learned from the training set were subsequently applied to validation and test data to prevent information leakage

The final outcome of the data preprocessing step resulted in three datasets: subjective, objective, and combined. These three curated baseline dataset forms the foundation for classification and explainability analyses within the SCOPE-PD framework.

2.3   Classification Algorithms for PD Prediction

To explore different machine learning strategies for differentiating PD from HC, five widely used supervised algorithms were implemented: Logistic Regression (LR), Support Vector Machine (SVM) with a radial-basis kernel, K-Nearest Neighbors (KNN), Random Forest (RF), and Extreme Gradient Boosting (XGBoost). All analyses were carried out in Python using the scikit-learn and XGBoost libraries.

Since all three datasets (subjective, objective, and combined) are imbalanced, as shown in Fig. 2, native weighting was used instead of data augmentation using SMOTE and ADASYN to avoid synthetic sample bias. For LR, SVM, and RF, the argument **class_weight=“balanced”** was applied to automatically adjust each class's contribution to the loss function in proportion to its frequency. For XGBoost, imbalance was managed using the parameter **scale_pos_weight** as shown in Eq. 1. This parameter scales the gradient and Hessian for the positive class during training, effectively increasing the contribution of the minority class to the loss function. When the dataset is highly imbalanced (i.e., $n_0 \gg n_1$), setting a larger scale_pos_weight ensures that misclassified positive samples receive a higher penalty. As a result, the model becomes more sensitive to the minority class and can achieve better recall and balanced performance across both classes. KNN, which does not include a native weighting mechanism, was evaluated directly on the stratified data.

$$\text{scale\_pos\_weight} = \frac{n_0}{n_1} \tag{1}$$

where

- n_0: Number of negative samples
- n_1: Number of positive samples

Hyperparameter tuning and model evaluation followed a 5-fold stratified cross-validation framework. The dataset was first divided into an 80/20 train–test split, stratified to preserve class ratios. The training subset (80%) was used exclusively for model development and hyperparameter optimization through a 5-fold cross-validation (GridSearchCV) procedure, which exhaustively searched predefined parameter grids for each algorithm using the F1-score as the optimization metric to balance sensitivity and precision. The best-performing hyperparameter configuration, identified by the highest mean cross-validated F1-score across the training folds, was then retrained on the entire 80% training data and evaluated once on the held-out 20% test set, which served as an independent set for unbiased performance assessment.

2.4   Model Explainability Using SHAP

Local explanations were reported with SHAP waterfall/force plots for individual patients (showing how their feature values shift the prediction from the baseline to the final probability) and global explanations by aggregating attributions across the cohort. Specifically, for each feature, mean absolute SHAP was computed for the HC and PD groups separately, and these were visualized as stacked bars (HC + PD). This class-conditional aggregation clarifies not only which features are most important overall, but also in which group their importance primarily arises. The concept of local interpretability of SHAP was utilized for precision clinical decision-making that may assist in formulating strategies for individualized treatment.

Overall, the use of SHAP enabled a comprehensive understanding of both global feature importance and individual-level predictions. By revealing the roles of subjective and objective features in model decision-making, this explainable AI framework strengthened the interpretability, clinical relevance, and trustworthiness of the proposed PD prediction system.

3   Results and Discussion

The results are divided into two sections: a. experimental results of ML models and b. interpretation of ML predictions for Parkinson's disease based on subjective, objective, and combined features.

3.1   Experimental Results of Machine Learning Models

Model performance varied across feature categories, revealing distinct predictive patterns between subjective, objective, and combined feature representations. The performance evaluation values are listed in Table 2. For subjective assessments, the RF model achieved the highest overall performance with an F1-score of 0.9786 ± 0.0016, ROC-AUC of 0.9875 ± 0.0032, and PR-AUC of 0.9970 ± 0.0008, indicating strong discriminative power and robustness in capturing nonlinear feature interactions from patient-reported measures. The SVM (RBF) and XGBoost models followed closely, with comparable precision and recall, while LR and KNN exhibited slightly lower but still competitive results, confirming that both linear and non-linear classifiers can effectively separate PD and healthy control groups when subjective scales are well standardized.

When using objective cognitive and motor features, performance improved further across most algorithms. RF again outperformed others with a mean accuracy of 0.9785 ± 0.0069 and recall of 0.9924 ± 0.0072, demonstrating excellent sensitivity to PD-related performance impairments. SVM and XGBoost achieved similar AUC values above 0.99, suggesting consistent generalizability and reliability across folds. These findings highlight that objective test metrics offer slightly superior predictive value compared to subjective measures.

When subjective and objective features were combined, model performance reached its peak across all metrics. The RF had the best overall accuracy (0.9866

± 0.0091) and F1-score (0.9917 ± 0.0055). The ROC-AUC (0.9992 ± 0.0006) and PR-AUC (0.9998 ± 0.0001) showed that it could almost perfectly tell the difference between the two groups. The SVM (RBF) and XGBoost followed closely, maintaining mean F1-scores above 0.988, while LR also performed strongly (F1 = 0.9871 ± 0.0045), confirming the complementary predictive value of integrating subjective and objective features. Although KNN achieved the highest precision (0.9978 ± 0.0050), its comparatively lower recall (0.8371 ± 0.0217) suggests limited sensitivity to PD cases despite its specificity toward healthy controls.

Although the reported accuracy is high, it was obtained under strict 5-fold nested cross-validation using class-level stratification, ensuring no overfitting to data noise. Thus, the high model performance reflects that the survey responses at the first visit can clearly distinguish PD from HC groups. Nevertheless, we recognize that external validation on independent cohorts is essential to confirm real-world performance. Overall, across all experiments, ensemble models such as RF and XGBoost consistently demonstrated superior and more stable performance than single estimators, emphasizing the importance of non-linear modeling and feature interaction learning in PD classification tasks. To further validate classification reliability and generalization, confusion matrices were generated for the best-performing models (Random Forest in all three datasets) using the Out-of-Fold (OOF) strategy (Fig. 3). The number of misclassified samples decreased from the models with the subjective features to the objective features and improved further with the combined features.

Table 2. Performance comparison across classification models for Parkinson's disease prediction using Subjective, Objective, and Combined features (mean ± standard deviation over 5 folds). SD: Standard Deviation.

Test Type	Model	Accuracy ± SD	Precision ± SD	Recall ± SD	F1 Score ± SD	ROC AUC ± SD	PR AUC ± SD
Subjective	Logistic Regression	0.9440 ± 0.0111	0.9851 ± 0.0034	0.9448 ± 0.0141	0.9645 ± 0.0072	0.9820 ± 0.0047	0.9956 ± 0.0013
	KNN	0.9010 ± 0.0150	0.9610 ± 0.0106	0.9143 ± 0.0121	0.9370 ± 0.0096	0.9525 ± 0.0148	0.9881 ± 0.0041
	SVM (RBF)	0.9563 ± 0.0043	0.9779 ± 0.0052	0.9676 ± 0.0098	0.9727 ± 0.0029	0.9778 ± 0.0123	0.9936 ± 0.0049
	Random Forest	**0.9655 ± 0.0027**	0.9755 ± 0.0074	**0.9819 ± 0.0062**	**0.9786 ± 0.0016**	**0.9875 ± 0.0032**	**0.9970 ± 0.0008**
	XGBoost	0.9609 ± 0.0079	**0.9846 ± 0.0101**	0.9667 ± 0.0101	0.9755 ± 0.0049	0.9831 ± 0.0037	0.9954 ± 0.0016
Objective	Logistic Regression	0.9663 ± 0.0099	0.9883 ± 0.0055	0.9695 ± 0.0079	0.9788 ± 0.0062	0.9928 ± 0.0013	0.9982 ± 0.0002
	KNN	0.9026 ± 0.0155	**0.9905 ± 0.0057**	0.8875 ± 0.0213	0.9360 ± 0.0109	0.9618 ± 0.0144	0.9859 ± 0.0061
	SVM (RBF)	0.9778 ± 0.0074	0.9857 ± 0.0058	0.9867 ± 0.0040	0.9862 ± 0.0046	0.9928 ± 0.0021	0.9980 ± 0.0008
	Random Forest	**0.9785 ± 0.0069**	0.9812 ± 0.0085	**0.9924 ± 0.0072**	**0.9867 ± 0.0043**	**0.9969 ± 0.0019**	**0.9991 ± 0.0007**
	XGBoost	0.9739 ± 0.0057	0.9858 ± 0.0109	0.9819 ± 0.0085	0.9838 ± 0.0035	0.9955 ± 0.0018	0.9988 ± 0.0006
Combined	Logistic Regression	0.9795 ± 0.0071	0.9931 ± 0.0055	0.9813 ± 0.0107	0.9871 ± 0.0045	0.9972 ± 0.0021	0.9993 ± 0.0005
	KNN	0.8673 ± 0.0150	**0.9978 ± 0.0050**	0.8371 ± 0.0217	0.9102 ± 0.0112	0.9744 ± 0.0059	0.9907 ± 0.0021
	SVM (RBF)	0.9818 ± 0.0045	0.9912 ± 0.0085	0.9863 ± 0.0094	0.9887 ± 0.0028	0.9984 ± 0.0006	0.9996 ± 0.0001
	Random Forest	**0.9866 ± 0.0091**	0.9875 ± 0.0137	**0.9961 ± 0.0041**	**0.9917 ± 0.0055**	**0.9992 ± 0.0006**	**0.9998 ± 0.0001**
	XGBoost	0.9826 ± 0.0045	0.9932 ± 0.0073	0.9853 ± 0.0092	0.9892 ± 0.0028	0.9988 ± 0.0006	0.9997 ± 0.0001

3.2   Interpretation of ML Predictions for PD

To provide transparent, clinically meaningful explanations of the tuned Random Forest, we used SHAP TreeExplainer, a tree-modelaware method that computes

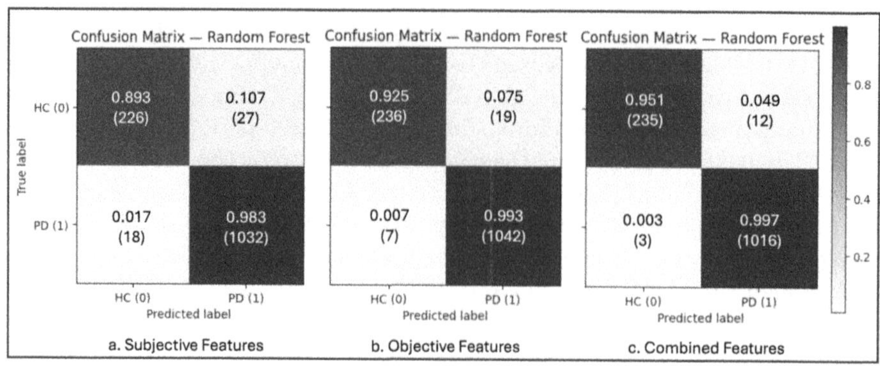

Fig. 3. Out-of-Fold (OOF) confusion matrix (normalized) for the best ML model. RF achieved the highest accuracy with all three types of features—subjective, objective, and combined.

exact, locally additive feature attributions for decision-tree ensembles [27]. TreeExplainer provides, for each subject, an attribution that indicates how much each feature increases or decreases the model's predicted probability of PD compared to a reference expectation ("baseline"). For binary classification, TreeExplainer returns one attribution vector per class. The PD channel (class = 1) was analyzed, where positive values push the prediction toward PD and negative values toward HC (class = 0).

Inputs to SHAP were the scaled features from the pipeline so that the attributions reflect exactly what the classifier sees. The contribution of features was analyzed in two aspects: global (contribution in predicting HC or PD) and local (contribution in a sample). The stacked bar plot (Fig. 4) shows the top 10 feature contributions towards class-conditional decisions for the features of subjective, objective, and combined tests. In the subjective dataset, 9 out of 10 top contributing features are from MDS-UPDRS II and one is from SCOPA-AUT (Fig. 4a). It is clear from this figure that the self-reported tremor feature (NP2TRMR) dominated the importance ranking by a wide margin within both HC and PD groups. The next tier comprised handwriting difficulty (NP2HWRT), dressing (NP2DRES), walking/balance (NP2WALK), speech (NP2SPCH), and eating/feeding (NP2EAT). These day-to-day activities showed mixed HC/PD splits: very low difficulty supports HC, whereas increasing difficulty contributes evidence toward PD. The SCOPA-AUT feature SCAU2 (In the past month, has saliva dribbled out of your mouth?) was found to be one of the most contributing features.

The top 10 features for objective tests are coming from MDS-UPDRS III test (Fig. 4b), where the top contributed feature (NP3BRADY) related to observing "spontaneous gestures while sitting, and the nature of arising and walking," which is a significant PD sign and central to diagnosis/staging. So, its dominance as a driver of PD is expected and provides the significance of this study. It also aligns with next-ranked items: NP3RTCON (rapid alternating movement/coordination), NP3FACXP (facial expression), and NP3FTAPR (finger tapping), which are closely related to bradykinetic/akinetic measures. While combining subjective and objective tests, eight features from MDS-UPDRS III and two features from MDS-UPDRS II appeared in the top 10 contributed features, as shown in 4c. The self-reported tremor impact (NP2TRMR) was found to be the most contributing in distinguishing PD and HC samples.

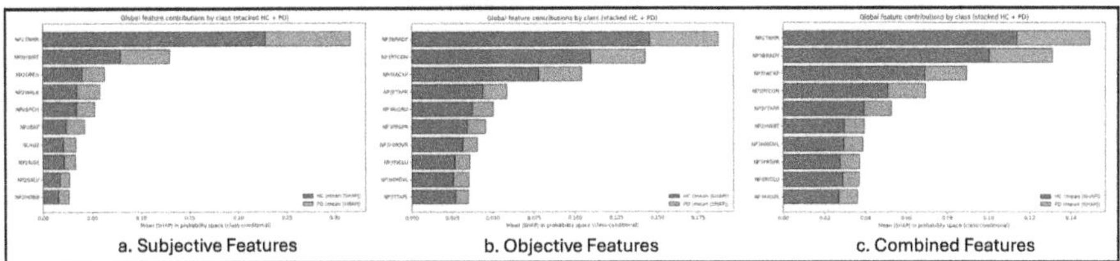

Fig. 4. Global feature contribution: Top 10 features contributing towards HC and PD cohorts. a. Subjective features. b. Objective features. c. Combined features.

To gain a deeper understanding of the model's decision-making process, we examined specific individuals diagnosed as PD patient, as shown in Fig. 5. We utilized SHAP's color-coded visualizations to identify the features that significantly influenced the model's prediction for each person. Red hues highlight features that strongly contribute in diagnosing PD, while blue hues indicate features aligned with HC prediction. The model's baseline probability of PD was 0.805 and rose to 0.963 once the patient's features were considered. The increase was driven mainly by tremor (NP2TRMR, +0.05) and bradykinesiarelated (NP3BRADY, +0.05) markers. Other positive features are finger tapping (NP3FTAPR, +0.02), hand movements (NP3HMOVR, +0.02), masked facies/facial expression (NP3FACXP, +0.02), upper-limb rigidity (NP3RIGRU, +0.02), posture (NP3PRSPR, +0.02), and dressing difficulty (NP2DRES, +0.01). Rapid alternating movements/coordination (NP3RTCON) countered the prediction (−0.05) meaning the PD patient had a slight issue with this activity. The remaining features had negligible effects. While SHAP predominantly highlighted established PD determinants such as tremor and bradykinesia, the framework also quantified their relative effect sizes across subjective and objective domains, offering clinicians quantitative thresholds for individualized interpretation. This transparency transforms well-known symptoms into measurable, patient-specific risk weights usable in clinical discussions.

3.3 Discussion

This study aimed to identify an effective model for predicting PD using self-reported and assessment-based tests and explain the contribution of features using SHAP based on the most effective model. Five individual models were evaluated: LR, SVM-RBF, KNN, RF, and XGBoost. RF attained the highest

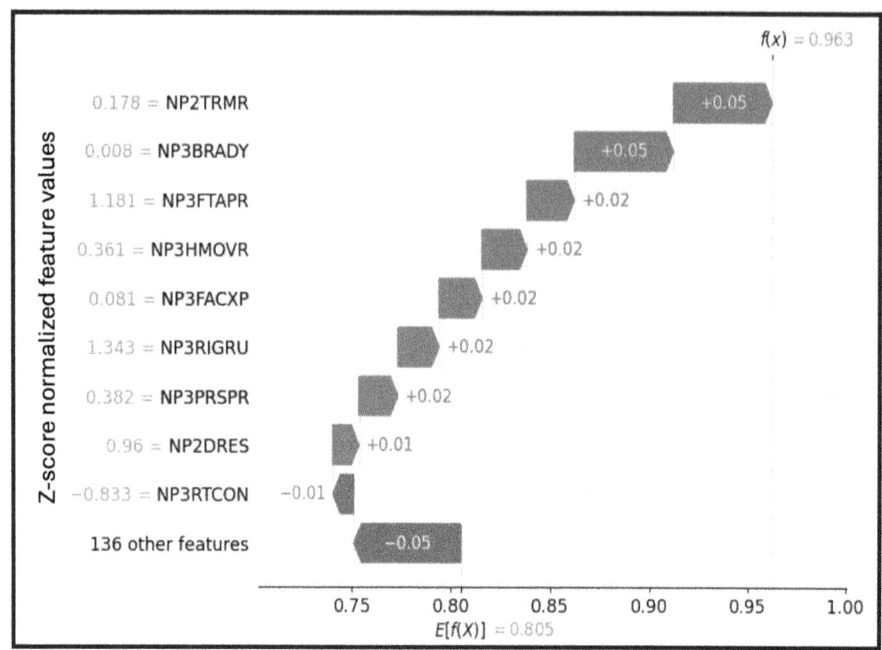

Fig. 5. Local feature contribution: Top 10 local features contributing to a specific PD patient.

accuracy rates of 96.55%, 97.85%, and 98.66% for subjective, objective, and combined tests, respectively. It positioned this study competitively within the existing literature to compare with other approaches as shown in Table 3. Grover et al. reported 81.67% accuracy by using only the speech features [6], while Templeton et al. achieved 92.6% accuracy by analyzing sensor-based metric [7]. Adebimpe et al. [8] achieved 93% accuracy with clinical (MoCA), cognitive (UPDRS), demographic, and lifestyle data. Dentamaro et al. [28] attained an accuracy of 96.6% by employing DenseNet on the detection of the prodromal stage by utilizing 3D MRI images and clinical data. Priyadharshini et al. [29] established a complete framework utilizing T2-weighted 3D MRI datasets and attained an accuracy of 96.8% through the integration of Gradient Boosting and SMOTE for data balancing. This study compares its findings with the existing literature by using multimodal self-reported and clinically assessed data from 1786 samples. To address the sample imbalance, this study utilized the default class-weight property for LR, SVM, and RF models, or the scale_pos_weight parameter for the XGBoost model, rather than applying SMOTE to generate synthetic data.

Limitations and Future Scopes: Only baseline data (first visit) were used to ensure the existence of multimodal data, and progression modeling was beyond the current scope. Incorporating longitudinal trajectories from follow-up visits is a next step toward progression and prognosis prediction. An expanded version of

Table 3. Comparison of the proposed SCOPE-PD model with existing studies on Parkinson's disease prediction.

Author	Data	Accuracy	Dataset	Classification
Grover et al. [6]	Voice recordings	81.67%	UCI ML Repository	Severe vs Not Severe
Templeton et al. [7]	Sensor-based metric	92.60%	Self-Collected	PD vs HC
Adebimpe et al. [8]	Subjective (UPDRS, MoCA)	93%	PPMI	PD vs HC
Dentamaro et al. [28]	3D MRI Image, Clinical	96.60%	PPMI	Prodromal stages
Priyadharshini et al. [29]	T2-weighted 3D MRI	96.80%	PPMI	PD vs Prodromal vs HC
SCOPE-PD Subjective	Subjective measures	96.50%	PPMI	PD vs HC
SCOPE-PD Objective	Objective measures	97.80%	PPMI	PD vs HC
SCOPE-PD Combined	Subjective + Objective	**98.60%**	PPMI	PD vs HC

SCOPE-PD will integrate genetic (e.g., GBA, SNCA) and MRI/fMRI imaging modalities from complementary cohorts to enable a full multimodal diagnostic framework. The absence of external validation datasets limits the ability to fully assess robustness and real-world applicability. Future efforts will incorporate multi-site data from PDBP and BioFIND for external validation. Deploying the model in clinical trials will facilitate the evaluation of its effects on diagnostic accuracy and patient care outcomes in practical environments.

4 Conclusion

The study SCOPE-PD aimed to develop and evaluate predictive models for PD applying ML techniques, corroborated by the explainable AI (XAI) method for better interpretability. Rather than claiming clinical deployment readiness, SCOPE-PD should be viewed as a methodological bridge linking subjective and objective markers through explainable modeling. Through a comprehensive evaluation of five ML models, this study found Random Forest with the highest performance, achieving accuracy above 96% across all three datasets. The novelty lies in unifying both modalities under a single interpretable framework, not in outperforming prior work. The results should therefore be interpreted cautiously until externally validated. The model could serve as a valuable screening tool for early PD detection, potentially identifying at-risk patients before significant symptom onset, as all data used here are collected at their first visit for diagnosis. The XAI analysis revealed that the MDS-UPDRS test features significantly contribute to the prediction of Parkinson's disease. The RF model's interpretability makes it appropriate for incorporation into clinical workflows, offering evidence-based diagnostic assistance. Moreover, the ability to identify individualized risk contributors opens opportunities for more personalized treatment planning and targeted interventions.

Data Acknowledgment: Data used in the preparation of this article were obtained [on May 3, 2025] from the Parkinson's Progression Markers

Initiative (PPMI) database (https://www.ppmi-info.org/access-data-specimens/download-data), RRID:SCR_006431. For up-to-date information on the study, visit http://www.ppmi-info.org. PPMI a public-private partnership is funded by the Michael J. Fox Foundation for Parkinson's Research and funding partners, including 4D Pharma, Abbvie, et al

Funding Information. This work was partially supported by the NIH/NCI R21CA290324 and NIH/NHGRI UG3HG013615. The content is solely the responsibility of the authors and does not necessarily represent the official views of the funding agencies.

Conflict of Interest Statement. All authors of this work declare that there are no conflicts of interest in the authorship nor publication of this contribution.

References

1. Luo, Y., Qiao, L., Li, M., Wen, X., Zhang, W., Li, X.: Global, regional, national epidemiology and trends of Parkinson's disease from 1990 to 2021: findings from the global burden of disease study 2021. Front. Aging Neurosci. **16**, 1498756 (2025)
2. Yaribash, S., Mohammadi, K., Sani, M.A.: Alpha-synuclein pathophysiology in neurodegenerative disorders: a review focusing on molecular mechanisms and treatment advances in parkinson's disease. Cell. Mol. Neurobiol. **45**(1), 1–16 (2025)
3. Chaudhuri, K.R., et al.: Economic burden of Parkinson's disease: a multinational, real-world, cost-of-illness study. Drugs-Real World Outcomes **11**(1), 1–11 (2024)
4. Siciliano, M., et al.: Correlates of the discrepancy between objective and subjective cognitive functioning in non-demented patients with Parkinson's disease. J. Neurol. **268**(9), 3444–3455 (2021)
5. Islam, M.M., Templeton, J.M., Poellabauer, C., Mondal, A.M.: Redone-pd: reflections of dopamine-related gene mutations on neurocognitive functions in healthy controls and Parkinson's disease. In: 2024 IEEE International Conference on Bioinformatics and Biomedicine (BIBM), pp. 6113–6120. IEEE (2024)
6. Grover, S., Bhartia, S., Yadav, A., KR, S., et al.: Predicting severity of Parkinson's disease using deep learning. Proc. Comput. Sci. **132**, 1788–1794 (2018)
7. Templeton, J.M., Poellabauer, C., Schneider, S.: Classification of Parkinson's disease and its stages using machine learning. Sci. Rep. **12**(1), 14036 (2022)
8. Esan, A.O., Olawade, D.B., Soladoye, A.A., Omodunbi, B.A., Adeyanju, I.A., Aderinto, N.: Explainable ai for Parkinson's disease prediction: A machine learning approach with interpretable models. Current Res. Transl. Med. 103541 (2025)
9. Pereira, C.R., et al.: A new computer vision-based approach to aid the diagnosis of Parkinson's disease. Comput. Methods Program. Biomed. **136**, 79–88 (2016)
10. Lundberg, S.M., Lee, S.-I.: A unified approach to interpreting model predictions. Adv. Neural Inf. Process. Syst. **30** (2017)
11. Sobhan, M., Islam, M.M., Trepka, M.J., Holt, G.E., Dimitroff, C.J., Mondal, A.M.: Tilda-x: Transcriptome-informed lung cancer disparities via explainable ai. Cancers **17**(21), 3454 (2025)
12. Tran, V.Q., Byeon, H.: Explainable hybrid tabular variational autoencoder and feature tokenizer transformer for depression prediction. Expert Syst. Appl. **265**, 126084 (2025)

13. Nalls, M.A., et al.: Diagnosis of Parkinson's disease on the basis of clinical and genetic classification: a population-based modelling study. Lancet Neurol. **14**(10), 1002–1009 (2015)
14. Johns, M.W.: A new method for measuring daytime sleepiness: the epworth sleepiness scale. Sleep **14**(6), 540–545 (1991)
15. Greenberg, S.A.: The geriatric depression scale (GDS). Best Pract. Nurs. Care Older Adults **4**(1), 1–2 (2012)
16. Goetz, C.G., et al.: Movement disorder society-sponsored revision of the unified Parkinson's disease rating scale (MDS-UPDRS): scale presentation and clinimetric testing results. Movement Disord. Official J. Movement Disord. Soc. **23**(15), 2129–2170 (2008)
17. Schenck, C.H., Bundlie, S.R., Ettinger, M.G., Mahowald, M.W.: Chronic behavioral disorders of human rem sleep: a new category of parasomnia. Sleep **9**(2), 293–308 (1986)
18. Visser, M., Marinus, J., Stiggelbout, A.M., Van Hilten, J.J.: Assessment of autonomic dysfunction in Parkinson's disease: the scopa-aut. Movement Disord. Official Jo. Movement Disord. Soc. **19**(11), 1306–1312 (2004)
19. Spielberger, C.D.: State-trait anxiety inventory for adults (1983)
20. Weintraub, D., Mamikonyan, E., Papay, K., Shea, J.A., Xie, S.X., Siderowf, A.: Questionnaire for impulsive-compulsive disorders in Parkinson's disease-rating scale. Mov. Disord. **27**(2), 242–247 (2012)
21. Benton, A.L., Varney, N.R., Hamsher, K.: Visuospatial judgment: a clinical test. Arch. Neurol. **35**(6), 364–367 (1978)
22. Benedict, R.H., Schretlen, D., Groninger, L., Brandt, J.: Hopkins verbal learning test-revised: normative data and analysis of inter-form and test-retest reliability. Clin. Neuropsychol. **12**(1), 43–55 (1998)
23. Wechsler, D.: Wechsler Adult Intelligence Scale WAIS-IV Canadian. Pearson, (2008)
24. Hobson, J.: The montreal cognitive assessment (MOCA). Occup. Med. **65**(9), 764–765 (2015)
25. Zarino, B., Crespi, M., Launi, M., Casarotti, A.: A new standardization of semantic verbal fluency test. Neurol. Sci. **35**(9), 1405–1411 (2014)
26. Smith, A.: Symbol digit modalities test. Los Angeles, CA (2013)
27. Lundberg, S.M., et al.: From local explanations to global understanding with explainable ai for trees. Nat. Mach. Intell. **2**(1), 56–67 (2020)
28. Dentamaro, V., Impedovo, D., Musti, L., Pirlo, G., Taurisano, P.: Enhancing early Parkinson's disease detection through multimodal deep learning and explainable AI: insights from the ppmi database. Sci. Rep. **14**(1), 20941 (2024)
29. Priyadharshini, S., et al.: A comprehensive framework for Parkinson's disease diagnosis using explainable artificial intelligence empowered machine learning techniques. Alex. Eng. J. **107**, 568–582 (2024)

Factorization-Driven Representation Learning Techniques for Protein Tertiary Structure Prediction: A Comprehensive Review

Kazi Lutful Kabir$^{(\boxtimes)}$ iD

George Mason University, Fairfax, VA 22030, USA
kkabir@gmu.edu

Abstract. Protein tertiary structures capture the complex three-dimensional (3D) arrangements of the constituent atoms that define how proteins function. Different strategies have been applied in the form of representation learning to interpret these high-dimensional structure spaces. Factorization-based techniques, such as matrix and tensor factorization, have also been proven to be an effective set of approaches due to their utility in capturing latent organizations from complicated structural spaces and have already been applied in the context of tasks like tracking conformational changes across structure ensembles of biomolecules, identifying the biologically-active tertiary structure(s) from the given computed protein models, analyzing biomolecular dynamics simulations, etc. This paper exhibits a comprehensive review of factorization-driven methods employed to learn representations of tertiary structures for the protein structure prediction task. It covers how protein structures are encoded, represented, analyzed, and evaluated via different factorization-based frameworks. Furthermore, it outlines major open challenges and offers prospective research directions.

Keywords: Factorization · Matrix · Protein Tertiary Structure · Tensor

1 Introduction

A protein molecule's tertiary structure describes the three-dimensional (3D) arrangement of its atoms and is inherently dynamic as well as complex, which adopts multiple conformations under physiological conditions. Recognizing this dynamic behavior needs organizing these structures into distinct structural states, a task well-suited to unsupervised representation learning. Proteins typically undergo rapid transitions within a given state and moderate transitions between different states, making such techniques an effective tool for summarizing conformational behavior and identifying states that correspond to cellular interactions.

K. L. Kabir et al. (Eds.): BICOB 2026, CCIS 2977, pp. 290–302, 2026.
https://doi.org/10.1007/978-3-032-26028-4_21

Factorization-based methods [9,11] provide an unsupervised framework for learning representations of protein structural ensembles by decomposing high-dimensional structural data into a small number of latent components. These methods aim to represent protein molecular structures as combinations of underlying basis patterns or latent factors, which can capture dominant structural features and structural variability. Factorization approaches differ in their data representations, choice of objective functions, and imposed constraints (e.g., non-negativity or orthogonality), and, like clustering, they lack universally accepted evaluation metrics. In computational biology, factorization methods have been applied to various problems [10] such as identifying clusters in protein-protein interactions [13], extracting sequence motifs [22], recognizing genomic sub-types [8], unraveling functionally-relevant genes [38], interpreting cancer sub-types [7], and so on. This article scrutinizes existing research in the area of protein structure prediction focusing on factorization-driven approaches, highlighting key findings and current limitations of these approaches for analyzing tertiary structures of proteins.

The remainder of the paper is arranged as follows. Section 2 identifies as well as outlines an area taxonomy and then epitomizes existing approaches along that taxonomy. The article wraps up with a synopsis of the prospective directions (in Sect. 3) and supplementary remarks in Sect. 4.

2   Comprehensive Review

2.1   Taxonomy-Based Survey

A taxonomy offers a structured framework for organizing the methods and techniques developed within a particular domain, and crucially helps reveal existing research gaps. However, no such attempt appears in the literature specifically on factorization-driven approaches for tertiary structure prediction of proteins. To address this, a taxonomy is presented here that is driven by a comprehensive review of the research landscape of this arena. Table 1 provides a snapshot of the existing methods, organized into four major categories: representation of structures into a form suitable for factorization, element composition (structure feature or proximity measure) of the representation, factorization technique applied, and the utility of the method in terms of structure selection.

2.2   Representation of Protein Structure Set for Factorization-Based Methods

From Cartesian Coordinates to Matrix/Tensor: The tertiary structure of a protein is typically described as an ordered sequence of three-dimensional coordinates of its constituent atoms. For a molecule containing N atoms, a straightforward representation treats the structure as a single point S in a 3N-dimensional Cartesian space:

$$S = (x_1, y_1, z_1, \ldots, x_N, y_N, z_N)$$

Table 1. Snapshot of the Factorization-driven Representation Learning Approaches for Protein Tertiary Structure Prediction

With respect to	Category	Literature Instances
Representation	Feature Matrix	Okun et al. [31], Jung et al. [16], NMF-Rank [3], NMF-MAD [5]
	Proximity Matrix	SNMF-DS [20], NMF-EL_QA [17]
	Tensor	NTF-REL [19]
Element	Structure Feature	Okun et al. [31], Jung et al. [16], NMF-Rank [3], NMF-MAD [5]
	Proximity of Structures	SNMF-DS [20], NTF-REL [19], NMF-EL_QA [17]
Factorization Type	Feature Matrix Factorization	Okun et al. [31], Jung et al. [16], NMF-Rank [3], NMF-MAD [5]
	Symmetric Proximity Matrix Factorization	SNMF-DS [20], NMF-EL_QA [17]
	Tensor Factorization	NTF-REL [19]
Structure Selection	Single Structure Selection	NMF-Rank [3], NMF-MAD [5], SNMF-DS [20]
	Group Selection	NMF-Rank [3], NMF-MAD [5], SNMF-DS [20]
	Quality Assessment	NTF-REL [19], NMF-EL_QA [17]

Structural databases like the Protein Data Bank (PDB) [6] store protein structures in this form, providing explicit lists of atoms with their corresponding 3D spatial coordinates.

Feature Matrix: One way to embed these structures into a form suitable for factorization-driven techniques is by constructing a feature matrix, M, where the features are extracted from the set of structures. In this matrix M, each row corresponds to a specific feature, while each column represents an individual structure. Thus, the entry $M(i, j)$ denotes the i-th feature value for the j-th structure. To construct a feature matrix, several types of structure-specific features can be considered. For instance, in [5], this feature matrix consists of 38 features from 3 categories (26 energy-based, 9 consistency-based, and 3 contact-based). Okun et al. [31] considered a 125-dimensional feature vector composed of six feature sets, and Jung et al. [16] utilized profile-profile alignment features to construct the same.

Proximity Matrix: A pairwise similarity (or distance) matrix provides an alternative way to encode a set of structures. In this representation, the matrix T is symmetric, and each entry $T(i, j)$ denotes the proximity between structures i and j within the given set of structures. In fact, this representation has been utilized in [20] using RMSD [18] as the proximity metric. Other similarity measures, such as TM-Score, GDT-TS, have been employed in [17] to form the symmetric similarity matrix.

Tensor: Different proximity measures for comparing two protein structures capture distinct characteristics and can provide complementary insights [32]. To incorporate this information, a tensor R was constructed in [19] by stacking the symmetric similarity matrices, each computed over the same set of structures. Each layer of the tensor can be represented as $R^k_{m,m}$ where k indexes the specific similarity metric. The entry $R(i, j)$ quantifies the similarity with respect to metric k between the structures indexed by positions i and j in the set of m structures.

2.3 Structure Features and Proximity Measures for Comparing Structures

Structure Features: Different types of features can be considered for constructing a feature matrix for a given set of protein structures. Akhter et al. [3] considered a total of 39 features, including consistency-based, energy-based, contact-based, Rosetta REF2015 energy terms, and Rosetta Score12 total energy. Among them:-

- **Energy Function-based Features**: In [3], out of 27 energy-based features; 18 are gathered from the Rosetta REF2015 as well as Score12 energy functions, including 17 raw REF2015 energy terms and the total energy values from both REF2015 and Score12. The remaining 9 features come from potential energy terms: RWplus [42], a side-chain orientationdependent potential; DFIRE [44], a distance-dependent potential; and dDFIRE [40], which incorporates orientation dependence. Three features are obtained from GOAP [40], an all-atom distance-dependent and orientation-dependent potential, DFIRE component, as well as angle-dependent term. In addition, OPUS-PSP [27] contributes three features: a packing energy term, a LennardJones repulsive energy, and the total energy score.
- **Structural Consistency-based Features**: NMF-MAD [5] and NMF-Rank [3] considered 9 features on the basis of structural consistency. At first, the secondary structure is predicted from the primary sequence using PSIPRED and extracted from each structure using DSSP. Matches between PSIPRED and DSSP assignments for β-sheets, α-helices, and coils—normalized by sequence length—formed 3 features, while the combined

PSIPRED confidence scores for matching elements constituted a 4th feature. Moreover, 5 additional features are derived from solvent accessibility [28]. Solvent accessibility is predicted from the sequence using RaptorX [39] and computed from the structures using DSSP. Residues were classified as buried or exposed using a cutoff. The normalized counts of matching buried and exposed residues contributed 2 more features, and the combined probabilities for mismatches form another. Two further features are obtained from the correlation (Pearson) and similarity (Cosine) between solvent accessibility and secondary structure patterns predicted from the sequence and those observed in the structures.

- **Residue Contact-based Features**: NMF-MAD [5] and NMF-Rank [3] utilized three features based on contact information. One feature is the relative contact order, interpreted as the average sequence separation between the contacting residue pairs, normalized by sequence length [34]. Two additional features are derived following the procedure in [41]. Contacts are predicted from the amino acid sequence using RaptorX-contact, and the top ten predicted contacts are treated as references. For each structure, true positives, false positives, as well as false negatives are determined by comparing its top ten contacting residue pairs against the reference set. Precision and recall, respectively, are used as the remaining two features.

Besides, Okun et al. [31] considered a 125-dimensional feature vector composed of six feature sets that include amino acids composition, secondary structure, hydrophobicity, polarity, polarizability, and normalized van der Waals volume. On the other hand, Jung et al. [16] utilized profile-profile alignment features.

Proximity Measures: On the other hand, instead of structural features, proximity measures can be considered to construct a symmetric matrix where each element represents the pairwise similarity or distance between each pair of structures in a given structure set. Out of several measures to compare the two tertiary structures of a protein molecule, the most widely utilized ones are:

- **RMSD**: The root-mean-square deviation (RMSD) between corresponding atoms is widely utilized as a similarity measure for comparing two protein tertiary structures after optimal superposition [18].

$$RMSD = \sqrt{\frac{1}{N} \sum_{1}^{N} \mid S_A^i - S_B^i \mid^2} \tag{1}$$

where, N is the number of atoms, S_A^i and S_B^i represent the coordinate vectors for the i-th atom of the structure A and structure B, respectively (after optimal superimposition). As RMSD is a distance metric, [20] considered the inverse of RMSD to construct the symmetric non-negative similarity matrix for factorization.

- **Template-Modeling (TM) Score**: TM-score calculates the global structural similarity between a protein structure and a reference structure based on the distances between corresponding residue pairs.

$$TM - Score = max\left[\frac{1}{N}\sum_{i=1}^{T}\frac{1}{1+(\frac{d_i}{d_0})^2}\right] \tag{2}$$

where N denotes the number of residues in the reference structure, T is the aligned residue length with respect to the reference structure, d_i represents the distance of the i-th pair of residues after alignment, and d_0 is a scaling parameter used to normalize the matching difference [43]. TM-Score was utilized in NTF-REL [19] and NMF-EL_QA [17].
- **Global Distance Test-Total Score (GDT-TS)**: It measures the similarity between two structures with their corresponding superimposed residues.

$$GDT - TS(S_A, S_B) = \frac{T_1 + T_2 + T_4 + T_8}{4} \tag{3}$$

where T_p represents the percentage of residues from the structure S_A to be superimposed with the corresponding residues from the structure S_B with distance threshold, p ($p \in \{1, 2, 4, 8\}$).

Besides, MaxSub score [36] and GDT-HA [32] have also been considered for constructing symmetric proximity matrices for a given set of protein tertiary structures.

2.4   Factorization Technique

Non-negative Feature Matrix Factorization: NMF-MAD [5] and NMF-Rank [3] factorized the non-negative feature matrices constructed from the set of structures under observation. In fact, non-negative matrix factorization (NMF) is a prominent unsupervised learning technique that estimates a given non-negative matrix $X \in \mathbb{R}_+^{F \times N}$ as the product of two low-rank, non-negative factor matrices $W \in \mathbb{R}_+^{F \times K}$ and $L \in \mathbb{R}_+^{K \times N}$, such that

$$X_{ij} \approx \sum_{s=1}^{K} W_{is} L_{sj} \tag{4}$$

Both factor matrices are constrained to be non-negative and possess a reduced dimension K [33]. NMF is commonly interpreted as a low-rank decomposition method that minimizes a chosen distance measure $\| \cdot \|_{\text{dist}}$, i.e.,

$$\min \|X - WL\|_{\text{dist}} \quad \text{subject to} \quad W_{is} \geq 0, L_{sj} \geq 0.$$

NMF is backed by a statistical framework in which the data features are modeled as superpositions of latent components, with the number of components equal to the reduced dimension K [9]. These components can be interpreted as latent features [12]. The optimization problem $\min \|X - WL\|_{\text{dist}}$ can be solved using a variety of approaches that rely on alternating non-negative least squares, where one factor matrix is held fixed while the other is updated at each iteration. For a given choice of distance measure, NMF optimization is equivalent to an expectation-maximization (EM) algorithm. Under this probabilistic interpretation, the observed variables represented by the columns $d_1, \ldots, d_N$ of the matrix X are generated from the latent variables $l_1, \ldots, l_K$, corresponding to the columns of L. And, each observation x_i is drawn from a probability distribution with mean

$$\langle d_i \rangle = \sum_{s=1}^{K} W_{is} l_s \tag{5}$$

where K denotes the number of latent variables [26]. The contribution of each latent variable l_s to the observation d_i is mediated through the basis patterns denoted by the columns $w_1, \ldots, w_K$ of the matrix W. In [2,5], these basis patterns can be interpreted as pseudo-structure whose linear combinations spread over the full structure ensemble space. Each structure can therefore be expressed as a linear combination of these pseudo-structures, having coefficients provided by the corresponding columns of the matrix L.

Symmetric Non-negative Proximity Matrix Factorization: SNMF-DS [20] and NMF-EL_QA [17] considered the decomposition of symmetric proximity matrices for a given set of protein tertiary structures. Actually, symmetric non-negative matrix factorization (symmetric NMF) is a specific case of NMF in which the factor matrices are identical and entirely positive [14]. For symmetric NMF [25], the cluster membership indicator matrix, $W \in \mathbb{R}_+^{n \times k}$ is obtained by solving

$$min_{W \geq 0} \ f(W) = ||S - WW^T||_F^2 \tag{6}$$

where, $S \in \mathbb{R}_+^{n \times n}$ is a symmetric non-negative proximity matrix. The objective function is subjected to minimization using the Frobenius norm, and typically, the number of clusters k is much smaller than the number of data points, n. The optimization problem in Eq. (5) is solved using alternating non-negative least squares (ANLS) along with block principal pivoting, which converge to stationary points [21]. In fact, symmetric NMF provides a graph-based clustering framework that captures non-linear manifold patterns in the data more effectively than classical NMF or distance-based clustering methods, which are primarily suited for linear data manifolds [25]. Traditional graph-based approaches, such as spectral clustering, require eigenvector-based embeddings of the graph Laplacian followed by an additional, initialization-sensitive clustering step (e.g., k-means). In contrast, symmetric NMF directly incorporates non-negativity constraints into the optimization process, thereby avoiding dependence on post hoc clustering and yielding stable stationary-point solutions [25].

Tensor Factorization: NTF-REL [19] factorized tensors constructed by aggregating the symmetric proximity matrices from a set of protein structures. It utilized the RESCAL tensor factorization framework [30] with automatic latent dimension selection [37]. RESCAL decomposes a tensor $T \in \mathbb{R}^{n \times n \times m}$, formed from m graphs over n nodes, into a factor matrix $A^{n \times k}$ and a core tensor $R^{k \times k \times m}$ where k denotes the number of latent groups. The factorization minimizes the reconstruction error

$$\operatorname{argmin}_{A,R} \|T - R \times_1 A \times_2 A\|_F^2 \tag{7}$$

Here, $\times_i$ denotes the mode$-i$ tensormatrix product [23]. Each column of A represents a latent group, while each slice of R captures interactions among groups for a given relational instance. To accommodate the non-negativity of the data, a non-negative RESCAL formulation [24] is adopted, leading to the following optimization problem:

$$\operatorname{argmin}_{A,R_m} \sum_m \left\|T_m - AR_m A^\top\right\|_F^2 \quad subject\ to \quad \sum_j A_{ij} = 1, for\, 1 \leq j \leq k;\, A,\ R \geq 0$$

2.5 Utility for Structure Selection

Single Structure Selection: In protein structure prediction, single structure selection refers to the task of extracting one representative, most plausible tertiary structure from a given ensemble of structures. The goal is to detect the best single structure, typically the one nearest to the biologically-active structure. NMF-Rank [3], NMF-MAD [5], and SNMF-DS [20] can select a single structure from a given set of protein tertiary structures.

Multiple/Group Structure Selection: Multiple structure selection (also known as group/ensemble selection) involves choosing a representative subset of structures from a large set, instead of a single best structure. NMF-Rank [3], NMF-MAD [5], and SNMF-DS [20] also offer the utility of having a group of structures as a prediction, instead of a single best structure.

Quality Assessment: A quality assessment (QA) framework evaluates and ranks protein structures by estimating their closeness to the biologically-active structure, either on a per-structure basis or at the ensemble level. NTF-REL [19] and NMF-EL_QA [17], each of them can serve as a QA framework at the ensemble level.

2.6 Comparison of Methods

Table 2 outlines the comparison of the existing approaches.

Table 2. Comparison of Different Approaches/Methods

Method	Key Characteristics	Comparison from Literature	Evaluation Criteria
Okun et. al. [31]	−125-dimensional feature vector −Feature sets include amino acid composition, secondary structure, hydrophobicity, normalized van der Waals volume, polarizability, polarity	N/A	Protein Fold Classification Accuracy
Jung et. al. [16]	Feature Matrix: profile-profile alignment features	N/A	Protein Fold Recognition Rate
NMF-Rank [3]	– Features are based on the potential energy, structural consistency, and residue contacts	– Better than potential energy landscape basin-based methods[4]	- Single structure selection: RMSD loss, TM loss, GDT-TS loss - Group Selection: Purity
NMF-MAD [5]	– Features are based on the potential energy, structural consistency, and residue contacts	Better than NMF-Rank in terms of both single and group structure selection	– Single structure selection: RMSD loss, TM loss, GDT-TS loss - Group Selection: Purity
SNMF-DS [20]	- Symmetric non-negative matrix using proximity measures - Eigen-gap heuristics to find the number of structural groups in a non-parametric manner	- comparable to NMF-MAD in terms of group selection - slightly better than NMF-MAD in terms of single structure selection	- Single structure selection: RMSD loss, TM loss, GDT-TS loss - Group Selection: Purity
NTF-REL [19]	- Tensor constructed with stacked symmetric proximity matrices - Silhouette statistics [35] to determine the cluster stability	- Better than NMF-MAD and SNMF-DS in single structure selection	- Single structure selection: RMSD loss, TM loss, GDT-TS loss - Quality Assessment (QA) Framework
NMF-EL_QA [17]	- Integrated Multiview by taking into account matrices composed using proximity metrics - Utilization of unsupervised structure pruning	- slightly better than NTF-REL in single structure selection - for some protein targets, performance is comparable to Alphafold 2 [15,29] and Alphafold 3 [1]	- Single structure selection: RMSD loss, TM loss, GDT-TS loss - Quality Assessment (QA) Framework

3   Discussion

Despite considerable prior research, several open challenges and opportunities for further investigation remain, including:

i. Most of the factorization-based approaches are susceptible to initialization of the factor matrices. SNMF-DS utilized non-negative double singular value decomposition (NNDSVD) for the initialization of the factor matrix. Finding an optimal way of initialization that supports the objective function remains an open question.
ii. Structures can be represented in terms of dihedral angles. In that case, how to apply the factorization-based techniques in that representation?
iii. Are structural features alone sufficient to uncover meaningful patterns, or is it necessary to incorporate additional molecular properties (such as energy-based features [2]) to improve the factorization mechanism?
iv. Factorization-based techniques have sufficient potential to capture the dynamics of biomolecules. This unveils another potential research direction.

4   Conclusion

This article aims to organize and synthesize the existing body of research on factorization-based approaches for protein structure prediction. As discussed, direct comparisons among current methods are often challenging due to the absence of standardized benchmark datasets and evaluation metrics. As mentioned above, several research directions hold promise for advancing current methodologies toward uncovering the underlying organization of biomolecular structures, thereby enabling the identification of structural micro- and macro-states as an important exploratory step in understanding the behavior of dynamic biomolecules.

References

1. Abramson, J., et al.: Accurate structure prediction of biomolecular interactions with alphafold 3. Nature 1–3 (2024)
2. Akhter, N., Chennupati, G., Djidjev, H., Shehu, A.: Improved decoy selection via machine learning and ranking. In: 8th International Conference on Computational Advances in Bio and Medical Sciences (ICCABS), pp. 1–1. IEEE (2018)
3. Akhter, N., et al.: Improved protein decoy selection via non-negative matrix factorization. IEEE/ACM Trans. Comput. Biol. Bioinf. **19**(3), 1670–1682 (2021)
4. Akhter, N., Shehu, A.: From extraction of local structures of protein energy landscapes to improved decoy selection in template-free protein structure prediction. Molecules **23**(1) (2018)
5. Akhter, N., Vangara, R., Chennupati, G., Alexandrov, B.S., Djidjev, H., Shehu, A.: Non-negative matrix factorization for selection of near-native protein tertiary structures. In: International Conference on Bioinformatics and Biomedicine (BIBM), pp. 70–73. IEEE (2019)

6. Berman, H.M., Bourne, P.E., Westbrook, J., Zardecki, C.: The protein data bank. In: Protein Structure, pp. 394–410. CRC Press (2003)
7. Brunet, J.P., Tamayo, P., Golub, T.R., Mesirov, J.P.: Metagenes and molecular pattern discovery using matrix factorization. Proc. Nation. Aca. Sci. **101**(12), 4164–4169 (2004)
8. Carrasco, D.R., et al.: High-resolution genomic profiles define distinct clinico-pathogenetic subgroups of multiple myeloma patients. Cancer Cell **9**(4), 313–325 (2006)
9. Cichocki, A., Zdunek, R., Phan, A.H., Amari, S.I.: Nonnegative Matrix and Tensor Factorizations: Applications to Exploratory Multi-way Data Analysis and Blind Source Separation. Wiley (2009)
10. Devarajan, K.: Nonnegative matrix factorization: an analytical and interpretive tool in computational biology. PLoS Comput. Biol. **4**(7), e1000029 (2008)
11. Donoho, D., Stodden, V.: When does non-negative matrix factorization give a correct decomposition into parts? In: Advances in Neural Information Processing Systems, vol. 16 (2003)
12. Févotte, C., Cemgil, A.T.: Nonnegative matrix factorizations as probabilistic inference in composite models. In: 2009 17th European Signal Processing Conference, pp. 1913–1917. IEEE (2009)
13. Greene, D., Cagney, G., Krogan, N., Cunningham, P.: Ensemble non-negative matrix factorization methods for clustering protein-protein interactions. Bioinformatics **24**(15), 1722–1728 (2008)
14. He, Z., Xie, S., Zdunek, R., Zhou, G., Cichocki, A.: Symmetric nonnegative matrix factorization: algorithms and applications to probabilistic clustering. IEEE Trans. Neural Networks **22**(12), 2117–2131 (2011)
15. Jumper, J., Evans, R., et al.: Highly accurate protein structure prediction with alphafold. Nature (2021)
16. Jung, I., Lee, J., Lee, S.Y., Kim, D.: Application of nonnegative matrix factorization to improve profile-profile alignment features for fold recognition and remote homolog detection. BMC Bioinform. **9**(1), 298 (2008)
17. Kabir, K.L.: Combining non-negative matrix factorization with molecular energy landscape analysis for structure quality estimation of proteins. In: International Conference on Bioinformatics and Computational Biology, pp. 177–188. Springer (2025)
18. Kabir, K.L.: Unsupervised learning for tertiary structure prediction of protein molecules: systematic review. In: International Conference on Computational Advances in Bio and Medical Sciences, pp. 40–53. Springer (2025)
19. Kabir, K.L., Bhattarai, M., Alexandrov, B.S., Shehu, A.: Single model quality estimation of protein structures via non-negative tensor factorization. In: The 11th International Conference on Computational Advances in Bio and Medical Sciences (ICCABS), pp. 3–15 (2021)
20. Kabir, K.L., Chennupati, G., Vangara, R., Djidjev, H., Alexandrov, B.S., Shehu, A.: Decoy selection in protein structure determination via symmetric non-negative matrix factorization. In: International Conference on Bioinformatics and Biomedicine (BIBM), pp. 23–28. IEEE (2020)
21. Kim, J., Park, H.: Fast nonnegative matrix factorization: an active-set-like method and comparisons. SIAM J. Sci. Comput. **33**(6), 3261–3281 (2011)
22. Kim, W., Chen, B., Kim, J., Pan, Y., Park, H.: Sparse nonnegative matrix factorization for protein sequence motif discovery. Expert Syst. Appl. **38**(10), 13198–13207 (2011)

23. Kolda, T.G., Bader, B.W.: Tensor decompositions and applications. SIAM Rev. **51**(3), 455–500 (2009)
24. Krompaß, D., Nickel, M., Jiang, X., Tresp, V.: Non-negative tensor factorization with RESCAL. In: Tensor Methods for Machine Learning, ECML Workshop, pp. 1–10 (2013)
25. Kuang, D., Yun, S., Park, H.: Symnmf: nonnegative low-rank approximation of a similarity matrix for graph clustering. J. Global Optim. **62**(3), 545–574 (2015)
26. Lee, D.D., Seung, H.S.: Learning the parts of objects by non-negative matrix factorization. Nature **401**(6755), 788 (1999)
27. Lu, M., Dousis, A.D., Ma, J.: OPUS-PSP: an orientation-dependent statistical all-atom potential derived from side-chain packing. J. Mol. Biol. **376**(1), 288–301 (2008)
28. Miller, S., Janin, J., Lesk, A.M., Chothia, C.: Interior and surface of monomeric proteins. J. Mol. Biol. **196**(3), 641–656 (1987)
29. Mirdita, M., Schütze, K., Moriwaki, Y., Heo, L., Ovchinnikov, S., Steinegger, M.: Colabfold: making protein folding accessible to all. Nat. Methods **19**(6), 679–682 (2022)
30. Nickel, M., Tresp, V., Kriegel, H.P., et al.: A three-way model for collective learning on multi-relational data. In: ICML, vol. 11, pp. 3104482–3104584 (2011)
31. Okun, O., Priisalu, H.: Fast nonnegative matrix factorization and its application for protein fold recognition. EURASIP J. Adv. Signal Process. **2006**(1), 071817 (2006)
32. Olechnovič, K., Monastyrskyy, B., Kryshtafovych, A., Venclovas, Č: Comparative analysis of methods for evaluation of protein models against native structures. Bioinformatics **35**(6), 937–944 (2019)
33. Paatero, P., Tapper, U.: Positive matrix factorization: a non-negative factor model with optimal utilization of error estimates of data values. Environmetrics **5**(2), 111–126 (1994)
34. Plaxco, K.W., Simons, K.T., Baker, D.: Contact order, transition state placement and the refolding rates of single domain proteins. J. Mol. Biol. **277**(4), 985–994 (1998)
35. Rousseeuw, P.J.: Silhouettes: a graphical aid to the interpretation and validation of cluster analysis. J. Comput. Appl. Math. **20**, 53–65 (1987)
36. Siew, N., Elofsson, A., Rychlewski, L., Fischer, D.: Maxsub: an automated measure for the assessment of protein structure prediction quality. Bioinformatics **16**(9), 776–785 (2000)
37. Truong, D.P., Skau, E., Valtchinov, V.I., Alexandrov, B.S.: Determination of latent dimensionality in international trade flow. Mach. Learn. Sci. Technol. **1**(4), 045017 (2020)
38. Wang, G., Kossenkov, A.V., Ochs, M.F.: LS-NMF: a modified non-negative matrix factorization algorithm utilizing uncertainty estimates. BMC Bioinform. **7**(1), 175 (2006)
39. Wang, S., Li, W., Liu, S., Xu, J.: Raptorx-property: a web server for protein structure property prediction. Nucleic Acids Res. **44**(W1), W430–W435 (2016)
40. Yang, Y., Zhou, Y.: Specific interactions for ab initio folding of protein terminal regions with secondary structures. Proteins Struct. Funct. Bioinform. **72**(2), 793–803 (2008)
41. Zaman, A.B., Parthasarathy, P.V., Shehu, A.: Using sequence-predicted contacts to guide template-free protein structure prediction. In: Proceedings of the 10th ACM International Conference on Bioinformatics, Computational Biology and Health Informatics, pp. 154–160. ACM (2019)

42. Zhang, J., Zhang, Y.: A novel side-chain orientation dependent potential derived from random-walk reference state for protein fold selection and structure prediction. PLoS ONE **5**(10), e15386 (2010)
43. Zhang, Y., Skolnick, J.: Scoring function for automated assessment of protein structure template quality. Proteins Struct. Funct. Bioinform. **57**(4), 702–710 (2004)
44. Zhou, H., Zhou, Y.: Distance-scaled, finite ideal-gas reference state improves structure-derived potentials of mean force for structure selection and stability prediction. Protein Sci. **11**(11), 2714–2726 (2002)

Author Index

GPSR Compliance
The European Union's (EU) General Product Safety Regulation (GPSR) is a set
of rules that requires consumer products to be safe and our obligations to
ensure this.

If you have any concerns about our products, you can contact us on

ProductSafety@springernature.com

In case Publisher is established outside the EU, the EU authorized
representative is:

Springer Nature Customer Service Center GmbH
Europaplatz 3
69115 Heidelberg, Germany